AI-driven Innovations in Physiotherapy and Oncology 3

AI-driven Innovations in Physiotherapy and Oncology 3

Edited by

Abhishek Kumar
Priya Batta
Sachin Ahuja
Pramod Singh Rathore

iSTE

WILEY

First published 2026 in Great Britain and the United States by ISTE Ltd and John Wiley & Sons, Inc.

ISTE Ltd
27-37 St George's Road
London SW19 4EU
UK

www.iste.co.uk

John Wiley & Sons, Inc.
111 River Street
Hoboken, NJ 07030
USA

www.wiley.com

Library of Congress Control Number: 2026934946

British Library Cataloguing-in-Publication Data
A CIP record for this book is available from the British Library
ISBN 978-1-83669-088-7

The manufacturer's authorized representative according to the EU General Product Safety Regulation is Wiley-VCH GmbH, Boschstr. 12, 69469 Weinheim, Germany, e-mail: Product_Safety@wiley.com.

Contents

Preface

AI-driven Innovations in Physiotherapy and Oncology 3 presents an advanced exploration of how artificial intelligence is reshaping rehabilitation sciences and cancer care.

This volume builds on prior editions by integrating recent breakthroughs in machine learning, medical imaging, predictive analytics and personalized treatment planning. It highlights clinically relevant applications, from intelligent physiotherapy assessment and robotic rehabilitation to AI-supported oncology diagnostics, prognosis and therapy optimization. Emphasizing interdisciplinary collaboration, this book bridges computational intelligence with evidence-based clinical practice.

Designed for researchers, clinicians and graduate students, this book aims to support innovation, improve patient outcomes and guide the responsible adoption of AI in modern healthcare systems.

February 2026

Preface written by Abhishek KUMAR, Priya BATTA, Sachin AHUJA and Pramod Singh RATHORE.

1

Reinforcement Learning Models for Adaptive Cancer Rehabilitation in Physiotherapy

In cancer rehabilitation, physiotherapy is very important in improving a patient's quality of life, physical function and mental health after treatment. While rehabilitation strategies have shown clinical value, they tend to score poorly with static, generic, cookie-cutter and one-size-fits-all frameworks that do little to help with a patient's individualized, evolving care. Given the complexity of fatigue, neuropathy and musculoskeletal issues that develop after treatment, the industry has rapidly evolved post-treatment complications and the demand for dynamic, patient-centered rehabilitation will only get stronger. The field of AI, specifically RL (reinforcement learning), has the potential to revolutionize physiotherapy through RL developing autonomous and self-adjusting treatment plans. RL works by receiving and processing patient information and feedback, and assessing treatment results to determine and modify plans in real-time. The real-time effectiveness of rehabilitative therapy is maximized through RL as disengagement, risk and ineffectiveness are systematically eliminated. This chapter will cover the use of reinforcement learning in flexible cancer rehabilitation, including currently used models, algorithm design, and the major clinical functional components and applications. It will also outline the major problems and provide solutions, such as data use, bias with algorithms and the need for clinical data understanding. The focus will be on combining wearable sensors, real-time data analysis and customizing to individual patients to develop smart physiotherapy solutions. Drawing upon machine learning, oncology, biomedical engineering and physiotherapy, this chapter outlines the complete strategy for the first clinical integration of RL-based systems for transforming rehabilitation of cancer and, ultimately, promoting survivorship.

Chapter written by Trupti YADAV and Sanjay BADJATE.

1.1. Introduction

Improvements in the detection and treatment of cancer have changed how we address this illness. The introduction of immunotherapy and other more personalized treatments have ensured that the more aggressive and targeted treatments are no longer the first line of treatment. The result of these changes has been the moving of more types of cancer from "fatal" to "manageable". The completion of treatment, however, is only the beginning of the journey for a cancer patient. There is a steadily growing number of cancer survivors who live with a diverse and complex assortment of long-term functional impairments such as fatigue, lymphedema, neuropathy and pain, as well as other conditions that are more complex and poorly understood clinically. Cardiopulmonary insufficiency is a consequence of cancer that is frequently overlooked and left untreated. (APTA Oncology 2025). Burnout, psychosocial impairment and inactivity are often other impairments that result from these sequelae. The integration of cancer rehabilitation with physiotherapy has been a positive response, but cancer rehabilitation also requires the integration of complex physical exercise as well as other identified components. It has been well established that such rehabilitation impacts comprehensive care and the costs associated with poorly managed healthcare.

What makes traditional rehabilitation programs ineffective is that they take a one-size-fits-all approach to all cases and completely disregard the intricacies of the recovery journey, including the recovery timeline, volume of exercise a patient can manage, other health issues and personal psychosocial factors. For example, considerations of fatigue, mobility and cardiovascular endurance can differ significantly, even among patients receiving the same cancer treatment (Keogh and MacLeod 2012). Furthermore, inconsistent motivation, pain and the risk of injury can make adherence to rehabilitation exercises difficult. Therefore, the need for dynamic, personalized care is apparent and necessary to optimize these cases. The use of artificial intelligence (AI), and more specifically, reinforcement learning (RL), can help avoid these pitfalls. Among machine learning paradigms, reinforcement learning is unique in that a learning agent (i.e. the decision-making algorithm) interacts with its environment and improves its performance over time based on feedback (AlMahamid and Grolinger 2021). In cancer rehabilitation, your digital therapy assistant is an RL agent, the environment is the patient and their body, and the reward is the feedback signal of the patient and various indicators that show a treatment response.

Rehabilitation exercises and their parameters, including intensity, duration and rest intervals, can all be configured depending on RL models (Liao et al. 2020). These models predict outcomes using real-time data from wearables, clinical documentation and data from the patients themselves. For example, a patient may exhibit training fatigue if they reveal a lowered step count, high heart rate and high muscle tremors. In this capture, an agent can lower the strain on exercises in the upcoming set and propose a pause. In contrast, if the patient performs exercises and reports pain ease, the agent will signal an increase in challenge to reinforce higher adaptive potential. There are multiple interlinked components that configure the tech framework for such an RL-abetted system. An RL engine, at the system's core, cognitively iterates the decision-making policy from each patient. Wearable smartwatches, EMGs, IMUs and pulse oximeters serve to capture the needed biosignal and biomechanical data. These data stream to cloud tech for real-time storage, processing and analysis. The outcomes are then used to shift the RL algorithm for current, real-time understanding of the patient's state.

Patients and therapists can use a mobile app or web interface to engage with the system, see progress, get feedback and modify goals. One of RL's greatest advantages is dealing with dynamic and nonlinear systems, such as cancer rehab. Recovery is not a straight line: it is a series of strides, plateaus and setbacks, with complications and other diseases that create regressions (Stavrev and Ginchev 2024). RL systems adjust predictively to these changes and can be further enhanced with transfer learning and other multi-agent techniques. With transfer learning, RL systems built on prior rehab cases can be used with patients from similar demographics to mitigate the cold-start problem. Multi-agent reinforcement learning (MARL) can address the different – but cooperative – health dimensions of strength, flexibility, cardiovascular endurance and mental health to co-adjust the rehab plan. Another powerful attribute of RL rehab is the potential for ongoing real-time adjustments to the rehab plan. In contrast, RL-gamified systems can keep patients motivated and compliant with ongoing therapy-scheduled clinic visits and self-reports. In rehab, these reports do not capture other dynamic changes. RL systems remove the guesswork. They adjust the therapy in real time as the patient engages in it.

Rewards can be given by point systems, badges and achievements to encourage patients to perform exercises, while an RL agent fine-tunes to maintain interest and reduce extreme frustration (Lopes and Lopes 2022). While there are many uses for RL, there are still considerable challenges regarding its use for rehabilitation. The primary issue is probably data privacy and security. Systems complying with HIPAA and GDPR would not be a full solution, given the sensitive nature of the data. Confidentiality and trust would still be at risk without robust encryption, anonymization and secure access. Secondly, there would be considerable and appropriate RL bias without the use of training and additional datasets. A model based on a closed, narrow group of patients would not be suitable for the more expansive, broader use. This is particularly true in oncology. Age, gender, disease, subtype and socioeconomic status are some characteristics that determine how individuals respond to rehabilitation and what is required. Therefore, there is a model gap without a purposeful, integrated approach from data scientists and clinicians from multiple fields – and even patients. Lastly, a model will be of little use if there is a lack of interpretability and transparency (Milani et al. 2024).

Developing and implementing XRL techniques could help alleviate the unwillingness of health practitioners and the public to use and trust black-box systems when health decisions are made. These systems provide reasoning that could facilitate trust and support joint decision-making. Lastly, before the systems are widely implemented, comprehensive clinical validation needs to be performed. Initial studies, pilot and preliminary research provide a good starting point, but definitive proof to support the use of RL-driven rehab systems will be obtained through large-scale, randomized, controlled studies that evaluate safety, effectiveness and costs. Integration, training and ongoing support are vital in achieving this goal.

1.2. Fundamentals of reinforcement learning (RL)

1.2.1. *Overview*

RL focuses on sequential decision-making within various domains. Inspired by behavioral psychology, the RL agent learns by trial and error and adjusts its decision-making based on the rewards or penalties received. The agent interacts with the environment by making observations and following policies that allows them to achieve optimal cumulative rewards over the long run (Rădulescu et al. 2020). The sequential decision-making can be framed within the Markov decision process (MDP), where the agent, environment and the state space, action space and reward function forms the core of the interaction:

– *Agent*: decision-maker learns from the environment.

– *Environment*: the external system the agent interacts with.

– *State (S)*: a representation of the current situation.

– *Actions (A)*: a set of choices available to the agent.

– *Reward (R)*: a scalar feedback signal indicating the effectiveness of an action.

The agent operates within an environment by moving from state to state based on the actions it takes and the environment's response to the actions. The agent optimizes its policy by receiving rewards that will be synchronized with the cumulative expected rewards. The agent refines policies over repeated episodes to improve outcomes.

Depending on the algorithm being used, learning may include value functions such as Q-learning, policy gradients or even actor–critic structures. Unlike supervised learning, where there is a labeled dataset, there are no labeled examples in RL; the learning signal is derived from rewards collected through interaction (Christiano et al. 2017). This is why RL is applicable in cases where there are unknown optimal solutions, and it is necessary to devise effective methods through exploration. It has been successfully applied to robotics, games and self-driving cars, and is now used more frequently in healthcare, such as in cancer rehabilitation, along with autonomous vehicles.

1.2.2. *Key components in cancer rehabilitation*

Utilization of RL in cancer rehab moves us away from traditional static therapy pain management to more versatile, tailored and individualized interventions. Here, every RL element can be customized to fit the rehab clinical scenario to design smart rehab systems that adjust to the individual patient's therapy needs:

– *Agent*: in a rehab scenario, the agent can be a machine learning model, virtual therapist or smart controller of a software system. This agent determines the optimal rehabilitation protocol based on the patient's state. The agent can be part of a digital health ecosystem that comprises a mobile app, a web portal or an integrated device in the rehab equipment. The agent captures the patient's responses and modifies the treatment in real time (Ianculescu et al. 2025).

– *Environment*: the patient's surrounding dynamic physical and mental states combined. Broadly defined, the environment consists of any and all factors that may influence results in rehabilitation, be it fatigue, emotional state, co-morbidities, medication, therapy adherence, etc. Healthcare environments, in contrast to the simulated environments of games or robotics, are noisy and non-stationary, and frequently only partially observable. Ethical and safety parameters also come into play, mandating a thoughtful calibration of the learning algorithm so that it cannot make dangerous recommendations (Xu et al. 2022).

– *State*: the physician believes that every patient they come across has a story to tell. They digest and synthesize different forms of storytelling, both coherent and abstract. These biographies include:

– *Biometric information*: information gathered through wearable devices which include heart rate, oxygen saturation, muscle activity (EMG), movement patterns, sleep data and mobility statistics.

– *Therapy knowledge*: exercise history, data from prior sessions, recovery patterns and medication compliance.

– *Patient-reported outcomes (PROs)*: pain, fatigue, emotional state and energy levels, which might be collected through surveys and/or voice recorders.

– *Contextual information*: temporal information, time of treatment (e.g. recent chemotherapy) or other situational attributes which may influence patients' performance level or outcomes.

These states, which involve high dimensionality and continuous data, therefore require function approximation methods, such as neural networks.

– *Action*: this domain of actions covers clinical decisions regarding the rehabilitation pathway and includes, but is not limited to, the following:

– *Modifying exercise parameters*: adjusting the intensity, duration, type or frequency of exercises.

– *Altering therapy modalities*: introducing new activities such as balance training, resistance therapy or mindfulness exercises.

– *Recommending breaks or rest days*: based on signs of overexertion or physiological fatigue.

– *Triggering alerts to clinicians*: clinicians could use the AI system to receive alerts about patients whose data indicate potential risks to their health.

Importantly, whatever is done must be clinically safe and be in accordance with therapeutic guidelines. Therefore, a human-in-the-loop mechanism is routinely applied, where healthcare professionals review or override the recommendations:

– *Reward*: the reward acts as a feedback signal and informs the agent of better treatment strategies. In the case of cancer rehabilitation, defining a good reward function is both important and difficult as it needs to reflect both short-term improvements and long-term outcomes (Yu et al. 2021). Some of the most common reward indicators are:

– *Functional gains*: increases in mobility, strength or endurance.

– *Symptom reduction*: decreases in reported pain, fatigue or psychological distress.

– *Adherence and engagement*: regular participation in therapy sessions.

– *Patient satisfaction:* positive subjective reports or high-quality-of-life scores.

– *Avoidance of adverse events*: preventing injury, overexertion or relapse.

Rewards can be immediate (e.g. reduced pain after a session) or delayed (e.g. improved mobility over weeks). This temporal – credit – assignment problem is well dealt with by RL algorithms that are geared towards maximizing long-term returns.

1.3. Cancer rehabilitation needs and challenges

1.3.1. *Functional deficits post-cancer treatment*

Cancer treatments surgery, chemotherapy, radiation therapy and immunotherapy are often lifesaving but can lead to various functional deficits, having a dramatic impact on a patient's quality of life.

Rehabilitation, therefore, becomes an important component of survivorship care, tackling the physical, neurological and psychological after-effects of treatment:

– *Fatigue and reduced endurance*: one of the most common side effects of cancer treatment is persistent fatigue, often disproportionate to the level of exertion. This fatigue can hinder participation in daily activities, delay return to work and reduced overall engagement in physical exercise, which is essential for recovery (Kumar et al. 2026a).

– *Musculoskeletal limitations*: when surgery lasts longer than planned, patients do not use some muscles and joints after a procedure; they may become weaker, stiffer, have decreased range of motion and imbalances in posture. These types of problems are visible particularly after major surgery for cancers of the breast, head-and-neck or bone.

– *Neuropathy and balance issues*: chemotherapy-induced peripheral neuropathy (CIPN) is the deadliest syndrome in cancer survivors and causes neuropathy and disturbed coordination when someone moves. This syndrome, especially with the added risk of falling adds to the situation, and makes balance training indispensable.

– *Psychosocial distress*: concerning the cancer scare, survivors bear a fourfold greater risk of psychosocial stress, especially anxiety, depression and body image issues. Physical recovery depends, in some cases, on the patient's psychological state, where emotional distress may lie at the root of non-compliance to prescribed physiotherapy, making emotional care imperative.

– *Cardiorespiratory deconditioning*: inactivity may be exacerbated by the use of certain chemotherapeutic agents with cardiotoxic effects, leading to further decline in cardiac and pulmonary function. Patients may present with symptoms such as dyspnea, tachycardia, and other signs of exercise intolerance, which must be carefully monitored during exercise to avoid risk (Kumar et al. 2026b).

Given the wide range and complexity of problems encountered in cancer rehabilitation, care must be individualized and patient-centered. A uniform, unchanging rehabilitation plan is unlikely to address these diverse needs effectively.

1.3.2. *Challenges in current rehabilitation protocols*

Even with the recognized potential benefits of oncology rehabilitation, the current clinical practices remain largely ineffective and inaccessible due to some systemic and technological bottlenecks:

– *Failing to personalize prescriptions*: many facilities are still providing generic prescribing instructions for exercise that fail to match the type of cancer, the stage of treatment and the functional level of patients. This approach can make therapy less effective and result in patients leaving therapy.

– *Limit adaptability based on patient progress*: traditional rehabilitation programs are typically updated at intervals, following planned assessments. This set schedule is unable to keep up with the unpredictable pace of recovery and thus sudden improvements or setbacks may not be quickly reflected in therapy intensity and focus (Strunk and Lichtwarck-Aschoff 2019).

– *High therapist to patient ratio*: the increase in the number of cancer survivors increases the need for rehab services; however, trained physiotherapists are limited, particularly in resource-constrained settings. As a result, patients may receive limited one-on-one attention, compromising the quality of care.

– *Subjective assessment tools*: to assess the degree of physical function and symptoms, the clinician and the patient complete questionnaires. While useful, these techniques are subjective and can introduce variability when deriving a diagnosis and formulating a treatment plan (Rothrock et al. 2019).

– *Lack of continuous data integration*: rehabilitation results are documented intermittently and, in many cases, infrequently. Not having continuous, real-time data from a patient's wearable devices, activity tracking systems in the home or mobile health applications impedes the ability to assess a patient's adherence and progress between the scheduled clinical appointments.

1.4. RL-based framework for adaptive rehabilitation

1.4.1. *System architecture*

A cancer rehabilitation system powered by RL could be described as an intelligent system designed as a closed-loop ecosystem that adjusts to the changing emotional and physical states of the patient. The system architecture includes different components that offer personalized, patient-centered and flexible data-based physiotherapy to patients. The most important components are the wearable sensors that help in collecting real-time data on a patient's physiology and biomechanical responses. These sensors, consisting of electromyography (EMG) devices, accelerometers, gyroscopes and wearable devices capture data on muscle activation, limb movements and positioning, and core body orientation throughout a range of movements the patient performs during a rehabilitation session.

The mobile application, or a web portal that functions as a user interface, represents the sensor system layer in its function. The interface provides patients with daily workout plans, instructional videos, performance feedback, motivational messages and progress tracking. Clinicians are using dashboards, which allow them to see how their patients are doing, approve changes to therapy and communicate with their patients. Information is shared so that patients can make the final decisions, although therapists continue to provide clinical oversight. The RL engine (which is an intelligent module) is constantly learning from sensor data and feedback from the patient in order to provide suggestions for real-time, adaptive therapy (Tetteroo et al. 2014).

The system modulates the intensity, duration and type of exercises, depending on the previous results of each person, so that everyone has a personalized rehabilitation plan. All interactions and outcomes are stored in a cloud-based system that maintains longitudinal patient data. This data is used to train and improve models of RL, large-scale analytics, cohort studies and retrospective analysis.

1.4.2. *Data sources*

The quality of the RL system depends upon the quality and variety of input data. Multiple streams of data provide a full picture on the state of the patient:

– *Physiological signals*: these are heart rate, oxygen saturation (SpO_2), respiratory rate and muscle activation measured by EMG. The signals show the condition of the patient during the therapy or detection of overexertion or fatigue (Rudroff 2025).

– *Kinematic data*: collected by motion sensors, this includes joint angles, step counts, range of motion and gait patterns. The RL agent will utilize this information to determine how to move, identify dysfunction and prevent overfocusing on only one task.

– *Patient feedback*: subjective measures such as pain level, emotions, motivation and the way we perceive exertion, are collected from the patient through questionnaires on the app, or voice inputs. These inputs personalize rehabilitation and help balance the physical and psychological outcomes of rehabilitation in the learning process (Olugbade et al. 2019).

– *Electronic health records (EHRs)*: in addition to rehabilitation data, there is the context of the EHR: cancer history and stage, treatment details, other conditions, medications and past injuries. The data allows the RL model to personalize their approach to each patient based on their respective risks and medical history. The integration of these different data sources allows the system to create a cohesive, multi-faceted view of the patient's health, which in turn fine-tunes and maximizes the safety of the therapy recommendations.

1.4.3. *Adaptive learning process*

The RL agent is incorporated within the trial and feedback system to continuously improve performance. This is the core of the system. The following outlines the entire flow of the process. The patient completes the prescribed exercises as the system guides them. Wearable sensors, the app, and other input and output devices provide, respectively, the physiological, kinematic and subjective data and then the system evaluates the signal of the data for the session. The RL agent assesses the data and determines the outcome of the session.

When a session ends positively – better mobility in the joints, less pain or improved endurance – a high reward is assigned. On the other hand, if the session ends with excessive fatigue, more pain, poor adherence or discomfort, a low or negative reward is given. With the given reward, the agent updates its policy for the session to refine its future recommendations. With refinement, the system aims to optimize therapy plans to maximize recovery and minimize discomfort and risks. This provides more adaptive personalized rehabilitation. This, in turn, translates to greater adherence, improved negativity and better clinical outcomes over time.

RL algorithm/model	Core principle	Application in cancer rehabilitation	Key advantages	Potential limitations
Q-learning	Value-based method that learns optimal actions by estimating future rewards	Adjusting exercise intensity or frequency based on real-time performance and fatigue data	Simple to implement; effective for discrete actions	Inefficient in large or continuous state spaces
Deep Q-network (DQN)	Combines Q-learning with deep neural networks for high-dimensional input	Adaptive progression of exercise levels using sensor and motion data	Handles complex data from wearables; learns nonlinear relationships	Prone to instability without proper tuning

Policy gradient methods	Directly optimizes the policy function without value estimation	Personalized exercise scheduling and sequencing	Suitable for continuous control problems	Requires large data; slower convergence
Actor–critic model	Uses separate networks to evaluate actions and update policy	Balancing intensity and recovery in endurance or strength training	Stabilizes learning and speeds convergence	Computationally demanding
Multi-agent RL (MARL)	Multiple agents collaborate for multi-dimensional optimization	Coordinating physical, psychological and cardiovascular therapy aspects	Enables holistic, adaptive therapy plans	Complex to coordinate and train effectively
Inverse RL	Learns optimal strategies by observing expert behavior	Mimicking expert physiotherapist decisions for automated recommendation	Reduces need for labeled data; enhances realism	Requires high-quality expert demonstration data
Explainable RL (XRL)	Focuses on transparency and interpretability of decision-making	Provides understandable therapy recommendations to clinicians	Builds trust and improves clinical acceptance	Still evolving; may trade off performance

Table 1.1. *Reinforcement learning algorithms and applications in cancer physiotherapy*

1.5. Benefits of RL in cancer physiotherapy

The impact of ML and RL on the transformation of rehabilitation physiotherapy for oncology patients has been substantial. These technologies' most significant potential target has been the personalization of therapy programs. While traditional rehabilitation plans are static and do not take the current condition of the patient into account, RL systems are structured in a way that allows patients' rehabilitation physiotherapy systems to attain real-time adjustments of exercise intensity, duration, type and intersperse breaks based on fatigue, performance and physiological signals to monitor fatigue in patients' post-cancer treatment. This allows the treatment to be more effective and also considers the patients' cancer-related comorbidities (Kishor et al. 2025).

Moreover, the AI-gamification systems in physiotherapy rehabilitation increase adherence to therapy. Patients' engagement increases because artificial therapy system rehabilitation gamification includes virtual rewards, performance dashboards and interactive challenges and thus motivates them to complete therapy more often. This is critical for recovery. AI systems also increase patients' participation through visual goal attainment systems and rehabilitation progress feedback. This is crucial for cancer patients, who often suffer psychologically. These systems also increase operational efficiency in clinical rehabilitation systems.

Automating the assessments, monitoring and tracking of the progress of patients allows therapists to reduce the time spent on admin work and leaving more time for tailored patient work. By using intelligent scheduling systems, resources can be allocated in a more effective manner, thereby ensuring that high-risk and low-compliant patients receive interventions when needed (Nandan Prasad 2024). This type of optimization is vital in the healthcare sector when there is a shortage of staff and limited resources. In cancer rehabilitation, there is a major safety concern when patients are overexerted and there are adverse outcomes. Well-trained ML systems that can analyze and interpret the warning signs of fatigue, pain and distress caused by an overload of cardiovascular exertion to the heart, can – to an extent – recognize the safety threshold of a patient.

Flexible and adaptive exercise systems that re-design the exercise schedule in real-time can ensure that the patient's cardiovascular and pain thresholds are not breached and that the exercise stays within the safety limits. If abnormal patterns are detected, alerts and decision-support tools can trigger prompts for clinician intervention. Lastly, the objective and transparent nature of physiotherapy is enhanced by ML and RL technologies through data-driven decision-making. There is an enormous amount of data in each session that can be used to evaluate the effectiveness of treatment (Fusco et al. 2025).

Analytics dashboards help clinicians visualize the progress of patients in different cohorts and identify time-based trends. Treatment plans may then be adjusted using evidence-based practice rather than relying solely on clinician opinion. This is to improve clinical outcomes and contribute to the continuous improvement of programs.

1.6. Limitations and ethical considerations

1.6.1. *Privacy and security issues*

The application of RL in physiotherapy in oncology necessitates the gathering and analysis of very sensitive information related to health, such as biometric data, patient activity history and the patient's clinical history. Safeguarding sensitive information from unauthorized access and misuse requires the implementation of strong breach protection measures, such as end-to-end encryption in transit and safe storage (Chevrier et al. 2019). To diminish the chances of re-identifying data subjects, controls for anonymization and pseudonymization must be utilized whereby personal identifiers are either fully masked or replaced. Legal and honest practice requires regulation, including the US's Health Insurance Portability and Accountability Act (HIPAA) and Europe's General Data Protection Regulation (GDPR). Security audits, as well as breach reporting and data control governance, are additional facets of the secure digital environment required for RL-based rehabilitation.

1.6.2. *Data bias*

The neutrality of RL algorithms is determined by the neutrality of their training data. If datasets lack diversity with respect to age, sex, race, type of cancer or social class, RL systems will generate suggestions that will be ineffective or even detrimental to the health of the individuals in the dominant group. Such bias in algorithms creates inequities in opportunities for care and goes against the spirit of personalized medicine (Agarwal et al. 2023). It is thus important to have inclusive and representative datasets. Furthermore, performing fairness-enabled RL and periodically analyzing the outputs of the models will help to spot and mitigate the bias.

1.6.3. *Interpretability*

Several RL models are "black boxes" that make decisions that are difficult for clinicians to comprehend or accept. In high-stakes healthcare environments, it is not a luxury, but essential to healthcare clinical adoption – and it is not an afterthought. Without rationales for recommendations, clinicians may hesitate to adopt RL-guided recommendations. Techniques such as policy visualization, reward decomposition and saliency mapping can be used to create transparency to make the model's reasoning visible and easier to assess. Collaborative development with clinicians can also help to develop models that will provide actionable, interpretable outputs customized to real-world rehabilitation situations (Iyer et al. 2018).

1.6.4. *Dependency and dehumanization*

There are aspects of rehabilitation where RL systems can be highly effective in optimizing a rehabilitation protocol, but there is a risk of losing the human touch of care if we rely too much on automation. Emotional support, empathy and therapeutic rapport, which are important to recovery, may be weakened if digital systems replace (instead of aid) human therapists (Salil et al. 2025).

1.7. Future directions

1.7.1. *Multi-agent reinforcement learning (RL)*

With multi-agent RL, we can deploy several agents that will each focus on a different aspect of the oncology rehabilitation practice (e.g. cardiovascular, muscular or psychological rehabilitation). The agents learn how the patient adapts and then collaborate to develop a focused rehabilitation plan. One agent could, for example, help the patient optimize aerobic intensity while another modifies strength training at different or relative levels matched to real-time EMG (electromyography) data, and a third tracks psychological responses (mood or fatigue) to psychologically evaluate the patient (Xie et al. 2025). The coordinated effort of several agents allows the patient to engage more effectively in the rehabilitation process, while the healing milieu focuses on the psychological aspects of recovery. The level of collaboration is more advanced and adaptable.

1.7.2. *Transfer learning*

Transfer learning enables the adaptation of RL models developed for a certain patient group to be minimally revised for other patient groups. For example, a model developed for breast cancer survivors can be adapted for the head and neck cancer cohort with some additions of targeted clinical attributes (Krishna et al. 2024; Pai et al. 2024). This greatly decreases the time and effort spent in developing models from the ground up and addresses the cold start problem.

1.7.3. *Human-in-the-loop approaches*

The systematic control of self-learning RL loops by a human ensures that the processes remain safe, ethically compliant and compassionate. Based on their clinical evaluation, clinicians can intervene, validate or sit around the RL-generated recommendations. This collaborative model combines automation and human expertise and establishes a trust and acceptance by both patients and clinicians (Gilbert et al. 2022).

1.7.4. *Integration with electronic health records (EHRs)*

Integration of RL systems to EHRs will allow for context-aware decisions to be made by pulling longitudinal health data, treatment histories and co-morbidities. Such integration provides scale and clinical actionability to RL systems by improving the specificity and individualization of rehabilitation protocols (Perera et al. 2025; Yadav et al. 2024).

1.8. Conclusion

RL holds great promise for tailoring physiotherapy in cancer rehabilitation. With wearable sensors augmented by real-time data analytics and cloud computing, RL-based systems adjust treatment plans aligned with Ruben's recovery stage, which increases the therapy's impact while decreasing the risk. These systems gather feedback and learn the most efficient exercises so that the rehabilitation and recovery programs provide the most effective patient-centered rehabilitation programs. Numerous challenges pertaining to RL's integration into the clinic include, but are not limited to, algorithm transparency and explainability, data privacy, patient confidentiality, ethical issues and lack of clinic-grade evaluations. Trust in the RL systems documented in the medical literature to provide these graphic elucidations for interventional RCTs is not justified. Subsequently, such documentation is required as the primary driver for interventional RCTs in the RL literature, along with regulatory documents and interdisciplinary teams integrating oncology, physiotherapy and computing science. RL has the potential to be the primary driver in the integration of algorithm fine-tuning to provide rehabilitation therapy post-cancer treatment, improving the therapy's effectiveness, responsiveness and personalization. The integration of complex algorithms with RL technology in physiotherapy post-cancer treatment rehabilitation not only redefines the physiotherapy post-cancer treatment rehabilitation paradigm but also reinforces the promise of RL technology to provide human-centered care to these patients.

1.9. References

Agarwal, R., Bjarnadottir, M., Rhue, L., Dugas, M., Crowley, K., Clark, J., Gao, G. (2023). Addressing algorithmic bias and the perpetuation of health inequities: An AI bias aware framework. *Health Policy and Technology*, 12(1), 100702.

AlMahamid, F. and Grolinger, K. (2021). Reinforcement learning algorithms: An overview and classification. In *2021 IEEE Canadian Conference on Electrical and Computer Engineering (CCECE)*, 12 September, 1–7.

APTA Oncology (2025). Combined sections meeting (CSM) 2025 poster and platform presentations. *Rehabilitation*, 43(1), 38–66.

Chevrier, R., Foufi, V., Gaudet-Blavignac, C., Robert, A., Lovis, C. (2019). Use and understanding of anonymization and de-identification in the biomedical literature: Scoping review. *Journal of Medical Internet Research*, 21(5), e13484.

Christiano, P.F., Leike, J., Brown, T., Martic, M., Legg, S., Amodei, D. (2017). Deep reinforcement learning from human preferences. *Advances in Neural Information Processing Systems*, 30 [Online]. Available at: https://proceedings.neurips.cc/paper_files/paper/2017/file/d5e2c0adad503c91f91df240d0cd4e49-Paper.pdf.

Fusco, R., Granata, V., Setola, S.V., Pupo, D., Petrosino, T., Lamanna, C.P., Castaldo, M., Riga, M.G., Karaboue, M.A., Izzo, F. et al. (2025). Visual perception and pre-attentive attributes in oncological data visualisation. *Bioengineering*, 12(7), 782.

Gilbert, T.K., Dean, S., Zick, T., Lambert, N. (2022). Choices, risks, and reward reports: Charting public policy for reinforcement learning systems. arXiv preprint, arXiv:2202.05716, 11 February.

Ianculescu, M., Constantin, V.Ş., Guşatu, A.M., Petrache, M.C., Mihăescu, A.G., Bica, O., Alexandru, A. (2025). Enhancing connected health ecosystems through IoT-enabled monitoring technologies: A case study of the monit4healthy system. *Sensors*, 25(7), 2292.

Iyer, R., Li, Y., Li, H., Lewis, M., Sundar, R., Sycara, K. (2018). Transparency and explanation in deep reinforcement learning neural networks. In *Proceedings of the 2018 AAAI/ACM Conference on AI, Ethics, and Society*, 27 December, 144–150.

Keogh, J.W. and MacLeod, R.D. (2012). Body composition, physical fitness, functional performance, quality of life, and fatigue benefits of exercise for prostate cancer patients: A systematic review. *Journal of Pain and Symptom Management*, 43(1), 96–110.

Kishor, I., Goyal, P., Reddy, H.G., Goyal, P., Mamodiya, U. (2025). AI-driven haptic technologies revolutionizing patient rehabilitation. In *Integrating AI With Haptic Systems for Smarter Healthcare Solutions*, Mamodiya, U., Kalam, A., Bossoufi, B. et al. (eds). IGI Global, Hershey, PA, 47–84.

Krishna, S.R., Gudur, A., Jain, S., Deivasigamani, S., Tiwari, M., Venkatesan, K.G. (2024). Deep learning for automatic diagnosis of skin cancer using dermoscopic images. *Journal of Artificial Intelligence and Technology*, 4(2), 114–123.

Kumar, A., Batta, P., Ahuja, S., Rathore, P.S. (eds) (2026a). *AI-driven Innovations in Physiotherapy and Oncology 1*. ISTE Ltd, London and Wiley, Hoboken.

Kumar, A., Batta, P., Ahuja, S., Rathore, P.S. (eds). (2026b). *AI-driven Innovations in Physiotherapy and Oncology 2*. ISTE Ltd, London and Wiley, Hoboken.

Liao, Y., Vakanski, A., Xian, M. (2020). A deep learning framework for assessing physical rehabilitation exercises. *IEEE Transactions on Neural Systems and Rehabilitation Engineering*, 28(2), 468–477.

Lopes, J.C. and Lopes, R.P. (2022). A review of dynamic difficulty adjustment methods for serious games. In *International Conference on Optimization, Learning Algorithms and Applications*, 24 October, 144–159. Springer, Cham.

Milani, S., Topin, N., Veloso, M., Fang, F. (2024). Explainable reinforcement learning: A survey and comparative review. *ACM Computing Surveys*, 56(7), 1–36.

Nandan Prasad, A. (ed.) (2024). Regulatory compliance and risk management. In *Introduction to Data Governance for Machine Learning Systems: Fundamental Principles, Critical Practices, and Future*. Apress, Berkeley, CA, 485–624.

Olugbade, T.A., Singh, A., Bianchi-Berthouze, N., Marquardt, N., Aung, M.S., Williams, A.C. (2019). How can affect be detected and represented in technological support for physical rehabilitation? *ACM Transactions on Computer-Human Interaction (TOCHI)*, 26(1), 1–29.

Pai, S., Bontempi, D., Hadzic, I., Prudente, V., Sokač, M., Chaunzwa, T.L., Bernatz, S., Hosny, A., Mak, R.H., Birkbak, N.J. et al. (2024). Foundation model for cancer imaging biomarkers. *Nature Machine Intelligence*, 6(3), 354–367.

Perera, D., Habib, G., Xu, Q., Tan, D.J., He, K., Cambria, E., Feng, M. (2025). Beyond prediction: Reinforcement learning as the defining leap in healthcare AI. arXiv preprint, arXiv:2508.21101, 28 August.

Rădulescu, R., Mannion, P., Roijers, D.M., Nowé, A. (2020). Multi-objective multi-agent decision making: A utility-based analysis and survey. *Autonomous Agents and Multi-Agent Systems*, 34(1), 10.

Rothrock, N.E., Cook, K.F., O'Connor, M., Cella, D., Smith, A.W., Yount, S.E. (2019). Establishing clinically-relevant terms and severity thresholds for Patient-Reported Outcomes Measurement Information System®(PROMIS®) measures of physical function, cognitive function, and sleep disturbance in people with cancer using standard setting. *Quality of Life Research*, 28(12), 3355–3362.

Rudroff, T. (2025). Digital biomarkers and AI for remote monitoring of fatigue progression in neurological disorders: Bridging mechanisms to clinical applications. *Brain Sciences*, 15(5), 533.

Salil, R., Jose, B., Cherian, J., Sheeja P.R, Vikraman, N. (2025). Digitalized therapy and the unresolved gap between artificial and human empathy. *Frontiers in Psychiatry*, 15, 1522915.

Stavrev, S. and Ginchev, D. (2024). Reinforcement learning techniques in optimizing energy systems. *Electronics*, 13(8), 1459.

Strunk, G. and Lichtwarck-Aschoff, A. (2019). Therapeutic chaos. *Journal for Person-Oriented Research*, 5(2), 81.

Tetteroo, D., Timmermans, A.A., Seelen, H.A., Markopoulos, P. (2014). TagTrainer: Supporting exercise variability and tailoring in technology supported upper limb training. *Journal of Neuroengineering and Rehabilitation*, 11(1), 140.

Xie, N., Zhang, X., Lu, C. (2025). An exploratory framework for EEG-based monitoring of motivation and performance in athletic-like scenarios. *Scientific Reports*, 15(1), 26156.

Xu, M., Liu, Z., Huang, P., Ding, W., Cen, Z., Li, B., Zhao, D. (2022). Trustworthy reinforcement learning against intrinsic vulnerabilities: Robustness, safety, and generalizability. arXiv preprint, arXiv:2209.08025, 16 September.

Yadav, T., Varadharajulu, G., Gudur, R. (2024). Multimodal therapeutic exercises and their role in mitigating chemotherapy-induced peripheral neuropathy: A longitudinal study on patient outcomes. *Cureus*, 16(11), e74214. doi: 10.7759/cureus.74214.

Yu, C., Liu, J., Nemati, S., Yin, G. (2021). Reinforcement learning in healthcare: A survey. *ACM Computing Surveys (CSUR)*, 55(1), 1–36.

2

AI-Enabled Gait and Balance Assessment in Oncology Rehabilitation

Cancer survivors often experience changes in their walking pattern and struggles with their balance during and after their treatment which may include chemotherapy, surgery and radiation, and may leave them physically inactive for extended periods. These changes can result in falls, loss of independence and an overall lower quality of life. The Berg Balance Scale and Timed Up & Go test are useful measures in routine practice but provide only episodic information, are open to subjective interpretations and can fail to capture decline in a patient's balance and gait in a timely manner. The use of artificial intelligence (AI) in oncology rehab has the potential to shift the paradigm for balance and gait assessments from subjective intermittent assessments to continuous, objective and interpretable assessments. Markerless computer vision, telehealth systems, smart floors and wearable inertial measurement units (IMUs) provide real-time gait and balance testing in clinical and home environments. AI can assess a patient's rehab progress and personalize their rehab program, improving proactive care through earlier recognition of decline. Real-time analytics, enhanced by virtual reality, robotic systems and remote rehab systems, can improve patient compliance with their rehab. This chapter provides an in-depth overview of AI-assisted systems in balance and gait analysis for oncology rehabilitation, focusing on clinical rationale, technology, approaches for implementation, barriers to use and future research. The aim is to provide a blueprint for integration of smart mobility analytics to enhance functional rehabilitation, improve safety and enhance the quality of life of cancer survivors.

2.1. Introduction

As oncology rehabilitation develops, the need for assessing and monitoring patient mobility in a timely and precise manner becomes increasingly important in rehabilitation to assist patients in their recovery and improve the quality of life for cancer survivors. After cancer treatment, patients would experience post-surgical

---

Chapter written by Dhairyasheel PATIL and Pankaj THOTE.

complications, treatment-related neuropathies and prolonged inactivity, resulting in major interruptions to their gait and balance. The challenges to gait and balance increases the likelihood of falls, loss of autonomy, inactivity and participation in daily life (Rattanakrong 2021). These challenges create a serious problem in survivorship and quality of life.

Tools for evaluating gait and balance such as the Berg Balance Scale (BBS), Timed Up & Go (TUG) test and gait speed measurements have been used in rehabilitation for many years. Although these measures are useful in rehabilitation and easy to use, they could be subjective, fail to provide a comprehensive ongoing assessment and might be too coarse to capture early functional mobility changes. The patient becomes more vulnerable to serious injury and chronic loss of mobility before the deficit is detected. The use of AI in mobility assessment is the best approach to solve these issues.

Integrating wearable sensors, depth cameras, pressure-sensing insoles and even smartphone-based accelerometers with machine learning algorithms makes it possible to continuously monitor movements and provide real-time assessments of gait and balance performance. These systems provide the ability to assess the subtle deviations in symmetry of the gait, variability of the stride, timing of the steps and the transitions of posture. AI-enhanced gait analysis offers a number of real-time clinical applications. It improves the accuracy of rehabilitation planning since there are dynamic, real-time assessments of functional status that allow therapists to adjust exercise regimens on the fly and directly target specific impairments (Su et al. 2023). It also provides additional clinical assessments that allow the clinician to stay ahead of deteriorating conditions, such as falls, providing opportunities for timely preventative intervention.

AI systems encourage patient engagement through intuitive dashboards and feedback systems that allow users to track their progress and follow prescribed rehabilitation routines more closely (Reddy 2025a). Multiple AI techniques power these systems, for example, supervised machine learning and movement classification, unsupervised clustering for pattern discovery into mobility data and deep learning techniques such as recurrent neural networks (RNNs) and convolutional neural networks (CNNs) that accurately process time-series data. These systems are trained on diverse datasets of human movements, both pathological and during recovery, to identify and assess normal versus impaired gaits with impressive sensitivity and specificity.

There are several studies validating AI tools for assessing gait in rehabilitation for cancer. For instance, research on survivors of breast cancer shows that wearable sensor systems identify asymmetries in arm swing and stride length well before patients perceive functional restrictions. Likewise, post-radiation studies on head and neck cancer survivors have used AI motion capture to reveal and assess cervical spine mobility, postural control restrictions and provide data that physiotherapists have used for focused treatment planning.

There has been ongoing integration of such technologies in inpatient and outpatient oncology. A few rehabilitation facilities use robotic treadmills and AI-enabled feedback mechanisms that allow therapists to make real-time adjustments to the training intensity, while monitoring changes in gait metrics. Other facilities use home-based robotic systems where patients have lightweight sensors during daily activities and the data attached to the robotic systems (Boltaboyeva et al. 2025). These systems are positioned for safe monitoring of patients who are immunocompromised and live in remote regions to limit travel to highly congested hospitals.

These systems are highly effective for health monitoring. However, the potential in these systems is limited by several factors: the need for continuous monitoring raises data privacy and security issues, and the trust of clinicians can only be gained by ensuring algorithm transparency and clinical interpretability.

The need for system cost and availability will drive inequitable use of these systems in different health systems. The positive impact on AI rehabilitation in oncology is clear, and with the ongoing developments in sensor miniaturization, edge computing and federated learning, the systems will become even more adaptive and personalized, while ensuring data privacy.

AI will not only get better at evaluating mobility outcomes, but also predicting mobility outcomes, which will help personalize rehabilitation timelines even more (Tariq 2025). Furthermore, the combination of rehabilitation and real-time feedback on gait and balance outcomes may allow the use of VR and AR technologies during rehabilitation to gamify physical training and provide immersive experiences.

2.2. Clinical background: gait and balance in cancer survivors

Mobility limitations from cancer therapy develop due to intricate physiological, neurological and biomechanical interactions. For long or intensive treatment oncology patients, each treatment approach (chemo, surgery, radiation) adds specific problems to movement and functional independence.

Chemotherapy-induced peripheral neuropathy (CIPN) contributes to these problems, especially from drugs such as paclitaxel- and platinum-based compounds. These agents can injure sensory nerves and cause feet/hand numbness and tingling. The proprioceptive loss particularly affects feet, which has major challenges regarding balance, coordination and stable gait control, and walking and stair climbing become dangerous.

CID, as a consequence of cancer therapy, also impairs mobility. Systemic inflammation, cancer cachexia and prolonged bed rest cause significant loss of strength, muscle mass and endurance. Patients gain even less strength and cardiovascular fitness to perform basic activities, such as walking long distances, carrying groceries or chair rises. Cardiovascular fitness loss may even cause sedation. In severe CID cases, loss of strength will improve even basic activities. Functional movement may improve with surgical imbalance adjustments or reduction and reconstruction of weight-bearing structures.

For example, patients with sarcomas may undergo limb-sparing surgery which may lead to asymmetric loading during walking. Also, a mastectomy with axillary lymph node dissection may result in shoulder and upper body postural imbalance which may further alter the person's gait (Ribeiro et al. 2018). The compensatory movement strategies that patients adopt after surgery may become persistent and contribute to chronic pain and musculoskeletal disorder.

The effects of radiation therapy may further complicate the situation. Radiation-induced fibrosis may result in stiff tissues, contractures and impaired muscle elongation and contraction, which in turn may lead to a loss of the joint's active range of motion. These effects may be especially pronounced in the pelvis, neck and chest, which directly restrict stride length, posture and trunk rotation during movement. The combination of these effects is often greater than the sum of the individual effects on movement, which is especially true when multiple treatment modalities are utilized, as is common in multimodal oncology care (Bauder and Chien 2024).

Cancer-related fatigue related to cancer deteriorates the mobility issues. This constant exhaustion is not relieved by rest as is the case with normal fatigue. It decreases the motivation and tolerance of the patients toward physical exercises. Consequently, the speed of patients reduces, they move less and require more rest intervals. A combination of all of these effects forms the pattern of changed gait: reduced speed, reduced stride, increased variability of the step length and timing, decreased cadence. In most cases, these gait changes are accompanied with compromised posting and balance during standing and walking.

Due to this fact, risk of falls among cancer survivors is increased, which may lead to injuries, hospitalization and a lack of independence. The necessary steps to address mobility deficits include early detection, which should establish the direction of the treatment and to track the progress.

Clinical tools applied to this situation are the following:

– *Berg Balance Scale (BBS)*: a widely validated tool assessing static and dynamic balance through fourteen functional tasks. It is particularly effective in determining who is at risk of falls, but it can have ceiling or floor effects on some patients; it is either too easy or too hard.

– *Timed Up and Go (TUG)*: this is a tool that is easy to administer and quick and can assess mobility and dynamic balance; it involves timing the time it takes to stand up out of a chair, walking three meters, performing a turn, returning and sitting again. The tool is particularly useful in patients with frailty or who are old-aged and have cancer.

– *Ten-meter walk test*: this is a common gait speed measurement that predetermines functional deterioration, hospitalization and death among aging adults and even cancer patients.

Parameter	Conventional clinical tools	AI-enabled approaches	Advantages of AI integration
Assessment tools	Berg Balance Scale (BBS); Timed Up and Go (TUG); ten-meter walk test	Wearable IMUs; computer vision (pose estimation); smart floors, mobile-based systems	Enables objective, real-time and continuous assessment
Nature of data	Episodic and subjective	Continuous, quantitative and multi-dimensional	Captures subtle or early gait changes
Environment	Clinic-based	Clinic- and home-based (telehealth)	Remote monitoring and reduced hospital visits
Sensitivity to change	Limited, often detects only visible decline	High sensitivity to micro-deviations in gait or balance	Allows early intervention and personalized therapy
Patient engagement	Minimal	Interactive dashboards, gamified VR platforms	Improves adherence and motivation
Clinician feedback	Manual and periodic	Automated, AI-interpreted with explainable metrics	Supports evidence-based and adaptive decision-making
Cost and accessibility	Low-cost but limited insight	Initially higher setup cost scalable via mobile/consumer devices	Long-term cost-effective; enhances accessibility
Data utilization	Sparse clinical records	Big data analytics and predictive modeling	Supports prognostic insights and population-level research

Table 2.1. *Comparative overview of conventional and AI-based gait and balance assessment methods in oncology rehabilitation*

Clinical rehabilitation cannot be done without these tools; however, they are limited. Their primary restriction is that they are episodic. Evaluations are typically conducted at the time of clinic visits, and only a short period of the performance is measured, and in usually favorable circumstances. They are also manually administered and scored, which may be difficult in regard to patients who cannot travel or those with health complications. The tools can also fail to detect the minor changes in movement, and hence it can be hard to note the beginning of decline or minor, significant gains during treatment.

In order to close these gaps, we should have ongoing and technology-facilitated monitoring systems that provide objective detailed and real-time information regarding mobility. Cancer rehabilitation is embracing wearable sensors and AI analytics more to complement conventional tests and aid in the improvement of outcomes.

2.3. AI-enabled gait and balance assessment: overview of technologies

2.3.1. *Markerless computer vision/pose estimation*

Recent researchers on the topic of pose estimation in computer vision have achieved this through the use of CNNs. These techniques directly derive two- or three-dimensional skeletal landmarks using regular video. The videos might be the recordings of one camera or several cameras, and they might be recorded by some everyday gadgets such as smartphones, tablets or webcams. The systems monitor joint movements, position and gait parameters automatically. In comparison to other methods of traditional motion-capture systems that require costly cameras and reflective dots, markerless systems are non-invasive and cost-effective. Their measurements have been found to be equal to those of professional gait-analysis equipment, including pressure-sensitive walkways, in clinical studies. They demonstrate a significant inter-correlation in the identification of the gait speed, cadence, step time and stride length. Their portability and low costs allow them to be implemented in any hospital and clinic as well as in the remote or home environment. This means that patients can enjoy an increased and convenient monitoring of their mobility (e.g. cancer survivors).

2.3.2. *Wearable IMUs and smart sensing systems*

Wearable IMUs are made up of acceleration and gyroscopes. They are finding more extensive applications in gait analysis. These sensors are used on the trunk, thighs or ankles to measure a large set of temporals – spatial gait parameters: the step-time variability, trunk sway, stride asymmetry and double-support duration

(Rodrigues 2017). The biomechanical data can be further extended by adding force-sensitive smart insoles or instrumented floors. Gait patterns that are abnormal can be detected using machine learning algorithms that are trained using this data, rehabilitation progress can be evaluated and real-time feedback to implement corrective movement strategies can be provided. These systems are specifically useful in rehabilitation of oncology patients, in which small changes in gait due to neuropathy or fatigue before apparent functional impairment could be detected.

2.3.3. *Telehealth and mobile-based systems*

The increasing inclusion of gait analysis in telehealth systems and mobile health (mHealth) applications is a big breakthrough in the area of remote patient monitoring (Mazzà et al. 2021). Smart phones and tablets have internal sensors that detect acceleration and gyroscopes which can be used by the user to conduct short walking exercises at home. The information is uploaded to the cloud services or directly analyzed at the device level with AI algorithms identifying gait dysfunctions, identifying mobility patterns (e.g. antalgic, ataxic, shuffling) and their temporal changes. This makes the rehabilitation services more democratic, particularly to patients, especially in rural or underserved areas, and allows constant tracking of progress without the need to regularly visit the clinic.

2.3.4. *Learnable and privacy-protective AI*

Clinicians must have faith in AI-based systems to adopt them. Numerous new models thus use explainable AI (XAI) methods which indicate what gait characteristics, such as shortened arm swing or longer step time, were most relevant toward a diagnosis or recommendation (Özateş et al. 2024). Also, systems are being developed to run data on device (locally) to avoid sending sensitive video or sensor data over the air to other servers, thereby securing patient information. This is why balance of transparency, security, and performance is important in the ethical implementation of AI in healthcare.

2.3.5. *VR and rehabilitation robotics are transforming*

The gait training approach is run through providing immersive and adaptable environment. Robotic gait trainers use dynamic resistance, speed and support depending on the performance of the patient in real time. On the same note, VR platforms offer gamified balance and mobility activities which are both interesting and therapeutic. The AI constantly compares gait parameters and adjusts the complexity of tasks to achieve the maximum therapeutic intensity and encourage neuroplasticity, which improves the results in cancer and neurological rehabilitation (Reddy 2025b).

2.4. Evidence base and validation in the oncology setting

Specific oncology validation of AI gait analysis remains scarce; however, in other fields, such as neurology, geriatrics and general rehabilitation, there is evidence of clinical promise. In those aspects, AI analysis is as reliable as gold-standard laboratory equipment (motion capture, force plates), and it demonstrates a high level of test–retest reliability and sensitivity to small biomechanical alterations – it can often identify impairments before conventional tests. This early detection can be useful in cancer management, where gait problems related to the effects of nerve damage caused by chemotherapy, radiation fibrosis or post-surgery alterations are slow to manifest. AI technology is scalable, non-invasive and cost-effective, assisting in ongoing monitoring both in hospitals and at home. Individualized rehabilitation and adaptive planning of treatment can be implemented through the use of longitudinal data collection. Therefore, even though oncology-related research is becoming more prominent, cross-disciplinary results provide a strong basis regarding implementing AI gait analysis in cancer rehabilitation (Ratziu et al. 2024; Dhaouadi et al. 2025).

2.5. Core components of an AI-enabled oncology gait–balance platform

2.5.1. *Non-intrusive and accessible data capture*

Current AI-based mobility assessment systems are based on non-invasive, convenient methods of data collection, and hence can be relevant to clinical and home environments. Gait and balance analysis can be used cheaply and in large scale with motion captured by smartphones, tablets and webcams. Skeletal landmarks are read using computer vision algorithms that need no markers or special suits and enable patients to pass their tests at home. The IMUs, gyroscopes and acceleration sensors provide additional biomechanical data, including angles of joints, limb acceleration and force (Hurtado-Perez et al. 2025). These types of sensors are frequently implemented in smartwatches or wearables specific to a rehab.

Another potential new solution is smart floors with pressure sensors: they identify gait patterns, foot pressure, weight-shifting behaviors in the course of therapy, which gives them high-resolution information.

2.5.2. *AI models and analytics*

AI extracts complicated mobility statistics in the raw sensor or video data. The parameters of gait such as step length, cadence, walking speed, symmetry and variability of time are computed using machine learning models, which are trained over large datasets. Measures of balance, which include postural sway, center-of-mass motion, single-leg stance time and limits of stability are also measured. Such figures assist clinicians in identifying minor issues in patients receiving cancer care.

Recent AI applications include explainable analytics, which indicate the characteristics which contribute to each outcome or deviation (Li et al. 2023). This openness fosters trust and justification of clinical decisions as it comes out to explain why a patient can exhibit abnormal gait or balance.

2.5.3. *Dashboard and reporting interface*

The main element of AI-powered mobility systems is an interactive dashboard that displays up-to-date visualizations of important mobility indicators. The interface would compare the data of a patient against age and sex-related normative ranges and interpret the results in the form of color-coded indicators that can be easily interpreted. It further shows trend analysis, which allows the clinicians to monitor the progress of a patient over a given period of time and observe any indicators of decline at an early stage (Fong et al. 2024).

2.5.4. *Integration with rehabilitation prescription*

AI is now beginning to be applied in rehabilitation programs to inform interventions. Personalization is associated with creating customized exercise plans based on the profile of mobility of each patient and increasing or decreasing the difficulty (where necessary).

These adaptive plans keep track of the progress or backwardness. An example of this is that a patient who has just had radiation therapy, which caused fibrosis, should start with the exercises of low load and range of motion and then proceed with resistance training due to the improvement of mobility. The system will automatically re-assess and make rehabilitation lean with the functional objectives and course of recovery of the patient (Tamantini et al. 2025).

2.6. Implementation in oncology settings: use cases and workflow

2.6.1. *Baseline and ongoing assessment*

Every oncology patient gets a set distance ambulation, turning and standing sequence as part of their balance and walking rehabilitative assessment. These distances and times are captured through AI-wearable sensors, smart floor tiles and even smartphone cameras. AI watches over cadence, shift length, step symmetry, sway and turn times and keeps a balance of postural sway (Lim et al. 2022). This serves as a complete functional baseline profile over gait and balance. AI watches over baseline for every future visit and home assessment for distance and time. A set of measurements will reveal patterns of recovery, decline or new complications. With timely, pointed assessments, these systems justify evidence-required adjustments in rehabilitative interventions.

2.6.2. *Personalized intervention and progress monitoring*

After recognizing the specific issues someone might be facing, AI algorithms will suggest personalized recovery rehabilitation plans specific to the patients' gait, balance or other issues due to the condition or inactivity. For example, if someone has an issue with taking long steps, they might be suggested to practice cadence training to encourage quicker paces. Similarly, if a patient has issues turning, suggesting dynamic balance exercises will be of help. These interventions can be suggested through digital platforms and can be programmed adaptively to encourage improvement. Patients will need to follow real-time cueing, through any of the three possible styles, to correct errors of movement. To ensure that we are able to get the most out of rehabilitation, we keep on improving the effectiveness and efficiency of the plan by conducting constant evaluations (Langote et al. 2024).

2.6.3. *Fall risk stratification and safety monitoring*

Due to the presence of neuropathy, fatigue and deconditioning, cancer survivors are very vulnerable to falls. The fall risk stratification by using AI is done by considering the gait variability, turning instability and single-leg performance in the stance, which are usually overlooked during a standard clinical examination. High-risk patients are those getting automated messages to be evaluated further, balance training, assistive devices or environmental solutions, such as handrails or non-slip flooring (Kumar et al. 2026a). This constant observation will easily identify any acute loss of mobility, which are usually the antecedents to falls, hence improving safety in the clinical and home facilities.

2.6.4. *Remote telerehabilitation*

To reduce the necessity to visit the clinic frequently, AI-based gait and balance monitoring could be performed at home using domestic equipment such as a smartphone or tablet. Guided tasks are performed by the patients as the system records and analyzes motion data. Results are then abandoned to secure platforms to be accessed by clinicians to supervise and monitor progress remotely (Kumar et al. 2026b). The model of telerehabilitation will increase access (particularly in rural or underserved areas) and promote compliance (merging rehabilitation into daily life).

2.7. Challenges and considerations

2.7.1. *Data quality and clinical validation*

Gait and mobility analytics based on AI is effective in rehabilitation of cancer patients, and it requires high-quality and domain-specific data. A majority of the existing AI models are trained on neurology, geriatrics or generic rehabilitation samples. There is a lack of datasets representing the vast variety of types of cancers, treatment phases and functional impairments that oncology patients may have. The therapies of cancer may alter the gait mechanics in a manner that generic data fails to capture such as chemotherapy-induced neuropathy, post-operation compensation or radiation-induced fibrosis. Unless oncology cohorts are carefully clinically tested, models might be inaccurate or irrelevant. Thus, multi-centered studies, data association with electronic health records and prospective trials are necessary to enhance the diversity of the data and demonstrate clinical utility in cancer centers (Kumar and Joghee 2025).

2.7.2. *Bias and generalizability*

The quality of AI systems depends on the quality of data they are trained on. Unless the training set is diverse in terms of representation between genders, age, ethnicity and subtype of cancer and socioeconomic status, the models derived are subject to systematic bias. This bias may lead to incorrect prognoses, misplug risk or advising unequally, in particular, to underserved groups.

It is also important to ensure generalizability in order to provide fair care (Raza et al. 2024). Therefore, developers of the models need to focus on the variety of the datasets, apply fairness-conscious algorithms and constantly track the performance of groups of people. Having a well-established reporting of dataset composition and validation results creates a sense of trust with the clinicians and patients.

2.7.3. *Regulatory and integration hurdles*

The challenge of commercializing AI tools in the medical field involves hurdles entailed in local regulatory environments. The AI algorithms to be implemented in the assistance of clinical decisions should meet the requirements of safety, efficacy and explainability, which are established by the regulators, including the FDA, EMA or CDSCO. Moreover, they have to integrate easily within hospital IT systems – such as electronic health record (EHR) systems, rehabilitation systems and remote-monitoring solutions. This fit requires interoperability, a high level of cybersecurity and interfaces that are easy to use. Training of clinicians and adapting a workflow is necessary as well, in order to make sure the tools are adopted and successfully utilized. Lastly, the speed at which it will be adopted will depend on institutional preparedness, facilitating policies and reimbursement systems (Jones et al. 2021).

2.7.4. *Patient engagement and equity*

Engagement of patients is the key to the success of AI-enhanced rehabilitation tools. However, low-digital literacy and lack of Internet accessibility or cognitive difficulties might make such technologies difficult to use among people. The design solutions should be accommodative with multilingual interfaces, easy-to-understand guidelines and cheap hardware. Additional usability and trust can be achieved by engaging patients and caregivers in the design process. One of the needs to prevent the appearance of digital health disparities and also provide general access to these innovations in oncology care is equity-focused strategies, including community outreach, and subsidies.

2.8. Future directions and research needs

Further studies on AI-based physiotherapy in cancer patients should aim at achieving two primary objectives in order to increase clinical effectiveness and international reach. First, there is the necessity to conduct randomized controlled trials (RCTs) specially designed on oncology populations. The initial research in the general rehabilitation field, as well as neurology, has demonstrated that AI may enhance the assessment and treatment, though the findings cannot be extrapolated to cancer patients unless specific experiments are conducted.

AI algorithms will be tested against conventional treatments in cancer-specific RCTs, which will provide evidence of their effectiveness in various tumor types and different stages, as well as the outcomes in the long term, including preventing falls, prolonging life and providing functional independence (Lippi et al. 2024). They will also uncover the most benefiting groups of patients that can be assisted with AI-assisted rehab and provide the data needed to allow the regulating of the proposed approaches and using it on a regular basis.

Second, researchers ought to combine AI with the wearable robotics, as well as the virtual reality (VR) technology. Robotic-assistive devices such as exoskeletons and soft robotic exoskeletons assist patients in relearning to walk and build muscle strength, particularly after chemotherapy or surgery. When combined with AI, these devices can make changes in interventions on the fly, depending on the performance of a patient (Shen et al. 2025). Another level of engagement is the immersive VR that simulates daily activities that involve cognition and motion. AI can study the movement of patients in such VR environments to modify the difficulty level and guarantee the high level of therapy administration (Thomas and Yadav 2024). Combined, these hybrid platforms can make adherence and recovery faster and more motivating through gamification.

Lastly, there is a pressing need to develop explainable, low-cost AI tools to be applied in low-resource environments. Rural or under-served cancer patients do not have sophisticated physiotherapy or motion-analysis centers (Alharthi 2024). The high-quality care can be provided to these communities by introducing lightweight, smartphone-based AI applications that utilize single-camera video or low-cost sensors. Their trust will increase and adoption will grow by making the recommendations of the AI viewable, i.e. having clear and understandable metrics by both the clinicians and the patients (Patel et al. 2024).

2.9. Conclusion

Gait and balance assessment instruments are AI-based and transforming the oncological rehabilitation process by providing constant, objective and interpretive assessments of movement. The common gait and balance clinical tests that are usually subjective and rarely done could overlook minor but clinically significant alterations in cancer patients. Conversely, biomechanical parameters can be studied in real time with AI systems, which can use wearable sensors, automated video analysis or smart-home technology, and thus provide continuous monitoring. They identify early symptoms of motor weakness, neuromuscular issues associated with treatment and risk of falls that enable clinicians to prevent functional mobility by customizing rehabilitation programs to enhance outcomes. But there are a number of obstacles that have to be hurdled to clinical adoption. AI can be implemented in oncology practice, and it should be validated to guarantee reliability and accuracy regardless of the type of cancer, rehabilitation stages and baseline functional status. The development of algorithms must be directed by ethical standards across all aspects of data collection and patient privacy and bias that may have a disproportionate impact on marginalized populations. The systems must have smooth integration with the ongoing work processes to reduce friction. This involves easy-to-use clinician interfaces, smooth connectivity with EHRs and sufficient training to make inferences out of AI. Properly designed and implemented AI-gait and balance technologies have the potential to increase the accuracy and inclusivity of rehabilitation processes in cancer care and offer scalable solutions to support survivorship care in the healthcare system.

2.10. References

Alharthi, A.S. (2024). Interpretable machine learning comprehensive human gait deterioration analysis. *Frontiers in Neuroinformatics*, 18, 1451529.

Bauder, N. and Chien, L.C. (2024). Enhancing mobility in oncology: Evidence-based practices across the care continuum. In *Seminars in Oncology Nursing*. WB Saunders, Philadelphia, PA.

Boltaboyeva, A., Baigarayeva, Z., Imanbek, B., Ozhikenov, K., Getahun, A.J., Aidarova, T., Karymsakova, N. (2025). A review of innovative medical rehabilitation systems with scalable AI-assisted platforms for sensor-based recovery monitoring. *Applied Sciences*, 15(12), 6840.

Dhaouadi, S., Khelifa, M.M., Balti, A., Duché, P. (2025). Optical sensor-based approaches in obesity detection: A literature review of gait analysis, pose estimation, and human voxel modeling. *Sensors*, 25(15), 4612.

Fong, S., Pabis, K., Latumalea, D., Dugersuren, N., Unfried, M., Tolwinski, N., Kennedy, B., Gruber, J. (2024). Principal component-based clinical aging clocks identify signatures of healthy aging and targets for clinical intervention. *Nature Aging*, 4(8), 1137–1152.

Hurtado-Perez, A.E., Toledano-Ayala, M., Cruz-Albarran, I.A., Lopez-Zúñiga, A., Moreno-Perez, J.A., Álvarez-López, A., Rodriguez-Resendiz, J., Perez-Ramirez, C.A. (2025). Use of technologies for the acquisition and processing strategies for motion data analysis. *Biomimetics*, 10(5), 339.

Jones, L.K., Ladd, I.G., Gregor, C., Evans, M.A., Graham, J., Gionfriddo, M.R. (2021). Evaluating implementation outcomes (acceptability, adoption, and feasibility) of two initiatives to improve the medication prior authorization process. *BMC Health Services Research*, 21(1), 1259.

Kumar, R.M. and Joghee, S. (2025). A review on integrating breast cancer clinical data: A unified platform perspective. *Current Treatment Options in Oncology*, 26(1), 1–3.

Kumar, A., Batta, P., Ahuja, S., Rathore, P.S. (eds) (2026a). *AI-driven Innovations in Physiotherapy and Oncology 1*. ISTE Ltd, London and Wiley, Hoboken, NJ.

Kumar, A., Batta, P., Ahuja, S., Rathore, P.S. (eds) (2026b). *AI-driven Innovations in Physiotherapy and Oncology 2*. ISTE Ltd, London and Wiley, Hoboken, NJ.

Langote, M., Saratkar, S., Kumar, P., Verma, P., Puri, C., Gundewar, S., Gourshettiwar, P. (2024). Human–computer interaction in healthcare: Comprehensive review. *AIMS Bioengineering*, 11(3), 343–390.

Li, Z., Zhu, Y., Van Leeuwen, M. (2023). A survey on explainable anomaly detection. *ACM Transactions on Knowledge Discovery from Data*, 18(1), 1–54.

Lim, A.C., Natarajan, P., Fonseka, R.D., Maharaj, M., Mobbs, R.J. (2022). The application of artificial intelligence and custom algorithms with inertial wearable devices for gait analysis and detection of gait-altering pathologies in adults: A scoping review of literature. *Digital Health*, 8, 20552076221074128.

Lippi, L., de Sire, A., Folli, A., Turco, A., Moalli, S., Marcasciano, M., Ammendolia, A., Invernizzi, M. (2024). Obesity and cancer rehabilitation for functional recovery and quality of life in breast cancer survivors: A comprehensive review. *Cancers*, 16(3), 521.

Mazzà, C., Alcock, L., Aminian, K., Becker, C., Bertuletti, S., Bonci, T., Brown, P., Brozgol, M., Buckley, E., Carsin, A.E. et al. (2021). Technical validation of real-world monitoring of gait: A multicentric observational study. *BMJ Open*, 11(12), e050785.

Michalova, Z., Szekiova, E., Blasko, J., Vanicky, I. (2023). Prevention and therapy of chemotherapy-induced peripheral neuropathy: A review of recent findings. *Neoplasma*, 70(1), 15–35.

Özateş, M.E., Yaman, A., Salami, F., Campos, S., Wolf, S.I., Schneider, U. (2024). Identification and interpretation of gait analysis features and foot conditions by explainable AI. *Scientific Reports*, 14(1), 5998.

Patel, D.J., Sharma, P., Gudur, R., Banerjee, D., Singh, J. (2024). Prognostic study to detect ailments, including cancer, during solid organ transplantation to prevent postoperative complications using perioperative antibiotics therapy. *Onkologia i Radioterapia*, 18(9), 1–7.

Rattanakrong, N. (2021). Incidence of peripheral neuropathy, postural stability alteration, and factors associated with fall in patients with breast cancer during taxane-based chemotherapy. PhD Thesis, Chulalongkorn University.

Ratziu, V., Hompesch, M., Petitjean, M., Serdjebi, C., Iyer, J.S., Parwani, A.V., Tai, D., Bugianesi, E., Cusi, K., Friedman, S.L. et al. (2024). Artificial intelligence-assisted digital pathology for non-alcoholic steatohepatitis: Current status and future directions. *Journal of Hepatology*, 80(2), 335–351.

Raza, S., Shaban-Nejad, A., Dolatabadi, E., Mamiya, H. (2024). Exploring bias and prediction metrics to characterise the fairness of machine learning for equity-centered public health decision-making: A narrative review. *IEEE Access*, 12(99). doi: 10.1109/ACCESS.2024.3509353.

Reddy, K.J. (ed.) (2025a). Patient perspectives and outcomes. In *Innovations in Neurocognitive Rehabilitation: Harnessing Technology for Effective Therapy*. Springer, Cham, 111–135.

Reddy, K.J. (ed.) (2025b). Technological innovations in rehabilitation: Artificial intelligence. In *Innovations in Neurocognitive Rehabilitation: Harnessing Technology for Effective Therapy*. Springer, Cham, 73–91.

Ribeiro, A.P., Maniaes, T., Naomi Hamamoto, A. (2018). The effects of mastectomy and breast reconstruction on body posture and biomechanical aspects. In *Women's Health and Biomechanics: Where Medicine and Engineering Meet*, Brandão, S., Da Roza, T., Ramos, I., Mascarenhas, T. (eds). Springer, Cham, 57–69.

Rodrigues, P.L. (2017). Gait analysis and rehabilitation using inertial sensors. Master's Thesis, Universidade do Porto, Portugal.

Shen, Y., Jiang, L., Lai, J., Hu, J., Liang, F., Zhang, X., Ma, F. (2025). A comprehensive review of rehabilitation approaches for traumatic brain injury: Efficacy and outcomes. *Frontiers in Neurology*, 16, 1608645.

Su, K., Lin, X., Liu, Z., Tian, Y., Peng, Z., Meng, B. (2023). Wearable triboelectric nanogenerator with ground-coupled electrode for biomechanical energy harvesting and sensing. *Biosensors*, 13(5), 548.

Suchy, Y. (2015). *Executive Functioning: A Comprehensive Guide For Clinical Practice*. Oxford University Press, Oxford.

Sun, R., Lin, Z., Leng, S., Wang, A., Zhao, L. (2025). An in-depth analysis of 2D and 3D pose estimation techniques in deep learning: Methodologies and advances. *Electronics*, 14(7), 1307.

Tamantini, C., Umbrico, A., Orlandini, A. (2025). Automated planning and scheduling in robot-aided rehabilitation: A review. *Journal of NeuroEngineering and Rehabilitation*, 22(1), 180.

Tariq, M.U. (2025). Machine learning techniques for personalized rehabilitation: Enhancing patient outcomes through predictive algorithms. In *Predictive Algorithms for Rehabilitation and Assistive Systems*, Korupalli, V.R.K., Ranasinghe, P., Feldman, D. (eds). IGI Global, Hershey, PA, 221–252.

Thomas, M.S. and Yadav, T. (2024). Prevalence of upper cross syndrome in multipara women. *Journal of Ecophysiology and Occupational Health*, 24(1), 67–71.

3

Deep Learning-Driven Fatigue Monitoring in Cancer Physiotherapy Programs

Most people receiving or recovering from cancer treatment experience cancer-related fatigue (CRF), which is pervasive and causes frailty. It impacts quality of life and rehabilitation, and can persist for long periods of time, even after treatment. While CRF is something physiotherapy can help with, fatigue monitoring is still primarily based on self-reporting which is subjective and is of less value across time. In this chapter, we address how new deep learning (DL) technology can help with monitoring fatigue in real time or on an ongoing basis during cancer physiotherapy. We start with an overview of CRF in clinical oncology and discuss classical models of fatigue assessments. After this, we discuss CRF clinical monitoring through recent CRF monitoring technology and the use of wearable sensors that measure and collect health data on heart rate and variability, actigraphy and even thermal data. We discuss the use of DL, particularly convolutional neural networks (CNN), long short-term memory (LSTM) and other hybrid patterns to analyze fatigue over time and predict increases in CRF with great accuracy. DL models can detect fatigue in real time, predict future symptoms and adjust therapy in the present. Limited real-world use, heterogeneous groups and small sample sizes present considerable challenges to the large-scale use of these models. There are also issues of data integration, clinical use and predictive accuracy.

3.1. Introduction

Fatigue can be highly distressing for individuals or survivors of malignancies as they are undergoing or recovering from expensive and difficult treatments for their condition. For people affected by cancer, fatigue involves a consistent and overwhelming sense of tiredness in the body or mind or a combination of the two that is disproportionate to the degree of exertion; and in no way does restful sleep or prolonged periods of inactivity ease the condition. Fatigue, therefore, can be a major

Chapter written by Anand GUDUR and Faisal Hussain HUSSAIN.

barrier to an individual's performance of daily functioning, making it difficult to follow a prescribed rehabilitation program, and detracting from quality of life (Gruet 2018). It has been reported that more than 70% of cancer patients getting chemotherapy, radiotherapy or immunotherapy complain of moderate to severe fatigue.

Fatigue can affect cancer survivors long after their treatment has been completed, and in some cases, for an extended period of time. CRF has root causes in the cancer itself, its treatment, psychological stress and physical inactivity. As far as CRF is concerned, physiotherapy and structured exercises have the most evidence and are accepted as the best ways to alleviate the condition (Pedersen and Saltin 2015). While personalized physical activities are beneficial in improving cardiorespiratory fitness, strength and energy, their benefit can be lost if the activities are not matched to the person's condition and needs. Therefore, physiotherapy needs constant fine-tuning, and monitoring fatigue is paramount.

In the past, fatigue evaluation focused on the patient's point of view, using tools such as the Functional Assessment of Chronic Illness Therapy–Fatigue (FACIT-F), the Brief Fatigue Inventory (BFI) and visual analog scales. During physiotherapy sessions, fatigue is assessed using clinician-based tools. Although such tools capture personal feedback, their potential is limited because they are episodic, subject to recall bias, and heterogeneity of interpretation. Fast-paced clinical environments have a shortage of real-time, objective fatigue insights. This gap, however, is starting to be filled by the use of wearable devices enhanced by deep learning (DL) techniques. Patterns of movements, heart rate and changes in skin temperature, plus sleep–wake cycles capture fatigue proxies such as energy expenditure, physical exertion, stress and recovery, thus defeating stress proxies. DL is the best technique to use in the analysis of these signals due to the broad, intricate and complex-structured time-series data processed. While traditional machine learning applies handcrafted techniques, DL and neural networks derive the intrinsic patterns needed for analysis directly from the data (Dargan et al. 2020).

Spatial feature extraction from multivariate sensor data, such as thermal images and movement trajectories, is another domain where CNNs excel. RNNs, particularly LSTM and GRU, are geared toward sequential data. They are well suited for capturing time-related shifts in fatigue symptomatology. The application of DL to estimate fatigue in cancer and similar context populations is certainly plausible, as noted in the recent literature. In real time, fatigue levels have been predicted with high accuracy using LSTM architectures applied to streams of accelerometry and heart rate data. Performance in recognizing fatigue escalation has improved with hybrid architectures that integrate CNNs for feature extraction and LSTMs for time-series analysis. There are also studies on fatigue estimation that leverage multimodal sensor fusion, using diverse wearable technologies for enhancing robustness and context sensitivity (Li et al. 2023).

While computing in the cloud, privacy and security of sensitive health data must be addressed, particularly for unsupervised systems. Future studies should focus on building personalized models of fatigue that perform on adaptive baselines and track progression. Unsupervised and semi-supervised learning techniques can use unlabeled data; therefore, the need for extensive manual data labeling is lessened. Features such as medications within a certain time frame, sleep, emotions and cancer stage also improve the model. Finally, well-designed feedback systems such as visual dashboards and mobile alert applications will foster patient self-awareness and help clinicians to adjust therapy in real time (Hsu et al. 2025).

3.2. Cancer-related fatigue and physiotherapy

CRF is one of the most common and distressing symptoms someone with cancer has to go through. CRF is reported to the National Comprehensive Cancer Network (NCCN) as "a distressing, persistent, subjective sense of tiredness or exhaustion related to cancer or with cancer treatment that is not proportionate to recent activity and interferes with usual functioning". CRF is different from any other kind of fatigue, especially from those experienced by healthy individuals, and this definition shows the profound and debilitating complexity fatigue has. CRF can develop from the time the person is diagnosed, during treatment, or even in survivorship and can continue for several months or years after treatment has ended (Bower 2014).

The cancer and the side effects (or mental stress) caused by treatment with chemotherapy, radiotherapy and surgery, emotional and sleep disorders, nutritional deficiencies and other chronic inflammatory diseases are some of the reasons for developing CRF. Focus on CRF has shifted toward rehabilitation. Research shows that physiotherapeutic interventions and other forms of exercise are successful in the treatment of fatigue. Clinical guidelines and studies highlight the importance of physical activity in the management of fatigue at radiation as the primary means of non-pharmacological intervention (Hilfiker et al. 2018). Cancer survivors in early Cochrane reviews reported minimal fatigue after following structurally prescribed exercise programs.

These conclusions have been supported by subsequent studies of randomized control trials which not only report decreases in fatigue but also improvements in the overall quality of life, and physical and emotional well-being. Fatigue physiotherapy interventions are typically multimodal and personalized. Programs often include combinations of a range of motion exercises which include joint flexibility maintenance, strength training to counteract muscle atrophy, balance exercises to improve stability and fall prevention, and endurance exercises which improve cardiovascular fitness. The central philosophy focuses on progressively rebuilding resilience and functional independence but avoiding overexertion and the potential exacerbation of fatigue. A major difficulty in the fatigue management physiotherapy is striking a balance of exercise dose and the progression for the individual (Kumar et al. 2026a).

The exercise dose and progression must be individualized considering the individual basal fitness, treatment, comorbidity and fatigue level. In this situation, the management of fatigue in the immediate context is crucial. The combination of subjective fatigue score and objective monitoring helps to ensure that patients do not exceed the prescribed limits (Kumar et al. 2026b). The use of real-time exercise feedback enables dynamic monitoring and adjustment of activity levels to effectively manage fatigue and optimize exercise intensity, ensuring safe, personalized, and adaptive rehabilitation outcomes. Adjusting the exercise helps encourage safe participation while maximizing therapy outcomes and gaining voluntary adherence to the physiotherapy program.

3.3. Traditional and sensor-based monitoring

Fatigue in oncology has historically been evaluated using patient-reported outcome measures (PROMs), including the Functional Assessment of Chronic Illness Therapy-Fatigue (FACIT-Fatigue), Brief Fatigue Inventory (BFI), or numeric rating scales (NRS) (Fan et al. 2022). Although these measures are validated and used widely, they contain substantial and significant weaknesses. PROMs typically capture momentary and static assessments and are susceptible to subjective phenomena, including recall bias, emotional state variations and fatigue severity distortions. These weaknesses affect the sensitivity and reliability of assessments, particularly when evaluating fatigue over long periods, such as during cancer treatment and rehabilitation.

Now, there is interest in the use of objective measures. With the advent of wearables, including smartwatches, fitness trackers and biosensors, it is possible to collect and evaluate physiological and behavioral data in real time. These devices contain accelerometers, which measure gross and fine motor activity and classify activity into intensity zones (steps counting). Monitors record heart rates and signify exertional load and recovery. Actigraphy devices record sleep and characterize circadian rhythms, restlessness and patterns associated with fatigue. One compelling example is a pilot cohort study with breast cancer patients receiving radiotherapy (Ancoli-Israel et al. 2003).

Participants used commercial fitness trackers which captured their heart rate and step counts during their treatment. Incorporating a machine learning strategy by using repeated activity window (RAW) metrics which are temporal segments of data that reflect changes in activity and physiology over time, the researchers were able to describe patterns of fatigue due to radiation therapy. Approximately half the patients evaluated reported and were clinically recognized to have some level of fatigue. Most significantly, ML algorithms created to analyze the heart rate and step data accurately classified periods of fatigue and non-fatigue. This indicates the potential of constant passive monitoring to be a fatigue assessment to replace or minimize the need for an active subjective self-assessment (Price et al. 2019). Apart from fatigue studies, other works have also shown that the integration of data from wearables and other clinical data using DL models in oncology has substantive potential. For example, a study on survival prediction in palliative patients integrated 48 hours of wearable sensor data and clinical variables which included diagnosis, performance and lab results, alongside data capturing physical activity, sleep and movement.

Parameter	Traditional monitoring (self-report tools)	Sensor-based monitoring	Deep learning-integrated monitoring
Data source	Patient-reported outcomes (e.g. FACIT-F, BFI and VAS)	Physiological signals (HR, actigraphy and thermal data)	Multimodal sensor data (HRV, motion, temperature and skin conductance)
Measurement frequency	Episodic (at therapy visits)	Continuous (real-time capture)	Continuous with adaptive prediction windows
Objectivity	Subjective; prone to recall bias	Objective but limited by sensor accuracy	Highly objective and adaptive learning from data
Analytical approach	Statistical analysis and manual scoring	Descriptive/threshold-based analytics	CNN, LSTM and hybrid DL architectures for spatiotemporal analysis
Outputs	Fatigue severity scores	Physiological fatigue indicators	Real-time fatigue prediction, alerts and therapy adjustment
Strengths	Easy to administer and low-cost	Real-time insights; minimal recall bias	Predictive modeling, personalization and clinical decision support
Limitations	Subjective; inconsistent over time	Requires device adherence and calibration	Requires large datasets, ethical and privacy concerns

Table 3.1. *Comparative summary of traditional and deep learning-based fatigue monitoring methods in cancer physiotherapy*

Models that integrate multiple types of data from DL have achieved more than 92% accuracy in forecasting short-term survival outcomes, offering a stark improvement over classical prognostic assessments such as the Palliative Prognostic Index (PPI) or the Karnofsky Performance Status (KPS) scores. Although it did not focus specifically on fatigue, the study illustrates the growing possibilities of wearable devices coupled with DL analytics. For cancer physiotherapy, there are many potential benefits (Hoang 2025). Table 3.1 enlists a comparative summary of traditional and deep learning-based fatigue monitoring methods in cancer physiotherapy.

3.4. Deep learning models for fatigue monitoring

3.4.1. *Forecasting symptom escalation*

Being able to forecast the escalation of symptoms for cancer patients undergoing physiotherapy or treatment is vital, as patients go through a period of active treatment. Finkelstein et al. (2024) were the first to study symptom escalation through analyzing self-reported symptom logs of chemotherapy patients, noting symptoms such as fatigue, pain, nausea and emotional distress. For the study, the authors used DL models such as convolutional neural networks (CNNs), long short-term memory (LSTM) and gated recurrent units (GRU) on the logs. Surprisingly, CNNs performed well (79% accuracy) at predicting time-oriented reports of fatigue and pain, as they are used to time-series data common in image processing and can detect small patterns in the data. On the other hand, GRUs performed well (77% accuracy) on the reports of emotional and psychological symptoms, as they are designed to better capture temporal patterns in data than standard RNNs. The results from the various models highlight the multiple ways DL can be tailored to different symptom domains. For the first time, patient data forecasting symptom trajectories demonstrated the active management and control of fatigue episodes.

3.4.2. *Sensor-based and multimodal fatigue*

Estimation research has moved to collecting data through physiological sensors and has been implementing continuous measures of fatigue. Compared to self-reporting, which potentially has bias in recall or underreporting, these methods are more reliable and provide real-time data. One study applied thermal facial imaging and CNNs to measure fatigue induced by exercise. The thermal imaging model identifies and correlates fatigue markers around the eyes, forehead and cheeks, which have varying vascular and metabolic activities. Under a 15% error rate indicates high accuracy when estimating fatigue using the imaging technique. The integration of several physiological markers to identify varying types of fatigue is also pioneering new intersections in the field. One research group used a multimodal sensor and actuator array to gather data in a wearable device, which incorporated ECG and actigraphy data, and then applied LSTM models (Jeong et al. 2022).

3.4.3. *Assessing rehabilitation exercise*

Ensuring quality in rehabilitation exercise is critical, particularly with cancer treatment-induced fatigue. In the new wave of AI tools, particularly DL, it has become possible to automate the evaluation of exercise performance as it relates to rehabilitation movement capture and wearable sensor data. In one example, the use of autoencoder-recurrent neural networks (RNN) compresses and reconstructs a sequence of movement data to assess lapses in execution. They are able to pinpoint where performance and pattern of exercise deviate and provide fatigue feedback (Dorschky et al. 2023). These systems are primarily in real-time movement scoring, which has useful applications in remote rehabilitation. While it is true that the systems were not originally designed for oncology patients, the technology is more than appropriate for patients who are more commonly cancer survivors with chronic fatigue and require modified, and more frequent, supervision in the rehabilitation exercise.

A machine learning program specifically designed to recognize fatigue could assist healthcare professionals in modifying exercise plans to better suit their needs by identifying suboptimal movement patterns that may suggest overexertion.

3.5. Integration into cancer physiotherapy

Multimodal data integration is important for the implementation of DL within the physiotherapy component of the CRF programs. Continuous data streams from wearables need to be integrated, in addition to heart rate monitors, along with accelerometers that collect kinetic and kinematic data, patient-reported symptom scales, and clinical data, which consists of treatment timelines, comorbidities, medications, etc. (Cuomo et al. 2024).

These elements, in concordance, assess a patient's fatigue biomedically and subjectively. However, before a DL model can be trained, extensive data preprocessing is required. This involves the creation of uniform intervals of time which will need to be defined and structured within the parameters of temporal resampling, temporal data windows for consecutive segments of days to weeks, and filling in the gaps to sustain continuity and reliability for the entire dataset.

Several DL model architectures have been developed to predict fatigue. For instance, fatigue-predicting CNN models analyze spatial constructs while gait symmetry and joint kinematics data are integrated.

The RNN family, including LSTM and GRU frameworks, handles the heart rate variability or symptom trajectory data, which are temporal sequences. Hybrid models that integrate CNN and RNN are created in attempts to negate the individual short-comings of DL networks and to capture spatiotemporal features.

To help maintain the temporal resolution while expanding the training dataset, the training protocols utilize sliding window techniques. Balanced class representation with k-fold cross-validation mitigates overfitting and generalizability concerns (Zhang et al. 2025). Predictive accuracy and clinical relevance of the predictions are enhanced with techniques such as dynamic thresholding and patient-specific baselines, which are more personalized.

3.6. Applications and use cases

3.6.1. *Real-time fatigue monitoring*

Real-time monitoring of CRF with DL systems shows great promise. The systems use data from wearables – heart rate monitors, accelerometers, gyroscopes and skin temperature sensors. DL systems track physiologic monitoring of exertion and recovery. This means that during therapy sessions, fatigue can be modified dynamically (Pinto-Bernal et al. 2021).

CNNs analyze spatial patterns of movements, while RNNs, particularly LSTM architectures, analyze temporal patterns of fatigue data. DL systems can track CRF in real time. For instance, during resistance training in a physiotherapy session, the real-time system monitors for signs of fatigue and alerts the clinicians of exercising the CRF patient and suggests modifications to the training routine to include rest and recovery. Overexertion in the CRF patients can trigger complications and recovery delays. DL systems in optimized patient monitoring promotes safety and positive patient outcomes.

3.6.2. *Predictive warnings*

Aside from real-time monitoring, fatigue modeling using DL algorithms holds great potential. Rather than responding to fatigue levels in real time, DL algorithms can work with longitudinal datasets to look at historical sensor data, activity data, treatment plans, sleep data and medication data, and recognize patterns leading to critical fatigue episodes. Predictive modeling can issue alerts hours (or even days) in advance of an expected fatigue peak. This predictive modeling allows physiotherapists and oncologists to deploy strategies that help mitigate a peak.

For example, if a model predicts the likelihood of high fatigue on a certain day, clinicians can lower the exercise intensity, alter session times, shift medication and add nutritional elements. Patients can also be told to perform lighter, easy tasks and avoid demanding, active tasks to help cope with the fatigue. This is especially the case with chemotherapy and radiotherapy where fatigue patterns are predictable. Non-cyclic fatigue patterns associated with treatments are also predictable and are especially difficult to manage (Vercellini et al. 2024). DL can be used to track and learn patterns specific to an individual, identify early signs of an escalation and send notifications through a mobile app. Integration with an electronic health record (EHR) system to the mobile app is a great way to improve cross-team collaboration.

3.6.3. *Personalized therapy*

DL systems excel in personalization, and this becomes particularly beneficial in the case of CRF and its highly individual variation in symptom expression. While some patients may respond with considerable fatigue to slight exertion, others may be functionally present and active, even when high physiological strain is present. Conventional monitoring tools use general thresholds and thus tend to be unhelpful in accommodating individual variability. In contrast, DL models learn and adjust to each individual's fatigue pattern by drawing on archived personal baseline data over time (Adão Martins et al. 2021). The personalized aspect for this case is the adjustment of the DL system to understanding "abnormal" and "normal" fatigue for a specific individual with CRF.

For example, if a patient tends to tolerate elevated heart rates to a specific level and then reports no fatigue, the system learns to not flag those as fatigue. In contrast, if a patient has a highly limiting fatigue when there is a slight increase in skin conductance or a reduction in steps, then the model will adjust its sensitivity to limit normal ranges of fatigue for those even minor increments. Additionally, personalization in DL systems enhances the optimization of CRF-focused physiotherapy. DL systems can adjust to fatigue tolerance when recommending individual CRF training loads, rest periods and session lengths. CRF goals can be individualized to patient objectives for optimal motivation (Balkin et al. 2011).

3.7. Strengths, limitations and challenges

3.7.1. *Strengths of DL-driven fatigue monitoring systems*

DL models, especially with wearable devices, have considerable advantages when it comes to fatigue monitoring in cancer rehabilitation. Most importantly, there is continuous, objective monitoring. Fatigue assessment aids such as patient-reported outcome measures (PROMs) are subjective, and there are time lags between when symptoms are experienced and when they are reported; recall bias is also present (Solomon and Manea 2022).

However, DL models trained on streams of real-time sensors capturing data on heart rate, accelerometry and other parameters can monitor fatigue states without patients actively reporting. This kind of passive monitoring is greatly appreciated and invaluable to frontline clinicians, while burdening patients less. Another benefit is the alerts that DL models can issue right before fatigue is about to set in. By learning fatigue patterns, DL models can detect signs of a fatigue episode that a patient is unaware of, and probably the last thing the patient needs is a physiotherapy session that will exacerbate their symptoms.

When a DL model indicates a high likelihood of fatigue within the next 48 hours, clinicians can take steps, such as decreasing physical workload or prescribing longer rest intervals, which can help improve adherence and retention with the rehabilitation program. Another plus is the capacity for integration with remote care and telehealth. As rehabilitation for oncology services grows to incorporate hybrid and home-based care, the DL-monitored systems with fatigue detection wearables are becoming essential to delivering uninterrupted high-quality care (Pourpanah and Etemad 2024). These systems offer clinicians the convenience of remote consultations with automatic fatigue trend longitudinal dashboards, alerts and evidence-based fatigue trend documentation. The self-monitoring fatigue system also encourages engagement.

3.7.2. *Limitations and gaps in the current research*

While there is no denying the great potential DL-automated systems that monitor fatigue can offer, there are research-led constraints that disallow it from being ready for clinical use. One of the primary research deficits is small sample sizes. Most of the current datasets are pilot-scale and tend to be under 100 people, which contributes to the weakness and lack of statistical power in training and validating a model. This poses issues with generalizability of the results and creates potential overfitting and bias. The lack of physiotherapy-specific datasets is also a considerable constraint (García and Santos 2025).

In the wider field of oncology, prediction of fatigue has been investigated, but most studies do not focus their data collection and model building around a physiotherapy workflow, activity or treatment plan. Physiotherapy has different exercise profiles, different pacing and different therapeutic goals that may not be captured with the fatigue predictions that are generalized to oncology patients who are just resting or involved in standard activities of daily living. In this context, there should be physiotherapy data collection pipelines that integrate exercise sessions, exertion levels, session durations and recovery assessments after the sessions (Bettosi et al. 2025). Further, the potential for generalizing models across different types of cancer has not been sufficiently investigated.

3.7.3. *Ethical and operational challenges*

When implementing DL systems for fatigue monitoring, we must consider the ethical and operational perspectives of the use case. The most fundamental are consent and data protection issues. Continuous data collection through wearables raises ethical and legal issues regarding informed consent, ownership and the secure collection, storage and transfer of data.

The absence of legally binding agreements and the imposition of data protection laws, such as HIPAA and GDPR, would breach trust and disrespect the patient. The DL prediction of fatigue lacks an articulate explanation. Clinicians and patients must know the how and the why of a model's prediction. Trust and clinical adoption will suffer greatly if "black box" DL models are used (Alqudah and Moussavi 2025). Lastly, costs and availability of wearable sensors, as well as advanced devices with high-resolution physiological monitoring, may limit equitable deployment to well-funded institutions and urban centers. Fatigue monitoring technology has to be scalable, low cost and accessible to maximize their impact on cancer rehabilitation.

3.8. Future directions

In order to allow the use of DL in cancer rehabilitation to take the next step, some specific research areas need to be explored. First, large-scale research needs to be performed in structured physiotherapy settings with varying cancer-type patients. Most of the current research is made up of small sample populations and limited cancer types. This makes it difficult to generalize the model. Multicenter research will allow large sample populations to gain validation of the model (Wynants et al. 2019).

Second, the combination of different modalities such as physiological signals like heartbeat and respiration, thermal images, motion capture with inertial sensors and patient-reported outcome surveys will increase the fatigue model sensitivity. Fusion of such data will allow DL to capture the relationship between objective and subjective fatigue (Mane and Yadav 2025).

Third, creating explainable AI (XAI) is crucial for the integration of this technology into clinical practice. Clinicians need to understand why a model detects fatigue and alters therapy, for instance. Interpretability can be improved with attention mechanisms, SHAP (Shapley Additive Explanations) and counterfactual reasoning (Gudur et al. 2023). Finally, in order to derive individual predictions, movement models must incorporate individual patient factors – baseline fitness, comorbidities, treatment regimens and longitudinal fatigue data – to personalize predictions (Wilson et al. 2024). Adaptive learning as a model will allow the system to shift as the patient condition changes.

3.9. Conclusion

Monitoring and predicting cancer physiotherapy fatigue with DL technologies could eventually improve cancer care. Close monitoring of continuous streams of data with integration of wearable technologies (heart rate, motion, temperature and skin conductance) is feasible. These systems detect fatigue and differentiate behaviors associated with changes in physical activity. Clinically, relevant insights once impossible to extract during real-time and time-series analyses of data are now accessible with convolutional and RNNs. This real-time monitoring capability transforms fatigue monitoring during physiotherapy and, most importantly, enables forecasting fatigue episodes weeks ahead. Such forecasting capability enables clinicians to modify exercise rest intervals and treatment scheduling. More importantly, this promotes patient safety and therapy compliance. Additionally, individual patient baselines can further enhance fatigue prediction models. Still, obstacles such as integration in rest intervals, longitudinal data collection and patient privacy remain prior to clinical use.

These challenges need to be addressed to move DL tools for oncology rehabilitation to practical use. Nonetheless, the application of DL technology in cancer physiotherapy holds significant potential for improving the quality of life and functional outcomes, as well as providing more adaptable and patient-centered rehabilitation for cancer survivors.

3.10. References

Adão Martins, N.R., Annaheim, S., Spengler, C.M., Rossi, R.M. (2021). Fatigue monitoring through wearables: A state-of-the-art review. *Frontiers in Physiology*, 12, 790292.

Alqudah, A.M. and Moussavi, Z. (2025). A review of deep learning for biomedical signals: Current applications, advancements, future prospects, interpretation, and challenges. *Computers, Materials & Continua*, 83(3), 3753–3841.

Ancoli-Israel, S., Cole, R., Alessi, C., Chambers, M., Moorcroft, W., Pollak, C.P. (2003). The role of actigraphy in the study of sleep and circadian rhythms. *Sleep*, 26(3), 342–392.

Balkin, T.J., Horrey, W.J., Graeber, R.C., Czeisler, C.A., Dinges, D.F. (2011). The challenges and opportunities of technological approaches to fatigue management. *Accident Analysis & Prevention*, 43(2), 565–572.

Bettosi, C., Ballie, L., Shenkin, S., Romeo, M. (2025). Policy learning for social robot-led physiotherapy. *arXiv preprint*. doi: 10.48550/arXiv.2509.11297.

Bower, J.E. (2014). Cancer-related fatigue: Mechanisms, risk factors, and treatments. *Nature Reviews Clinical Oncology*, 11(10), 597–609.

Cuomo, S., Mulasso, A., Lupo, C., Rainoldi, A., Brustio, P.R. (2024). High-intensity interval training versus usual care during cancer prehabilitation: A systematic review and meta-analysis of the impact on cardiorespiratory fitness. *Sport Sciences for Health*, 20, S74.

Dargan, S., Kumar, M., Ayyagari, M.R., Kumar, G. (2020). A survey of deep learning and its applications: A new paradigm to machine learning. *Archives of Computational Methods in Engineering*, 27(4), 1071–1092.

Dorschky, E., Camomilla, V., Davis, J., Federolf, P., Reenalda, J., Koelewijn, A.D. (2023). Perspective on "in the wild" movement analysis using machine learning. *Human Movement Science*, 87, 103042.

Fan, V.S., Mehdipour, A., O'Grady, H.K., Richardson, J., Griffith, L.E., Campbell, K.L., Bosch, J. (2022). Systematic review: Patient-reported outcome measures of fatigue in inflammatory bowel disease. *Fatigue: Biomedicine, Health & Behavior*, 10(2), 60–82.

Finkelstein, J., Smiley, A., Echeverria, C., Mooney, K. (2024). AI-driven prediction of symptom trajectories in cancer care: A deep learning approach for chemotherapy management. *Bioengineering*, 11(11), 1172.

García, V. and Santos, O.C. (2025). Towards intelligent assessment in personalized physiotherapy with computer vision. *Sensors*, 25(11), 3436.

Gruet, M. (2018). Fatigue in chronic respiratory diseases: Theoretical framework and implications for real-life performance and rehabilitation. *Frontiers in Physiology*, 9, 1285.

Gudur, A., Pati, P., Garg, P., Sharma, N. (2023). Radiomics feature selection for lung cancer subtyping and prognosis prediction: A comparative study of ant colony optimization and simulated annealing. *International Journal of Intelligent Systems and Applications in Engineering*, 12(3s), 553–565.

Hilfiker, R., Meichtry, A., Eicher, M., Balfe, L.N., Knols, R.H., Verra, M.L., Taeymans, J. (2018). Exercise and other non-pharmaceutical interventions for cancer-related fatigue in patients during or after cancer treatment: A systematic review incorporating an indirect-comparisons meta-analysis. *British Journal of Sports Medicine*, 52(10), 651–658.

Hoang, M.L. (2025). A comprehensive review of machine learning, and deep learning in wearable IoT devices. *IEEE Access*, 13. doi: 10.1109/ACCESS.2025.3573937.

Hsu, S.H., Lu, L.J., Chou, P.C. (2025). Preliminary efficacy of a gamified mobile app for promoting self-health management among nurses in the post-COVID era: Single-group pre-post study. *JMIR Serious Games*, 13(1), e66262.

Jeong, H., Jeong, Y.W., Park, Y., Kim, K., Park, J., Kang, D.R. (2022). Applications of deep learning methods in digital biomarker research using noninvasive sensing data. *Digital Health*, 8, 20552076221136642.

Kumar, A., Batta, P., Ahuja, S., Rathore, P.S. (eds) (2026a). *AI-driven Innovations in Physiotherapy and Oncology 1*. ISTE Ltd, London and Wiley, Hoboken, NJ.

Kumar, A., Batta, P., Ahuja, S., Rathore, P.S. (eds) (2026b). *AI-driven Innovations in Physiotherapy and Oncology 2*. ISTE Ltd, London and Wiley, Hoboken, NJ.

Li, R., Gao, R., Suganthan, P.N. (2023). A decomposition-based hybrid ensemble CNN framework for driver fatigue recognition. *Information Sciences*, 624, 833–848.

Mane, A.S. and Yadav, T. (2025). The short-term efficacy of mulligan traction straight leg raise on low back pain associated with hamstring tightness in young adults. *Cureus*, 17(3), e80215. doi: 10.7759/cureus.80215.

Pedersen, B.K. and Saltin, B. (2015). Exercise as medicine–evidence for prescribing exercise as therapy in 26 different chronic diseases. *Scandinavian Journal of Medicine & Science in Sports*, 25, 1–72.

Pinto-Bernal, M.J., Cifuentes, C.A., Perdomo, O., Rincón-Roncancio, M., Múnera, M. (2021). A data-driven approach to physical fatigue management using wearable sensors to classify four diagnostic fatigue states. *Sensors*, 21(19), 6401.

Pourpanah, F. and Etemad, A. (2024). Exploring the landscape of ubiquitous in-home health monitoring: A comprehensive survey. *ACM Transactions on Computing for Healthcare*, 5(4), 1–43.

Price, E., Moore, G., Galway, L., Linden, M. (2019). Towards mobile cognitive fatigue assessment as indicated by physical, social, environmental, and emotional factors. *IEEE Access*, 7, 116465–116479.

Solomon, N.L. and Manea, V. (2022). Quantifying energy and fatigue: Classification and assessment of energy and fatigue using subjective, objective, and mixed methods towards health and quality of life. In *Quantifying Quality of Life: Incorporating Daily Life into Medicine*, Wac, K. and Wulfovich, S. (eds). Springer, Cham, 79–117.

Vercellini, P., Buffo, C., Viganò, P., Somigliana, E. (2024). Update on medical treatment of endometriosis: New drugs or new therapeutic approaches? *Gynecologic and Obstetric Investigation*, 90(6), 535–559. doi: 10.1159/000542947.

Wilson, M.K., Strickland, L., Ballard, T., Griffin, M.A. (2024). The next generation of fatigue prediction models: Evaluating current trends in biomathematical modelling. *Theoretical Issues in Ergonomics Science*, 25(1), 21–43.

Wynants, L., Kent, D.M., Timmerman, D., Lundquist, C.M., Van Calster, B. (2019). Untapped potential of multicenter studies: A review of cardiovascular risk prediction models revealed inappropriate analyses and wide variation in reporting. *Diagnostic and Prognostic Research*, 3(1), 6.

Zhang, Y., Muniyandi, R.C., Qamar, F. (2025). A review of deep learning applications in intrusion detection systems: Overcoming challenges in spatiotemporal feature extraction and data imbalance. *Applied Sciences*, 15(3), 1552.

4

Predictive Modeling of Lymphedema Risk Using AI in Oncology Physiotherapy

Lymphedema is a persistent disorder, which often has devastating effects on quality of life. It is common among cancer survivors, particularly those who have had their lymph nodes removed or gone through radiation treatment. Preventing damage in the long term relies on early detection and treatment. Physiotherapists engage in preventative treatment and lymphedema management as members of an oncology care team, but the current methods of risk assessment are still inaccurate and too general. Advanced clinical decision-making can be assisted by predictive analytics, in particular, machine learning (ML) and deep learning (DL).

This chapter discusses how artificial intelligence (AI) can be further used to forecast the lymphedema risk in cancer survivors and implement AI into their physiotherapy treatment process. It includes the clinical context of lymphedema, critiques the conventional predictive models and surveys the AI methodologies, such as logistic regression, decision trees, support vector machines (SVMs), neural networks and ensemble methods. Considerable attention is given to the processes of model development, including data acquisition, feature selection, model training and validation. The chapter describes the performance evaluation metrics of predictive models and discusses explainability as a major obstacle toward clinical implementation. Furthermore, it outlines how AI applications are incorporated into physiotherapy practices for personalized treatment, efficient resource allocation and early preventive care assistance. Issues of data quality, model interpretation, ethical issues and integration are challenged, as well as advanced hybrid modeling, wearable tech and mobile technologies as future opportunities. Predictive analytics is the most promising of all possible applications of AI in oncology physiotherapy and will provide the opportunity for proactive, data-driven rehabilitation planning for cancer survivors.

Chapter written by Rashmi GUDUR and Mrudula NIMBARTE.

4.1. Introduction

Secondary lymphedema is a common and often chronic complication that affects a significant proportion of cancer survivors, particularly those who have undergone treatments involving lymphatic system disruption, such as surgery, radiation therapy or lymph node dissection. It manifests as an abnormal accumulation of protein-rich lymphatic fluid in the interstitial tissues, typically in the arms, legs or torso, leading to progressive swelling, discomfort, restricted mobility and, in many cases, psychological distress such as anxiety, depression and body image concerns (Preedaa et al. 2025). Lymphedema has not only physical functional implications, but also severe implications on the overall quality of life, social involvement and long-term health outcomes.

In physiotherapy of oncology, lymphedema prevention and early detection are extremely important. Early treatment measures in the form of exercise, manual lymphatic drainage, compression therapy and patient education may help a lot in reducing the severity and progression of symptoms. However, the lymphedema is very hard to detect in early or subclinical stages. Previous attempts to assess lymphedema risk used passive clinical markers, clinical gestalt and rudimentary risk models. These strategies do not consider volumetric integration or real-time tailoring to patients, and thus, are not useful in precision medicine.

The development of AI and more specifically, ML and DL, can reshape lymphedema prediction and lymphedema care (Eldaly et al. 2022). AI can characterize and classify dimensions of data to detect clinically important correlations among risk factors. This function allows for the creation of granular, individualized prediction models, which can be used in oncology physiotherapy for early lymphedema intervention and personalized treatment planning.

This chapter focuses on AI-based predictive modeling for lymphedema risk in the context of cancer rehabilitation. It starts with the clinical aspects of secondary lymphedema, its pathophysiology, key risk factors and its prevalence in various cancers such as breast, gynecological, prostate and brain and neck cancers. Among the primary reasons for the development of lymphatic dysfunction are the tumor burden near lymphatic structures, the surgical removal of lymph nodes and radiation-induced fibrosis (Deng et al. 2019). Additional factors such as obesity, age, number of lymph nodes removed and the presence of infection or inflammation also contribute to an increase in risk. After presenting the clinical factors, the lack of focus on the limitations for predictive systems is described.

Most systems are created using logistic regression or decision-tree frameworks and are working on small and shallow feature datasets. Many of these systems are vulnerable to overfitting and will lack generalizability in systems, and also cannot track the interactions of physiological and behavioral systems over time. In addition to these gaps, the literature also lacks systems that use longitudinal, real-time data and dynamic systems, particularly to measure how risk changes over time in the cancer continuum.

Contrarily, the literature on AI systems, specifically predictive systems using supervised machine learning (ML), random forests, support vector machines (SVMs) and gradient boosting, has predictive systems for healthcare that are superior to the predictive systems described above. Advanced deep learning (DL) systems such as convolutional neural networks (CNNs) and recurrent neural networks (RNNs) perform exceptionally well on time-series data and complex images and can track limb measurements and levels of physical activity, as well as biosensor data from wearables and other devices (Krittanawong et al. 2023).

To illustrate this, CNNs can be trained on imaging studies from the time when a patient had an operation done to learn how to recognize lymphedema lymph node changes in the flow of lymph post-operatively. Also, RNNs can be used to analyze the time-sequenced slices of life from clinical data to identify predictive lymphedema lymph node changes. Predictive modeling comprises several steps from data acquisition to model training, evaluation and validation. These steps can be taken from electronic health records, imaging and even patient assessments and sensor data during physiotherapy, or through wearables.

When integrated across numerous organizations, data standardization, quality and completeness can be particularly problematic, along with the raw data that needs to be preprocessed. The models of AI that are available in the literature have outperformed traditional estimates in predicting lymphedema, since they are more specific and sensitive. Notably, explainable AI (XAI) tools, including SHAP and LIME, enhance the intelligibility of the output and thus induce better trust in clinicians in making decisions.

The adoption of predictive models into physiotherapy processes requires technical feasibility and acceptance of the models as well as clinical feasibility. AI tools must be integrated into electronic health records or physiotherapy software in such a way that they complement and do not interfere with clinical practices. As part of the point of care, physiotherapists can be directed by risk scores, visual alerts and tailored recommendations in assessing and planning interventions and educating patients. In addition, dynamic risk stratification, which is constantly provided on the basis of real-time patient data by predictive models, can be offered, allowing just-in-time interventions to be performed (Avin et al. 2015).

Several challenges persist. Confidentiality and security of data, in particular sensitive health data, is of topmost importance. To achieve model generalizability and ethical implementation, regulatory approval and validation must be carried out in a variety of populations. Use of algorithmic bias may occur when the training data are not reflective of the overall population. As a result, the AI model development and deployment should be based on transparency, accountability and fairness (Khatiwada et al. 2024).

4.2. Clinical background: lymphedema in oncology

Lymphedema is a progressive and chronic condition that is as a result of an excessive accumulation of lymph fluid in the tissues in the case of lymph injury. It usually attacks the limbs but may extend to the torso, face or the genitals. The disorder is caused either by defects of the lymphatic system at birth (primary lymphedema), or by external trauma to lymph vessels or lymph nodes (secondary lymphedema). Secondary lymphedema is much more widespread, particularly among cancer survivors who have gone through aggressive forms of treatment.

In oncology, lymphedema is most commonly observed in patients treated for breast, cervical, prostate and brain and neck cancers. These patients frequently undergo surgical procedures such as lymph node dissection or radiation therapy, which can disrupt normal lymphatic circulation. Radiation-induced fibrosis and tumor-related lymphatic obstruction further contribute to the development of this condition. The fundamental abnormality of lymphedema is one concerning the dysfunction of lymph transport capacity, which, if left untreated, ultimately leads to fluid stagnation, chronic inflammation and fibrosis of the affected tissues (Klein and Jodocy 2021).

In lymphedema, swelling of the limbs, heaviness and tightness, discomfort, and pain, reduced range of motion, and hypertrophy, hyperkeratosis and fibrosis of the skin in the advanced stages are the noticeable signs. In lymphedema, tissue swelling can follow straight after treatment or develop months or even years later, which is why long-term surveillance is necessary for patients. In the early stages, the condition lymphedema is most reversible; however, the chronic stages of lymphedema are serious and are marked by hardened tissues and painful, recurrent and chronic infections such as cellulitis and lymphangitis.

In addition to its physical impacts, lymphedema has major negative psychosocial effects. Patients suffer from discomfort, deformed limbs and are thus, due to the condition, uncomfortable, often causing them to become socially withdrawn (Peng et al. 2024). Quality of life is greatly reduced due to negative psychosocial effects from lymphedema. This is why risk factors for secondary lymphedema need to be addressed.

Factors include the number and extent of lymph nodes removed, cumulative radiation dose, higher body mass index (BMI), type and stage of cancer, chemotherapy regimens, age of the patient and the presence of comorbid conditions such as hypertension or diabetes mellitus. Obesity, in particular, is strongly correlated with poor lymphatic function and increases the likelihood of chronic lymphedema. The problem may also be exacerbated by recurrent infections or trauma of an affected limb.

Lymphedema treatment in physiotherapy is a major aspect of the overall cancer therapy. Early detection, by regular screening and assessment of limb volume, and patient education, is very beneficial. Physiotherapists can prevent lymphedema progression by managing it with a multimodal intervention of manual lymphatic drainage (MLD), compression bandage or garments, skin care, therapeutic exercises and self-management education. MLD is a special massage that helps to increase the lymph circulation and decrease the amount of fluid stagnation. Comfortable pressure therapy also maintains the drainage effect and prevents fluid reflux (Zhao et al. 2024).

Once dismissed, exercise becomes an important part of the lymphedema treatment. Resistance and aerobic exercises that are carefully planned increase the lymphatic activity, augment the pumping action of the muscles and improve fitness in general. The use of teaching patients on skin care, injury prevention and lifestyle change gives the patient enough power to control the condition at home.

4.3. Rationale for predictive modeling

Predictive modeling is able to identify patients that are more likely to have lymphedema, particularly during cancer management. Outcomes can be improved through early intervention. Conventional instruments such as clinical scoring mechanisms and logistic-regression nomograms give risk estimates of variables selected. While these methods are useful, they introduce risks, due to assumptions of linearity and independence. Nonlinear factors that influence the risk of lymphedema are, however, absent in these tools (Vargo et al. 2024).

New health data including electronic records, medical images, genomics and readings of wearable sensors allow new modeling methods. ML and AI have the ability to examine a vast amount of diverse data and identify intricate risk relationships that are not present in traditional statistics. They will be able to identify the symptoms of advanced diseases at early stages and provide individual and real-time rehabilitation plans.

These models are a combination of clinical measures, patient-reported and real-time sensor data to accelerate the risk assessment. The use of AI in the dynamics of physiotherapy allows clinicians to be proactive. As an example, they will be able to initiate early treatment and refer high-risk patients to special programs.

This enhances the patient outcomes and enables clinics to utilize resources more effectively. Precision medicine: predictive modeling predicts the right time to treat a lymphedema patient in the right way. AI models supersede the one-size-fits-all approach and transforms oncology rehabilitation into proactive instead of reactive care and enhances the life of cancer patients.

4.4. AI and ML overview

AI refers to computerized procedures that replicate the behavior of people such as learning, thinking and making decisions. In oncology physiotherapy, AI assists in predictive modeling for lymphedema by analyzing multidimensional data and helping to identify lymphedema at-risk patients (Peng and Lu 2024). Several ML algorithms of differing strengths and weaknesses are used in the field to enhance predictive power and clinical decision support.

In clinical prediction tasks, one of the most basic and fundamental algorithms is logistic regression, which predicts the likelihood of a binary outcome (e.g. lymphedema is present/absent) and the other input variables such as patient age, BMI, type of surgery and lymph nodes involved. It is computationally efficient and easy to interpret but struggles with model complexity, nonlinear relationships and interactions.

Decision trees and their more advanced versions, random forests, are much more flexible. They ease the model complexity by splitting the data into subsets at various reasonable thresholds of specific features. Random forests improve generalizability and reduce overfitting by using bagging, creating an ensemble of decision trees (Halabaku and Bytyçi 2024). These algorithms are particularly useful with datasets containing a mix of variable types and incomplete records.

SVMs produce admirable results when it comes to determining the optimal hyperplane necessary to segregate various classes in high-dimensional feature spaces. These classifiers work particularly well with structured datasets which have a clear margin of separation between classes and are particularly resistant to overfitting when adequately fine-tuned.

Gradient boosting machines (GBMs) are an advanced class of ensemble models which include XGBoost, LightGBM and CatBoost. GBMs are modeled gradually with successive learners, correcting the errors of the previous one and tend to be superior to other algorithms when applied to structured data because they are effective in dealing with complex relationships.

Artificial neural networks (ANNs) are analogies of the human brain that model a simplified representation of it through connection of numerous artificial neurons in layers. They extract nonlinear and complicated patterns of large datasets.

ANNs, together with CNNs and RNNs, are extended to DL, which allows them to deal with high-dimensional inputs, including a medical image and time-series data. Ensemble techniques involve multiple models that enhance the predictive performance by using different advantages of the models.

Ensemble techniques involve bagging, boosting and stacking, which is a combination of predictions of various algorithms to enhance reliability and accuracy (Rane et al. 2024). They come in handy when different individual models can be used to capture various aspects of the data. XAI is needed in clinical practice to understand the decision of a model.

The following tools are SHAP, LIME and Grad-CAM, which assist clinicians with interpreting algorithm outputs and deciding how significant this or that predictor is, which will contribute to gaining trust in AI-assisted decisions.

4.5. Model development approaches

Combining clinical knowledge, statistics and ML techniques, predictive models for lymphedema risk are easier to develop. The idea is to anticipate the patients who are likely to get lymphedema and, thus, it is possible to intervene earlier and shape the cases. In this case, the development phase is outlined in five directions, namely: collection, preprocessing, feature construction and selection, model training and evaluation.

4.5.1. *Data collection*

The predictive model is as good as the data on which the model is modeled. Much in-depth quality data is essential in developing accurate models that can be generalized. A perfect dataset contains a broad range of applicable characteristics that may contain:

– *Demographic data*: age, sex, ethnics, BMI, smoking, alcohol intake and other lifestyle-related variables.

– *Medical and clinical history*: preexisting conditions such as diabetes, cardiovascular diseases and hypertension, which may affect the lymphatic system or influence recovery (Preedaa et al. 2025).

– *Cancer characteristics*: type of cancer (e.g. breast, gynecological, melanoma, etc.), tumor size, tumor location, tumor histological grade and nodal involvement.

– *Treatment modalities*: surgical modalities (mastectomy vs lumpectomy), amount and extent of lymph nodes removed, radiation therapy protocols (dose and field) and chemotherapy.

– *Symptomatology and physical examination findings*: the evaluation of the presence of early lymphedema symptoms encompasses the evaluation of leg circumference, skin texture, assessment of skin changes and limb tightness or heaviness as well as range of motion.

– *Imaging and sensor data*: data from diagnostic imaging such as MRI and ultrasound, as well as quantitative assessments such as bioimpedance spectroscopy. Wearable devices can also provide real-time data on limb movement, temperature and swelling (Groenendaal et al. 2021).

4.5.2. *Feature engineering and selection*

The second stage after collection of data is ML algorithm preparation. This includes the cleaning of data, handling of missing data, continuous normalization and encoding of categorical variables, one-hot or the ordinal encoding.

Feature engineering is the addition/creation of new variables which can be more predictive than the raw data. Examples include composite risk scores, features of time-series data in the form of interaction terms and trend-based features.

The feature selection process is a very important phase in order to enhance the performance of the model, minimize overfitting and minimize the time necessary to train the model.

Common techniques include:

– *Univariate feature selection*: statistical tests (e.g. chi-square, ANOVA, etc.) to determine the relevance of individual features to the target outcome.

– *Recursive feature elimination (RFE)*: an iterative method that builds a model and removes the weakest features until the optimal set is identified.

– *Regularization methods*: techniques such as LASSO (least absolute shrinkage and selection operator) and Ridge regression penalize irrelevant or redundant features by shrinking their coefficients.

– *Genetic algorithms*: these are inspired by natural evolution and simulate the process of selecting the best feature combinations over successive iterations.

This stage significantly influences the interpretability and effectiveness of the final model.

4.5.3. *Model training and validation*

The data is then split into training and testing sets after the most informative features have been selected. The model is developed using the training set, and the performance of the model is tested using the testing set.

To improve reliability, cross-validation methods are used, including the k-fold cross-validation methods such as division of the data into k groups, the rotation of the training and the validation to view all of the data points (Bhagat and Bakariya 2025).

Hyperparameter optimization plays a very important role in optimizing model performance. This is the process of optimizing algorithm-specific parameters with the help of such techniques as grid search (exhaustive search over a predefined parameter space), random search (random selection of parameters) or Bayesian optimization (probabilistic modeling of performance).

A variety of metrics is used in determining model performance:

– *accuracy*: overall correctness of the model;

– *precision*: proportion of true positives among predicted positives;

– *recall (sensitivity)*: ability to correctly identify actual positives;

– *F1-score*: harmonic mean of precision and recall, balancing both metrics;

– *AUC–ROC* (area under the receiver operating characteristic curve): this indicates how the model distinguishes the classes with various threshold values.

An exemplary model will exhibit strong performance in all of these measures, which implies that it can be used in the practical clinical work in the field of oncology physiotherapy.

4.6. Performance metrics and model comparisons

Different AI models have been found to be highly effective at the risk prediction of lymphedema. The following are some of the areas of observation:

– *Logistic regression models*: these are generally good at the base level performance with moderate accuracy and high interpretability.

– *Random forests and gradient boosting*: these can model nonlinear relationship and interactions, which also makes them perform well compared to simple models. These models lead to high values of the AUC and equal sensitivity and specificity.

– *SVMs*: these prove to be particularly useful in situations with complex boundaries and high-dimensional feature space. They are capable of high accuracy and sensitivity when optimized (Gnana et al. 2016).

– *Artificial neural networks*: these are suitable for large datasets, especially when incorporating imaging or sensor data. DL models have shown superior performance in image-based classification tasks.

– *Ensemble approaches*: these combine models using techniques such as stacking or voting and generally improves predictive power and reduces variance.

Model comparison should consider not only performance metrics but also factors such as interpretability, computational efficiency and ease of integration into clinical workflows. Table 4.1 shows a comparison of AI models for lymphedema risk prediction.

Model type/algorithm	Data source(s)	Key features used	Performance metrics (typical range)	Clinical relevance/remarks
Logistic regression	Clinical and demographic data	Age; BMI; surgery type; lymph node count	Accuracy: 70–80%; AUC: 0.75	Simple, interpretable baseline model; limited for nonlinear patterns
Decision tree	EHR; clinical data	Treatment modality; comorbidities	Accuracy: 75–85%	Easy visualization; prone to overfitting on small datasets
Random forest	Mixed clinical and sensor data	Demographics + imaging + limb volume	Accuracy: 85–92%; AUC: 0.88–0.93	High robustness and generalizability; widely used in clinical AI
Support vector machine (SVM)	Structured physiotherapy datasets	BMI; age; motion metrics	Accuracy: 80–90%; sensitivity: 0.82	Effective in high-dimensional data; requires tuning
Gradient boosting (XGBoost, LightGBM)	EHR + imaging data	Clinical + treatment + radiation details	Accuracy: 88–94%; AUC: 0.95	Strong performer in structured data tasks; less interpretable
CNNs	Imaging (MRI; ultrasound)	Lymphatic morphology; tissue texture	AUC: 0.90–0.96	Excels in image-based prediction; needs large datasets
RNNs/LSTM	Time series from wearable sensors	Limb circumference; bioimpedance; mobility data	Accuracy: 85–93%	Ideal for longitudinal monitoring; higher computational cost
Hybrid/ensemble models	Multimodal data	Combined clinical + imaging + sensor inputs	AUC: 0.94–0.98	Best overall performance; supports personalized physiotherapy

Table 4.1. *Comparison of AI models for lymphedema risk prediction*

4.7. Explainability and clinical integration

Among the factors impeding the implementation of AI in clinical practice is the lack of understanding of how recommendations are derived as most AI solutions are black-boxed. Clinicians are unwilling to trust recommendations of AI solutions when they are unable to understand the rationale behind them (Kumar et al. 2026a). XAI provides a way to overcome this challenge by providing clarity in how AI predicts outcomes.

There are several XAI methods designed to enhance the interpretability of the tools. For example, in the analysis of structured data, clinicians can use suggested value plots, SHAP (SHapley Additive exPlanations) scores and LIME (Local Interpretable Model-agnostic Explanations) to ascertain the most important predictors in a given prediction. These methods are designed to quantify the reasoning of the AI to ensure the user can understand and trust the AI's prediction. For image-based AI, such as diagnostic imaging and post-operative imaging, AI tools such as Grad-CAM (Gradient-weighted Class Activation Mapping) provide image overlay focused on areas of the image that were most important in a prediction. This has applications in physiotherapy of risk stratification and oncology. To use AI in clinical practice, trust and transparency are key.

In any case, clinicians will rely on and apply AI models in the situations that give straightforward and easily interpretable information. Good clinical decision support implies that AI predictions become part of the clinician's workflow in their EHR. To use the same case in point, a physiotherapist can get an automated notification when a patient is at the risk of lymphedema that is foreseen to exceed a certain threshold and take the necessary action to correct lymphedema prophylaxis. Lymphedema rehabilitation plans are more automated and can also be tailored to rehabilitation risk profiles. The risk prediction model will be optimized as time goes by due to real-time monitoring and gathering information on what a patient is doing during their daily activities. The more data is gathered, the more accurate the approximations are (Kumar et al. 2026b).

The interdisciplinary cooperation is necessary to be successful in its implementation. The data scientists need to collaborate directly with clinicians and health IT professionals to create tools not only technically but also a clinical workflow and being usability-compatible. The ease of use, adjustable thresholds and connection with existing clinical pathways make it possible to ensure that AI-based tools can improve (as opposed to slowing down) healthcare delivery (Elhaddad and Hamam 2024). With the evolution of XAI technologies, they will have an essential part in reducing the gap between the performance of the algorithms and the clinical trust in the system, which will culminate in making patient care more informed, equitable and more effective.

4.8. Role in oncology physiotherapy

The implication of predictive modeling on physiotherapy oncology is important:

– *Early intervention*: high-risk patients can be detected shortly after they are treated and be placed into preventive programs, including compression therapy and manual lymph drainage (Donahue et al. 2023).

– *Individualized rehabilitation plans*: AI allows tailoring the exercise regimen, the frequency of follow-ups and educational resources in accordance with the risk profile of each person.

– *Resource optimization*: with the stratification of the patients, the resources of physiotherapy can be determined more efficiently to make sure that the neediest patients receive the necessary care as soon as possible.

– *Patient engagement*: predictive insights can be shared with patients to encourage adherence and self-management. Mobile apps and wearable devices can support home-based monitoring and feedback (Calthorpe et al. 2020).

– *Remote monitoring*: in settings where in-person visits are limited, AI-enabled tools can support tele-rehabilitation and virtual physiotherapy consultations.

Predictive modeling integration also enables the physiotherapists to be proactive in providing care rather than being reactive to enhance the outcomes and efficiency.

4.9. Challenges and limitations

Nevertheless, AI-based predictive modeling has a number of problems:

– *Data quality and availability*: incomplete, inconsistent or biased data can impair model performance. Large, diverse datasets are essential for training robust models.

– *Class imbalance*: lymphedema may occur in a small proportion of patients, leading to skewed class distributions. Techniques such as oversampling, under sampling or cost-sensitive learning are needed to address this (Cè et al. 2024).

– *Generalizability*: models trained on data from a specific institution or population may not perform well elsewhere. External validation is critical to assess applicability.

– *Interpretability versus performance*: more complex models often offer better accuracy but lower transparency. Finding the right balance is key for clinical acceptance.

– *Ethical and legal considerations*: issues include informed consent, data privacy, algorithmic bias and accountability for errors in prediction.

– *Integration challenges*: technical and operational barriers can hinder implementation in real-world settings. User training, workflow alignment and interoperability with existing systems must be addressed (Mohammadhossein et al. 2025).

The way to resolve these issues is through interdisciplinary collaboration, strong validation and the desire to make AI practices ethical.

4.10. Future directions

Some of the promising lines are under investigation in order to enhance predictive modeling in oncology physiotherapy:

– *Hybrid models*: mechanistic knowledge (e.g. fluid dynamics, lymphatic anatomy) can be improved with information-derived methods to make them more understandable and accurate (Mao et al. 2024).

– *Imaging integration*: DL on MRI, ultrasounds or 3D imaging are able to detect the early signs of lymphedema before clinical symptoms appear (Patil et al. 2024).

– *Wearable technology*: real-time data can be obtained through continuous monitoring by smart garments, bioimpedance and accelerators to provide the risk assessment which is always dynamic (Appelboom et al. 2014).

– *Mobile health applications*: apps that are connected with AI models can provide patients with self-assessment, exercise and progress tracking.

– *Prospective clinical trials*: longitudinal studies evaluating AI tools in real-world physiotherapy settings are essential to demonstrate effectiveness and safety.

– *Regulatory frameworks*: establishing standards for validation, transparency and approval will support safe deployment and public trust (Ranjitsingh and Rao 2025).

With the evolving nature of the field, the partnership of clinicians, researchers, technologists and policymakers will play a significant role in making use of the potential of AI in enhancing cancer survivorship care (Gudur et al. 2024).

4.11. Conclusion

AI-driven predictive modeling is a development in the oncological physiotherapy, namely the management of lymphedema. The introduction of ML and DL algorithms provides physicians with the opportunity to predict and analyze the risk of lymphedema by projecting and analyzing multiple datasets (clinical histories, images, sensor data and patient-reported outcome measures). It is on the basis of such projections of periods at risk that risk-specific early therapeutic interventions are formulated and implemented, thus enhancing the likelihood of successful therapy, and reducing the likelihood of the patient advancing to an elevated state of illness. The AI custom predictive models make it easier to create a proactive therapy where a clinician can make the correct, evidence-based, rational decisions by simplifying the therapy processes and redistributing the resources to rehabilitation. In this case, predictive modeling is directed toward converting reactive treatment in the field of oncology rehabilitation to proactive predictive treatment. It is with this in mind that ongoing issues, including the paucity of quality data, algorithm prejudices, patient data confidentiality and the absence of smooth clinical workflow connectivity, must be transcended. The cross-discipline teamwork and well-documented modeling work will contribute to the achievement of this. There will also be the need of exhaustive modeling in different demographics of patients to have proper predictive modeling. Easy-to-use interfaces and XAI will be needed by the clinical community to create trust and adoption of predictive modeling as well. Predictive modeling is aimed at transforming reactive treatment in oncology rehabilitation to proactive predictive treatment.

4.12. References

Appelboom, G., Camacho, E., Abraham, M.E., Bruce, S.S., Dumont, E.L., Zacharia, B.E., D'Amico, R., Slomian, J., Reginster, J.Y., Bruyère, O. et al. (2014). Smart wearable body sensors for patient self-assessment and monitoring. *Archives of Public Health*, 72(1), 28.

Avin, K.G., Hanke, T.A., Kirk-Sanchez, N., McDonough, C.M., Shubert, T.E., Hardage, J., Hartley, G. (2015). Management of falls in community-dwelling older adults: Clinical guidance statement from the Academy of Geriatric Physical Therapy of the American Physical Therapy Association. *Physical Therapy*, 95(6), 815–834.

Bhagat, M. and Bakariya, B. (2025). A comprehensive review of cross-validation techniques in machine learning. *IJSAT-International Journal on Science and Technology*, 16(1). doi: 10.71097/IJSAT.v16.i1.1305.

Calthorpe, R.J., Smith, S., Gathercole, K., Smyth, A.R. (2020). Using digital technology for home monitoring, adherence and self-management in cystic fibrosis: A state-of-the-art review. *Thorax*, 75(1), 72–77.

Cè, M., Chiriac, M.D., Cozzi, A., Macrì, L., Rabaiotti, F.L., Irmici, G., Fazzini, D., Carrafiello, G., Cellina, M. (2024). Decoding radiomics: A step-by-step guide to machine learning workflow in hand-crafted and deep learning radiomics studies. *Diagnostics*, 14(22), 2473.

Deng, J., Wulff-Burchfield, E.M., Murphy, B.A. (2019). Late soft tissue complications of head and neck cancer therapy: Lymphedema and fibrosis. *JNCI Monographs*, 2019(53), lgz005.

Donahue, P.M., MacKenzie, A., Filipovic, A., Koelmeyer, L. (2023). Advances in the prevention and treatment of breast cancer-related lymphedema. *Breast Cancer Research and Treatment*, 200(1), 1–4.

Duncan, P.W., Zorowitz, R., Bates, B., Choi, J.Y., Glasberg, J.J., Graham, G.D., Katz, R.C., Lamberty, K., Reker, D. (2005). Management of adult stroke rehabilitation care: A clinical practice guideline. *Stroke*, 36(9), e100–e143.

Eldaly, A.S., Avila, F.R., Torres-Guzman, R.A., Maita, K., Garcia, J.P., Serrano, L.P., Forte, A.J. (2022). Artificial intelligence and lymphedema: State of the art. *Journal of Clinical and Translational Research*, 8(3), 234.

Elhaddad, M. and Hamam, S. (2024). AI-driven clinical decision support systems: An ongoing pursuit of potential. *Cureus*, 16(4), e57728. doi: 10.7759/cureus.57728.

Gnana, D.A., Balamurugan, S.A., Leavline, E.J. (2016). Literature review on feature selection methods for high-dimensional data. *International Journal of Computer Applications*, 136(1), 9–17.

Groenendaal, W., Lee, S., Van Hoof, C. (2021). Wearable bioimpedance monitoring: Viewpoint for application in chronic conditions. *JMIR Biomedical Engineering*, 6(2), e22911.

Gudur, A.K., Kale, S.R., Gudur, R.A., Bhosale, S.J., More, A.L., Datkhile, K.D. (2024). Single nucleotide polymorphisms in ape1, hogg1, rad51 genes and their association with radiotherapy induced toxicity among head and neck cancer patients. *Asian Pacific Journal of Cancer Prevention*, 25(8), 2645.

Halabaku, E. and Bytyçi, E. (2024). Overfitting in machine learning: A comparative analysis of decision trees and random forests. *Intelligent Automation & Soft Computing*, 39(6), 1–10.

Khatiwada, P., Yang, B., Lin, J.C., Blobel, B. (2024). Patient-generated health data (PGHD): Understanding, requirements, challenges, and existing techniques for data security and privacy. *Journal of Personalized Medicine*, 14(3), 282.

Klein, G. and Jodocy, D. (2021). Symptoms and symptom management in survivorship patients. In *Survivorship Care for Cancer Patients: A Clinician's Handbook*, Rauh, S. (ed.). Springer, Cham, 145–202.

Krittanawong, C., Omar, A.M., Narula, S., Sengupta, P.P., Glicksberg, B.S., Narula, J., Argulian, E. (2023). Deep learning for echocardiography: Introduction for clinicians and future vision: State-of-the-art review. *Life*, 13(4), 1029.

Kumar, A., Batta, P., Ahuja, S., Rathore, P.S. (eds) (2026a). *AI-driven Innovations in Physiotherapy and Oncology 1*. ISTE Ltd, London and Wiley, Hoboken, NJ.

Kumar, A., Batta, P., Ahuja, S., Rathore, P.S. (eds). (2026b). *AI-driven Innovations in Physiotherapy and Oncology 2*. ISTE Ltd, London and Wiley, Hoboken, NJ.

Mao, L., Wang, H., Hu, L.S., Tran, N.L., Canoll, P.D., Swanson, K.R., Li, J. (2024). Knowledge-informed machine learning for cancer diagnosis and prognosis: A review. *IEEE Transactions on Automation Science and Engineering*, 22, 10008–10028.

Mohammadhossein, N., Richter, A., Richter, S., Thatcher, J. (2025). Navigating augmented reality: A practice-oriented guide to identifying and overcoming implementation obstacles. *Business Horizons* [Online]. Available at: https://www.researchgate.net/publication/389943442_Navigating_augmented_reality_A_practice-oriented_guide_to_identifying_and_overcoming_implementation_obstacles.

Patil, D.S., Yadav, T., Gudur, A. (2024). To compare the effect of swallowing and non-swallowing exercises on dysphagia in post operative oral cancer individuals: A comparative study. *Asian Pacific Journal of Cancer Prevention*, 25(8), 2945.

Peng, X. and Lu, Z. (2024). Development and validation of upper limb lymphedema in patients after breast cancer surgery using a practicable machine learning model: A retrospective cohort study. *International Journal of General Medicine*, 2(17), 3799–3812.

Peng, S.H., Huang, C.Y., Shih, C.M., Tsai, P.Y., Yang, J.C., Hsieh, C.H. (2024). Effects of lymphaticovenous anastomosis on quality of life, body image, and spiritual health in lymphedema patients: A prospective cohort study. *Healthcare*, 12(14), 1419.

Preedaa, M.G., Parkunan, T., Bharti, M.K., Mahapatra, A., Kumar, M. (2025). Clinical physiology of lymphatic disorders. In *Fundamentals of Veterinary Pathophysiology*, Rana, T. (ed.). CRC Press, Boca Raton.

Rane, N., Choudhary, S.P., Rane, J. (2024). Ensemble deep learning and machine learning: Applications, opportunities, challenges, and future directions. *Studies in Medical and Health Sciences*, 1(2), 18–41.

Ranjitsingh, L.M. and Rao, T.S. (2025). Establish legal and regulatory standards for the testing and validation of AI systems to ensure their reliability and safety in operational environments. *International Journal of System Assurance Engineering and Management*, 16, 1–6.

Vargo, M., Aldrich, M., Donahue, P., Iker, E., Koelmeyer, L., Crescenzi, R., Cheville, A. (2024). Current diagnostic and quantitative techniques in the field of lymphedema management: A critical review. *Medical Oncology*, 41(10), 241.

Wu, C., Wang, R., Lu, S., Tian, J., Yin, L., Wang, L., Zheng, W. (2025). Time-series data-driven pm2. 5 forecasting: From theoretical framework to empirical analysis. *Atmosphere*, 16(3), 292.

Zhao, H.Y., Zhou, Y., Che, C.C., Chong, M.C., Ji, Y, Zhu, Y.T. (2024). *Breast Cancer Treatment: An Interdisciplinary Approach*. Springer, Cham.

5

AI-Based Movement Quality Scoring for Post-Chemotherapy Rehabilitation

Cancer survivors require post-chemotherapy rehabilitation in order to get back to functional mobility, strength and quality of life. The persistence of neuropathy, fatigue, muscle weakness and poor balance can continue well beyond the cessation of chemotherapy and have an impact on everyday living. In conventional physiotherapy, subjective movement quality and progress monitoring of patients by clinicians results in poor assessments and discrepancies. AI-driven movement quality scoring systems offer a new objective means of contributing to the rehabilitation process. This chapter concentrates on the post-chemotherapy patient market and the current developments in computer vision, machine learning and wearable sensors in motor performance evaluation automation. We describe the quality of movement assessment with the help of AI methods such as deep learning models, kinematic analysis and pose estimation. Real-time feedback can be received by patients and clinicians, rehabilitation programs can be tailored to the requirements of the cancer survivors and the assessment can be conducted remotely. We point to such issues as privacy of patient data, absence of transparency in algorithms, intermittent flow of patients and limited number of labeled data related to cancer rehabilitation. This chapter includes the integration of AI into the workflow of physiotherapy, ethical issues, real-life examples and also a pilot study. The integration of AI movement quality scoring technology into the rehabilitation process will transform the sphere as it will provide both patients and clinicians with personalized, objective and real-time feedback. The inclusion of AI in individual rehabilitation routines can help enhance recovery, enhance patient interaction and contribute to the overall health of cancer patients to a significant degree.

5.1. Introduction

In as much as chemotherapy is inevitable in the management of most types of cancer, patients are usually left to deal with physical setbacks. Some of the most widespread and disabling of these include chemotherapy-induced peripheral

---

Chapter written by Trupti YADAV and Rahul PETHE.

neuropathy (CIPN), fatigue, muscle atrophy and loss of mobility (Lee et al. 2024). These side effects may greatly affect the performance of a patient during daily tasks, such as walking, standing, dressing or even balancing. The increasing demand includes the need to have systematic and personalized rehabilitation programs, which aim to restore physical functioning and improve the quality of life for cancer survivors.

Conventionally, measurement of physical recovery in the rehabilitation process has been largely subjective. The physiotherapists and clinicians merely observe and use subjective rating scales, such as the Berg Balance Scale, the Functional Independence Measure (FIM) and the Timed Up and Go (TUG) test. Although this has proven to be helpful, the subjective methods are essentially constrained due to variability of human observer, prejudice and the difference in ratings. Moreover, the changes that, although insignificant, can be the beginning of improvement – or of worsening – will not be detected by the conventional methods of evaluation.

The latest suggestion and application of artificial intelligence (AI) in the rehabilitation process has implemented a change in the situation. The objective quantification of the movement of people can be a sophisticated solution through AI paradigms, which are based on machine learning, deep learning and computer vision (Manakitsa et al. 2024). The AI-based solutions offer clinicians crucial objective movement information and meaningful statistics to comprehend the mobility and recovery phases. These innovations are most remarkable, particularly the AI-based movement quality scoring systems.

Such systems are connected to different types of resources, such as portable inertial measuring devices, video recordings or depth cameras such as Microsoft Kinetic or Intel RealSense cameras. The AI is able to ascertain such significant elements of biomechanics as the angle of the joints, posture, length of strides, balance and alignment. The systems apply the recognition tools to scan and identify abnormal movement, monitor CIPN-related impairments and provide feedback.

Motion data can be recorded by the sensors so that the full analysis is possible, even when AI models are used to compare thousands of gait datasets to define which biomechanics are abnormal. The evaluation of gait data provides significant data on the degree of dysfunction, the risk of falls, rehabilitation data, dynamic therapy and treatment gaps over time (Montero-Odasso and Speechley 2018). Most importantly, such devices provide real-time rehabilitative data to guide therapists and construct questions concerning their treatment. The second AI-related rehabilitation advantage is that there is a greater ability to customize an impairment.

AI is capable of processing data in bulk and training various patient profiles to offer individualized exercise prescriptions to the patient, depending on their impairments and rehabilitation objectives. Customization based on the data will be critical to this group of patients, since the physical response of post-chemotherapy patients will vary according to age, type of cancer, length of the treatment process and comorbidities (Tariq 2025). Moreover, it is possible to add AI-based systems to home-based rehabilitation models and then rehabilitate outside of the clinic. Through remote monitoring systems built into smart phones or other smart devices, patients can perform their specified exercises at home. The AI systems will be able to assess the quality of patient movement, give the required feedback and inform the therapist of any problematic movement pattern trends (Lobo et al. 2024). This will make care more accessible and give the patients more ownership and control over their rehabilitation.

This chapter examines AI's role in post-chemotherapy rehabilitation, especially in assessing the quality of movement. The chapter first discusses the nature and extent of the chemotherapy-induced physical impairments and then more thoroughly the AI technologies to assess, track and aid in the rehabilitation process. Clinicians can augment or replace traditional observation with AI, making their approach to cancer rehabilitation more accurate, truly centered around the patient and also supported by data. The goal is to enhance the functional outcomes and help maintain the overall health of cancer survivors in the long-term.

5.2. Impact of chemotherapy on physical function

5.2.1. *Common side effects affecting mobility*

Caring for cancer patients, while providing clinical benefits, facilitates the emergence of several complications that impact the patient's physical functioning and mobility. One of these complications includes chemotherapy-induced peripheral neuropathy (CIPN), which is caused by the neurotoxic effects of some chemotherapy drugs (paclitaxel, vincristine and cisplatin). CIPN causes peripheral nerve damage and may lead to immobilizing hyperesthesia, polyesthesia and burning pain, as well as to loss of proprioception, poor balance and unsteady gait. These complications can cause falls and disable patients from performing even the most simple instrumental activities of daily living (Wasilewski and Mohile 2021).

Fatigue, which is associated with cancer, and the physical decline that is suggested by a low level of activity, contribute to muscle atrophy and generalized weakness, which is primarily a loss of lower extremity muscle. This loss of muscle mass has an adverse impact on basic mobility, such as walking, sitting to standing, or providing stair climbing. Patients may also become prone to joint stiffness, primarily due to the side effects of inactivity, certain medications (e.g.

corticosteroids) and poor muscle activity control. Stiffness and loss of flexibility may increase the loss of range of motion and postural control.

These issues become progressively worse when people have been deconditioned from prolonged bedrest or sedentary behavior. It leads to a cycle of reduced movement, which then leads to even more impairments and a decrease in quality of life. Unless addressed properly and in time, the autonomy of the patient will be significantly affected and the rehabilitation process will be rendered useless.

5.2.2. *Limitations of traditional movement assessment*

During rehabilitation, physiotherapists would be sensitive to standardized methods of assessment, including the Timed Up and Go test, the Berg Balance Scale, the six-minute walk test and the oncological functional performance assessment of the goniometric assessment of the joints to gauge the mobility and balance of the patients. Although these tools are relevant in terms of information provided, they are still very far from providing an insightful view of the rehabilitation process of the patient (Weiss and Merlo 2020).

Consequently, subtle and often elusive shifts in mobility, balance, and coordination tend to manifest beyond the clinical setting, dispersed and difficult to discern across everyday contexts. Also, these assessments lean heavily into the clinical expertise of the assessing physiotherapist, which effectively lessens the reliability of the assessors and the available longitudinal data. Conditions such as the patient's pain, fatigue and enthusiasm can significantly overshadow the assessment result, while other components, such as the multidimensional movement patterns of rehabilitation that need to be assessed qualitatively, also remain absent.

Real-time assessment and feedback are also unavailable, which limits the capacity to monitor the rehabilitation process. Without detailed, objective quantitative data, optimizing rehabilitation strategies remains a challenge for physiotherapists, particularly during complex or variable rehabilitation scenarios.

5.2.3. *Need for advanced assessment tools*

Post-chemotherapy patients seek various assessments for mobility weaknesses, which are complex and variable. There is a lack of sophisticated methods that can record high volume data and movement assessments over a length of time; real-time analysis and interpretation relative to a baseline are needed for assessing mobility weaknesses and fully personalized interventions. AI has the potential to fill the gaps where mobility assessments are needed.

AI tools interfaced with wearable movement sensors and computerized gait analysis can assess and quantify various clinical features of interest, such as the degree of gait deviation for balance and control and compensatory movement associated with some patterns of dysfunction (Han et al. 2025). Machine learning of large datasets can characterize normal and pathological movement patterns and provide standardized and objective metrics to assist clinicians.

AI tools can automate routine clinician tasks, such as progress monitoring with real-time adjustment of rehabilitation exercises. AI tools interfaced with telehealth provide seamless monitoring to patients when they are not able to attend a clinic. Such instruments could aid clinicians with regard to mobility measurements and give objective information of movement patterns.

5.3. AI technologies for movement quality assessment

5.3.1. *Motion capture systems*

Motion capture (MoCap) technologies are an important component of the contemporary rehabilitation environments because they allow accurate tracking and quantifying of human motion. These systems can be widely divided into two major groups: optical motion capture systems and wearable sensor-based systems, each of which has its unique benefits, considering the use of kinematic data.

Optical systems, such as Vicon, Qualisys and OptiTrack, use multiple infrared cameras to track reflective markers placed on anatomical landmarks of the patient's body. These systems reconstruct three-dimensional movement by calculating the spatial position of each marker in real-time (Han et al. 2025). Optical systems are highly accurate and considered the gold standard in movement analysis. However, their high cost, need for controlled environments and reliance on marker placement by trained personnel limit their use outside of clinical or research settings.

Wearable IMUs (inertial measurement units) provide a flexible and more economical choice. IMUs have integrated accelerometers, gyroscopes and magnetometers, which help measure the linear and angular movements, and the orientation of the body segments. IMUs can also be attached to smart garments, wristbands, as well as other devices, which makes them more suited to continuous and ambulatory monitoring. During remote and home-based rehabilitation programs, the IMU systems also maintain good value as rehabilitation programs, since they have good portability and less spatial resolution than optical systems.

Systems capture a considerable variety of data on movements, which includes joint angles, stride length, gait cycles and angular velocity. As data is collected, it can be processed, analyzed and compared to normative patterns to identify deviations, quantify rehabilitation progress and personalize exercise interventions (Tsiakiri et al. 2025). These findings allow physiotherapists to prepare more personalized rehabilitation plans based on data.

5.3.2. *Pose estimation and computer vision*

Unrestricted motion capture hardware is no longer necessary because of the advancements made in deep learning and computer vision, resulting in the creation of markerless pose estimation technologies. Programs such as OpenPose, MediaPipe, PoseNet and DeepLabCut can take standard video recordings and, in real-time, extract 2D or 3D video joints.

These tools make use of convolutional neural networks (CNNs) centered on large training databases to find anatomical points of interest (elbows, knees, ankles, etc.) in video frames and draw skeletal representations of the human body by interlinking these points. OpenPose, for example, can analyze one image and find 135 points of interest including those of the face, limbs and the rest of the body, while MediaPipe is designed for real-time pose estimation on mobile devices as a lightweight alternative (Garg et al. 2023).

The estimation of poses can be used to extract advanced kinematic features from inexpensive devices such as a webcam or smartphone in order to make motion analysis available to clinics and patients. Most importantly, it facilitates remote and resource-limited areas to gain access to motion analysis. Pose estimation technology has the potential to assess body alignment, movement efficiency and motor coordination.

The oncology rehabilitation field benefits most from this technology. Patients often have issues with posture and gait as a result of chemotherapy-related neuropathy or atrophy of the muscles (Cristian 2020). By combining pose estimation with AI, health professionals can assess patients remotely as they complete exercises, identify inappropriate or compensatory movements patients may be taking and validate their attendance to rehabilitation sessions.

5.3.3. *Machine learning and deep learning models*

The next step after capturing motion data via computer vision or sensors is the classification and evaluation of the quality of the recorded movements. At this stage, machine learning and deep learning techniques are used. These models learn to identify different movement patterns and assess the quality of movements using labeled data.

For the classification of activities, various supervised learning models, such as random forests, support vector machines (SVMs), decision trees, and k-nearest neighbors (k-NN), are successfully applied. These models work well when the feature space is clearly defined, which is the case with parameters such as joint angles, step frequency or symmetry indices (Cunningham and Delany 2021). For instance, time-series joint data could be used to train an SVM to identify and differentiate healthy and impaired gait patterns.

However, the dynamic and temporal nature of human movement suggests that deep learning models designed for sequence tasks may be more suitable. RNNs, LSTMs and GRUs are examples of time-series models used to evaluate the progression and fluidity of movement in human activities. These models have the ability to identify and read the changes in movement which are not visible to the human eye in most cases. Transformer-based models have recently started to be used in motion analysis tasks.

In contrast to RNNs, transformers have attention mechanisms that help us to study the sequence as a whole. This assists in the capturing of intricate associations in the movement patterns. Transformers are currently being used to measure rehabilitation efforts in cancer patients – sit-to-stand, walking or reaching; and in the process, give an elaborate scoring and prediction of the recovery patterns (Han et al. 2025).

The AI models that are developed by using large datasets are trained to determine the movement deviations that are associated with particular conditions, including foot drop due to neuropathy, postural imbalance or asymmetrical gait. This provides clinicians with high-resolution information regarding the functional limitations of a patient.

5.3.4. *Scoring metrics and interpretability*

One of the most important values of AI-based movement analysis is that it can provide quantitative marking of rehabilitation performance. The AI systems can evaluate exercises based on the following measures:

– *Range of motion (ROM)*: measures the level of the movement of the joints.

– *Normal*: measures the balance between the right and the left sides of the body.

– *Smoothness*: this is a measure of fluidity and coordination of motion.

– *Timing*: this is an examination of the rhythm, speed and lag of succession in motion phases.

– *Accuracy*: compares the movement that has been executed with a set standard.

These metrics are objective in nature and hence can be normalized across patients, therefore allowing the tracking of progress of rehabilitation, as time goes by (Colombo et al. 2005).

Moreover, XAI systems justify a particular score with respect to particular motion characteristics (e.g. no knee flexion or slow arm swing coordination) that result in a poor score. This understandability develops confidence with the physiotherapists and patients.

To illustrate, AI is able to provide explanations of its score by visualizing the attention and saliency maps to reveal the most relevant joints and movements used in the task. This will enable clinicians to confirm the AI score and manipulate therapy according to their skills. Therefore, enhancing patient engagement. There is a better understanding of which movements should be reinforced, and patients will be able to improve and become more motivated.

5.3.5. *Real-time feedback systems*

Real-time feedback is one of the most exciting solutions to how AI can be utilized during the rehabilitation process. Such systems combine AI solutions with convenient interfaces, such as mobile apps, smart mirrors or AR devices, therefore allowing users to receive immediate corrective guidance during workout.

For example, a patient doing knee extension exercises at home could see indicators on a smartphone displaying whether the leg is at too low, too fast or at the wrong angle.

To help with form correction, the patient can be assisted by voice instructions, haptic vibrations or visual overlays, mimicking the presence of a physiotherapist (Baqai et al. 2019).

In rehabilitation oncology, exercises can be done at home and guided by a specialist, which is a great way to avoid clinic fatigue. Patients are motivated and their consistent adherence is tracked by real-time feedback systems.

AI technique/technology	Core components	Function in movement assessment	Clinical application in rehabilitation	Expected benefit/outcome
Motion capture systems (optical and IMU-based)	Infrared cameras, reflective markers; wearable inertial sensors (accelerometers and gyroscopes)	Captures 3D kinematics; joint angles, stride and posture data	Assessment of gait, limb motion and coordination in post-chemotherapy patients	Objective motion quantification; identification of compensatory movements
Pose estimation and computer vision	Deep learning models (OpenPose, MediaPipe and PoseNet); depth cameras, video footage	Markerless tracking of joint landmarks and body alignment	Real-time evaluation of posture, gait and task performance from standard video	Low-cost, accessible assessment tool for clinical and home settings
Machine learning models (SVM, random forest and k-NN)	Feature extraction from motion data (joint angles, timing and symmetry)	Classifies movement as normal, impaired or compensated	Detects chemotherapy-induced gait or coordination impairments	Automated movement classification with high sensitivity
Deep learning architectures (LSTM, GRU and transformer)	Time-series data; attention mechanisms; recurrent layers	Models' sequential movement dynamics; predicts deviations	Long-term tracking of recovery progress and motion fluidity	Captures subtle, temporal movement variations for precise scoring
Scoring and interpretability frameworks	Range of motion (ROM), symmetry, smoothness, timing and explainable AI (XAI) visualization	Generates quantitative movement quality scores with interpretability	Provides therapists with detailed feedback on which movements need correction	Transparent, standardized and explainable evaluation metrics
Real-time feedback interfaces (AR/VR, mobile apps and smart mirrors)	Haptic/visual feedback; wearable integration; AI-driven guidance	Provides immediate corrective and motivational cues	Home-based guided exercises and remote physiotherapy sessions	Enhanced engagement, adherence and safety in rehabilitation

Table 5.1. *Core AI technologies and their applications in post-chemotherapy movement quality assessment*

5.4. Clinical applications in post-chemotherapy rehabilitation

5.4.1. *Gait analysis*

Gait analysis is an integral part of the rehabilitation process for patients, especially for patients with cancer recovering from CIPN, muscle weakness or joint dysfunction. These conditions may lead to a change in an individual's gait pattern, resulting in asymmetry, limited range of motion and an elevated risk of falling. There is a notable increase in the use of AI to identify and quantify abnormal gait parameters (Lim et al. 2022).

AI systems, whether using wearable inertial sensors or vision systems such as depth cameras and smartphone applications, can capture and analyze critical gait characteristics, including stride length, cadence, foot clearance, step width and step symmetry. Machine learning applications are able to identify the presence of abnormal gait patterns that indicate walking issues such as foot drop or loss of balance.

This objective information attempts to lend greater objectivity to what is otherwise a largely qualitative assessment of gait severity. It also monitors progress, records progress and customizes rehabilitation. As an example, patients with greater time in the double-support and shorter strides could experience strength training and proprioceptive exercises as an advantageous effect (Williams 2020). On the other hand, individuals with a problem with foot clearance could require dorsiflexion-oriented exercises. AI also helps in analyzing gait temporally to detect regression, allowing for timely intervention at the clinic.

5.4.2. *Upper limb function assessment*

Chemotherapy and cancer treatments have an impact on the upper limb function, especially when it comes to surgeries, radiation therapy and chemotherapy-induced motor neuropathies involving breast cancer. These conditions can affect an individual's ability to control fine and gross motor, joint and muscle functions, as well as grip, which could make reaching, writing and lifting tasks more difficult.

Systems based on AI and pose estimation and sensor-based tracking technology can monitor upper limb kinematics in real-time. Such systems analyze smoothness of movements, joint alignment, reaction times and accuracy of task completions across several standard rehabilitation exercises and functional tasks, which can include reaching exercises. For example, an AI can monitor the arm's reach trajectory, detect compensatory movements of the trunk and measure joint angles to assess whether the movements are expected (Wang et al. 2022).

Tools for computer vision, such as OpenPose and MediaPipe, can analyze 2D or 3D joint landmarks, and can be used with a simple smartphone or webcam. This expands the accessibility of sophisticated movement assessment tools, which can be especially useful in low-resource environments, remote rehabilitation and situations when a patient with movement limitations must be assessed.

To ensure that the therapy is not too difficult or too easy, AI can give real-time feedback by describing movements as normal, compensated or impaired to alter the activity, making the process adaptive and dynamic (Gultner et al. 2019).

5.4.3. *Balance and fall risk detection*

Balance remains an issue of concern in oncology rehabilitation and particularly in patients who have CIPN, dysfunction of the vestibular apparatus or deconditioning of muscles. These patients typically tend to fall, and this may result in injuries, hospitalization and loss of confidence in their mobility.

The AIs evaluate balance based on the postural sway, changes of the center of pressure (CoP) and reaction to the perturbations (Wantanajittikul et al. 2022). In tests, they gather the data in the form of a standing position, dynamic sit-to-stand activities and gait activities in the form of walking a balance beam, force platforms, camera-based or wearable IMUs.

In regulated AI learning, the clinician will be the one who gives the abnormality test. In unsupervised models, abnormal balance is identified through the AI by referring to normative datasets. Predictive models assess the risk of falls, helping therapists to determine appropriate assessments, such as proprioceptive training, resistance exercises and the provision of assistive devices.

5.4.4. *Remote rehabilitation and telehealth*

Two of the most innovative uses of AI in cancer rehabilitation is in remote monitoring and telehealth. Due to cancer-related fatigue or immunosuppression, patients may struggle with the logistics of frequently coming into the clinic. AI-enabled rehabilitation systems assist patients in guided exercises and assessments for cancer rehabilitation in their home. These systems use wearable sensors, smartphones or webcams to track movements during therapy (Devi et al. 2025).

During therapy sessions, AI systems analyze movement performance, form and adherence to exercises in real-time, offering feedback to the patient. Clinicians have online access to dashboards that track patient progress and are able to make remote therapy plan adjustments and communicate with the patient during scheduled teleconsultations, which are pitched in the AI systems as needed. Gamification, goal tracking and reminders that are built into the systems enhance motivation and consistency, helping adherence to rehabilitation.

Furthermore, home session data collected during their rehabilitation will contribute to longitudinal records on the patient and assist the clinician in spotting trends, adjusting interventions or predicting relapses (Sprint et al. 2017). Remote AI systems are particularly useful in rural or remote, underserved regions with little to no access to oncology rehabilitation specialists.

This remote access to comprehensive rehabilitation services transcends geographical barriers to clinic location, ensuring equitable access to rehabilitation services for all patients.

5.5. Challenges and limitations

5.5.1. *Data limitations*

Quality and diversity are essential characteristics of the datasets that machine learning models are trained on. It is unfortunate that the datasets available for the post-chemotherapy rehabilitation field have these features. Most available movement datasets target the general population, or focus on specific neurological conditions, such as stroke or Parkinson's disease (Wang et al. 2021). Datasets that focus on movement rehabilitation for cancer survivors are rare. Those that are available are only loosely related, as cancer survivors have specific movement limitations due to CIPN, fatigue or atrophy of muscles that need to be factored in. Supervised data training is also impeded by poorly documented or non-standardized annotations in many datasets. This is compounded by issues in sample sizes, absence of longitudinal datasets and the inability to predict functional decline. As for overfitting, it is inevitable for models to train on inadequate datasets and perform optimally on training data while struggling with unseen or new cases. This will need dedicated multi-center initiatives for large integrated datasets, as well as the creation of standardized cancer rehabilitation datasets.

5.5.2. *Interpatient variability*

Cancer survivors are a very diverse population in part because of the many differences within each of the influencing factors, such as age, type of cancer, treatment received, disease or illness state, and other coexisting medical conditions. Consider the differences in recovery mobility of a young breast cancer survivor with that of an older prostate cancer survivor who also has diabetes, arthritis and other conditions. The variations we see between different individuals make it difficult for AI models to provide accurate predictions. This is because AI models rely on consistent patterns to make predictions. For instance, a movement scoring model uses a uniform scoring system, but does not consider different patients' needs and starting points. Hence, the case for personalized AI systems that can tailor assessments and suggestions is compelling. More sensitive models can indeed consider dynamic baselines, which scores patients on their historical performances rather than comparative scoring to statistical averages of the population (Su et al. 2018). Unfortunately, the absence of fine-grained, continuous and extensive data points increases the computational complexity of personalized models and is currently a huge limitation.

5.5.3. *Sensor and environment constraints*

Wearable sensors and optical motion capture systems provide accurate and substantial biomechanical data, but with tradeoffs. For useful data to be collected, wearable IMUs (Inertial Measurement Units) must be continuously worn. Unfortunately, some older patients and patients with neuropathy, along with other conditions, may be uncomfortable with the sensors, may have skin sensitivity and could find the devices difficult or cumbersome to place correctly – and therefore to recharge and maintain. On the other hand, optical devices such as depth cameras and infrared motion capture systems need to be set up in controlled lighting and fixed positions. This is more applicable to a lab or clinical setting rather than in-home use (Jean et al. 2022). The quality of data and effectiveness of models may be compromised by certain environmental factors such as lighting, background clutter and furniture arrangement. Such constraints make it difficult to use AI-powered movement analysis tools for rehabilitation in practical scenarios, particularly for monitoring in the patients' homes or during tele-rehabilitation.

5.5.4. *Privacy and ethical issues*

The application of AI to healthcare raises ethical issues, particularly regarding privacy and patients' rights. Motion capture and wearable sensors collect sensitive data, including video, biometrics and geolocation data. It is necessary to protect the confidentiality and integrity of data that is sensitive, and guarantee secure transportation. The compliance has to include HIPAA in the US and GDPR in the EU. To ethically use it, patients must be told what is planned to be done with their data, how it is going to be stored and shared. It has to be informed consent, but this may be challenging in the case of vulnerable patients, such as older cancer patients, who are less technologically available and digitally literate (Kumar et al. 2026a).

5.5.5. *Clinical integration*

AI has proven its ability to enhance rehabilitation, and yet the actual test is to adopt it in the daily clinical routine. The current healthcare systems continue to use non-interoperable electronic health records that are still fragmented, thus making the challenge of integrating AI tools into the current workflow more difficult. Clinicians are reluctant to implement AI due to the fear of losing control, workload and quality of data. To be efficient, AI tools must be designed through collaboration with clinicians, have easily understandable designs, give actionable insights and facilitate collaborative decision-making (Kumar et al. 2026b). The development or modification of training content, clinical policies and guidelines should also be motivated by AI in oncology rehabilitation to make its use ethical, safe and effective.

5.6. Future directions and opportunities

5.6.1. *Individualized rehabilitation programs*

AI can be used to generate very individualized rehabilitation programs. It constantly evaluates the deficits in movement of a patient, their treatment history and recovery history. These AIs are real-time adaptable, changing the parameters of exercise (intensity, duration and complexity) according to performance information. Individualization ensures that the program is not so difficult that it causes demoralization. This maximizes recovery results and lowers the risk of injury.

5.6.2. *Multi-modal data fusion*

The combination of various data types in new rehabilitative approaches is now in place. These comprise kinematic information about the motion analysis system and physiological information about the heart rate, electromyography (EMG) and galvanic skin response in real-time (Kumari et al. 2024). Inclusion of patient-reported outcomes – fatigue, pain and emotional states – gives a more holistic view of the recovery compared to clinical data. The method of data fusion enhances evaluation of the quality of movement and allows comprehensive treatment planning.

5.6.3. *Federated learning and data sharing*

Federated learning deals with the issue of data privacy and accessibility. It enables hospitals and clinics to cooperate in training AI models without revealing details about the patients. This will enable more powerful models without the loss of patient data, and it will maintain ethical standards of conducting research.

5.6.4. *Gamification and engagement*

Motivation opens the platform of gamified rehabilitation. AI-based feedback on patients is provided in the form of interactive games, which are compatible with the motivation and rehabilitation goals of the patient in real-time (Gudur et al. 2024). This would assist them in adhering to the regimen given to them, therefore enhancing consistency, engagement and functional outcomes.

5.6.5. *Cross-domain application*

AI systems were made with the objective of integrating rehabilitation technology in the field of cancer. However, the technology is easily transferable to other areas such as stroke recovery, rehabilitation for Parkinson's and orthopedic rehabilitation. With the potential to transform a variety of sectors, the systems are likely to gain widespread use in practice.

5.7. Conclusion

AI movement quality scoring systems offer a more advanced approach to physical rehabilitation. AI systems offer the potential for more objective rehabilitation post chemotherapy. AI systems using machine learning and computer vision integrated with wearable sensors enable the systems to quantify human movements. AI systems provide real-time critiques, remote progress monitoring and intervention modifications. This makes the systems helpful to both patients and rehabilitation professionals. However, for wide-ranging use, the issues of dataset diversity, model explainability, patient autonomy and real-world clinical applicability must be solved. This might be the first of many clinical applications of AI, and it will work best when engineers and data scientists cooperate with medical professionals and ethicists. Integrative AI systems will not substitute physical therapists but, under proper regulatory frameworks, AI will change the paradigm of movement assessments and enhancement of survivors of cancer; the systems provide physiotherapists with real-time critiques. AI will also level the core rehab systems to real-time advanced movement assessments. Integrative AI systems will promote autonomy, recovery and the overall quality of life for post-cancer rehabilitation.

5.8. References

Baqai, A., Memon, K., Memon, A.R., Shah, S.M. (2019). Interactive physiotherapy: An application based on virtual reality and bio-feedback. *Wireless Personal Communications*, 106(4), 1719–1741.

Colombo, R., Pisano, F., Micera, S., Mazzone, A., Delconte, C., Carrozza, M.C., Dario, P., Minuco, G. (2005). Robotic techniques for upper limb evaluation and rehabilitation of stroke patients. *IEEE Transactions on Neural Systems and Rehabilitation Engineering*, 13(3), 311–324.

Cristian, A. (ed.) (2020). *Breast Cancer and Gynecological Cancer Rehabilitation*. Elsevier, Amsterdam.

Cunningham, P. and Delany, S.J. (2021). K-nearest neighbour classifiers-a tutorial. *ACM Computing Surveys (CSUR)*, 54(6), 1–25.

Devi, B.R., Neela, V., Ashwitha, A., Reddy, C.S., Balaji, P., Sudhakara, M. (2025). AI-driven predictive models for personalized rehabilitation and assistive systems. In *Predictive Algorithms for Rehabilitation and Assistive Systems*, Korupalli, V.R.K., Ranasinghe, P., Feldman, D. (eds). IGI Global, Hershey, PA, 87–114.

Frangoudes, F., Matsangidou, M., Schiza, E.C., Neokleous, K., Pattichis, C.S. (2022). Assessing human motion during exercise using machine learning: A literature review. *IEEE Access*, 10, 86874–86903.

Garg, S., Saxena, A., Gupta, R. (2023). Yoga pose classification: A CNN and MediaPipe inspired deep learning approach for real-world application. *Journal of Ambient Intelligence and Humanized Computing*, 14(12), 16551–16562.

Gudur, R.A., Bhosale, S.J., Gudur, A.K., Datkhile, K.D. (2024). Effect of cyp1a1, cyp1b1 and cyp2c gene polymorphisms on doxorubicin and paclitaxel. *Bioinformation*, 20(10), 1244.

Han, X., Guffanti, D., Brunete, A. (2025). A comprehensive review of vision-based sensor systems for human gait analysis. *Sensors*, 25(2), 498.

Jean, S., Burnham, C.A., Chapin, K., Garner, O.B., Pant Pai, N., Turabelidze, G., Butler-Wu, S. (2022). At-home testing for infectious diseases: The laboratory where you live. *Clinical Chemistry*, 68(1), 19–26.

Kumar, A., Batta, P., Ahuja, S., Rathore, P.S. (eds) (2026a). *AI-driven Innovations in Physiotherapy and Oncology 1*. ISTE Ltd, London and Wiley, Hoboken.

Kumar, A., Batta, P., Ahuja, S., Rathore, P.S. (eds) (2026b). *AI-driven Innovations in Physiotherapy and Oncology 2*. ISTE Ltd, London and Wiley Hoboken.

Kumari, V.S., Vijila, J., Balu, R. (2024). Decision making biomedical support system. In *Artificial Intelligence-Based System Models in Healthcare*, Anand, A.J., Kalaiselvi, K., Chatterje, J.M. (eds). Wiley, Hoboken.

Lee, K.T., Bulls, H.W., Hoogland, A.I., James, B.W., Colon-Echevarria, C.B., Jim, H.S. (2024). Chemotherapy-induced peripheral neuropathy (CIPN): A narrative review and proposed theoretical model. *Cancers*, 16(14), 2571.

Lim, A.C., Natarajan, P., Fonseka, R.D., Maharaj, M., Mobbs, R.J. (2022). The application of artificial intelligence and custom algorithms with inertial wearable devices for gait analysis and detection of gait-altering pathologies in adults: A scoping review of literature. *Digital Health*, 8, 20552076221074128.

Lobo, P., Morais, P., Murray, P., Vilaça, J.L. (2024). Trends and innovations in wearable technology for motor rehabilitation, prediction, and monitoring: A comprehensive review. *Sensors*, 24(24), 7973.

Manakitsa, N., Maraslidis, G.S., Moysis, L., Fragulis, G.F. (2024). A review of machine learning and deep learning for object detection, semantic segmentation, and human action recognition in machine and robotic vision. *Technologies*, 12(2), 15.

Montero-Odasso, M. and Speechley, M. (2018). Falls in cognitively impaired older adults: Implications for risk assessment and prevention. *Journal of the American Geriatrics Society*, 66(2), 367–375.

Sprint, G., Cook, D., Weeks, D., Dahmen, J., La Fleur, A. (2017). Analyzing sensor-based time series data to track changes in physical activity during inpatient rehabilitation. *Sensors*, 17(10), 2219.

Su, T.L., Jaki, T., Hickey, G.L., Buchan, I., Sperrin, M. (2018). A review of statistical updating methods for clinical prediction models. *Statistical Methods in Medical Research*, 27(1), 185–197.

Tariq, M.U. (2025). Machine learning techniques for personalized rehabilitation: Enhancing patient outcomes through predictive algorithms. In *Predictive Algorithms for Rehabilitation and Assistive Systems*, Korupalli, V.R.K., Ranasinghe, P., Feldman, D. (eds). IGI Global, Hershey, PA, 221–252.

Tsiakiri, A., Plakias, S., Giarmatzis, G., Tsakni, G., Christidi, F., Papadopoulou, M., Bakalidou, D., Vadikolias, K., Aggelousis, N., Vlotinou, P. (2025). Gait analysis in multiple sclerosis: A scoping review of advanced technologies for adaptive rehabilitation and health promotion. *Biomechanics*, 5(3), 65.

Wang, A.B., Housley, S.N., Flores, A.M., Kircher, S.M., Perreault, E.J., Cope, T.C. (2021). A review of movement disorders in chemotherapy-induced neurotoxicity. *Journal of Neuroengineering and Rehabilitation*, 18(1), 16.

Wang, X., Fu, Y., Ye, B., Babineau, J., Ding, Y., Mihailidis, A. (2022). Technology-based compensation assessment and detection of upper extremity activities of stroke survivors: Systematic review. *Journal of Medical Internet Research*, 24(6), e34307.

Wantanajittikul, K., Wiboonsuntharangkoon, C., Chuatrakoon, B., Kongsawasdi, S. (2022). Application of machine learning to predict trajectory of the center of pressure (COP) path of postural sway using a triaxial inertial sensor. *The Scientific World Journal*, 2022(1), 9483665.

Wasilewski, A. and Mohile, N. (2021). Meet the expert: How I treat chemotherapy-induced peripheral neuropathy. *Journal of Geriatric Oncology*, 12(1), 1–5.

Weiss, H.M. and Merlo, K.L. (2020). Affect, attention, and episodic performance. *Current Directions in Psychological Science*, 29(5), 453–459.

Williams, D. (2020). Investigating how the double support period affects fall risk in bipedal gait. PhD Thesis, The Pennsylvania State University.

6

Virtual Reality and AI for Pain Management in Cancer Physiotherapy

Cancer treatment presents pain and movement disorders that impact the quality of life, everyday activities and the emotional condition of survivors. Pain management is a pioneer in the multidisciplinary, personalized approach required for physiotherapy-integrated oncology care. The advancements in the field through digital innovations (particularly virtual reality (VR) and artificial intelligence (AI)) are promising for evidence-based contributions to pain management and functional rehabilitation. VR helps patients forget about their pain and anxiety and take part in rehabilitation, immersing them in the world of interaction using three-dimensional reality. Gamified therapy exercises are developed and challenge patients and enhance compliance, movement, neuroplasticity and rehabilitation adherence. AI simplifies position analysis and uses proprietary predictive algorithms in real time to dynamically prescript exercise programs, which allows responsive rehabilitation programs. Real-time analytic AI tools and predictive algorithms match the objectives of patients and rehabilitation.

In this chapter, we conduct a review of VR and AI application in physiotherapy of cancer pain management and outline the state of the art of pain management VR applications, the use of AI in rehabilitation and the neurophysiological pathways that mediate pain at the point of interaction between these technologies. By combating issues of technological access, cost management, patient-friendliness and the absence of compatibility across health systems, we find opportunities to further study the overlap of the technologies to promote oncology care. The chapter also emphasizes the role of new and emerging advanced technologies, such as AI and VR, in oncology physiotherapy. Such innovative technologies are pain-optimized, safer and more mobile, more customized and individualized and give sustainable care, to help patients in the post-cancer rehabilitation process.

Chapter written by Dhairyasheel PATIL and Abhay KASHETWAR.

6.1. Introduction

Pain is common and debilitating for cancer patients. It can stem from surgical procedures, radiation and chemotherapy, or even the tumors themselves, thus arising from both the treatments and the disease. Its ramifications can reach emotional pain, lack of sleep and ultimately quality of life. Opioids and NSAIDs, even with older techniques of physiotherapy such as TENS, still do not seem to offer the pain relief patients seek (Alzeer 2023).

Unwanted side effects, possible addictions and varying effectiveness from patients all suggest the need for other options. Integrative approaches seem to be the ideal next step.

Technologies such as virtual reality (VR) and artificial intelligence (AI) are becoming reality in pain self-management and can transform the ways patients seek pain management with adaptive and even immersive evidence-based care (Cascella et al. 2023). VR in physiotherapy can help cancer patients and empower them mentally. VR can help create and simulate environments to help distract patients to aid in pain management.

Neuroimaging weights and VR track and retrain the attention of the user with the aid of millisecond shifts of focus, creating the condition of distraction analgesia by lowering, and then sustaining, the activity of the insula, the anterior cingulate cortex and the somatosensory cortex. Immersive VR with fabricated, and then guided, meditative VR practices recreate peaceful, nature-filled scenarios, then turns toward guided meditative VR practices (Prabhu et al. 2020).

For the purpose of rehabilitation facilitation, VR-supported physiotherapy expands and enhances existing methods and techniques. Instead of monotonous exercises in a clinical setting, patients can navigate through a forest while stretching, complete tasks in a game-like interface that requires limb movement, or practice balance and gait in a simulated environment (Sun 2022).

Other remaining factors are irrelevant distractions. Adherence, as a focal factor, with adaptive therapy regimens as the latest distraction in contrast to interactive VR environment therapy, is a relevant distraction (Meier et al. 2024).

AI algorithms examine data concerning movement, pain and engagement, enabling real-time adjustments for optimal therapy session customization. Algorithms can evaluate whether a patient's movement is compensating incorrectly and can intra-session adjust the difficulty of a virtual exercise. This is important to enhance personalization to ensure that the therapy is not too harsh or too lenient. AI systems can feed on predictive pain flare-ups and come up with preemptive responses rather than reactive ones. AI is used to identify patterns that indicate whether a patient will undergo increased pain or not, depending on the biometrics, self-reported pain, posture and their symptoms (Shi and Wu 2023).

This helps clinicians modify the objectives of therapy sessions to reduce pain through advance exercises and coping skills. AI and VR technologies are combined to support remote and home rehabilitation programs. This is vital in the surmounting of the technology and mobility hurdles to access rehab, which is prevalent in cancer patients and individuals with hard-to-travel circumstances.

Virtual assistants, which are based on AI, will guide patients through the exercises at home and monitor the adherence to exercise. They gather and transmit pertinent information to the caregivers and facilitate the smooth flow of care and lessen the workload in healthcare centers (Kalusivalingam et al. 2021). VR-based tele-rehabilitation platforms integrating AI are scalable and provide flexible solutions that are equally applicable in the city and countryside.

The biopsychosocial model of pain management is also supported by these technologies. Emotional distress, anxiety and helplessness usually increase pain among cancer patients. Psychological pain may grow, even in the case of failure of opioid therapy. VR environments can be used to enhance the mood and decrease the levels of anxiety since the patient gets a strong sense of being in a virtual world. AI examines the behavior, speech and tone of patients to identify psychological pain and prescribe mental-health treatments and mindfulness meditation.

The holistic approach is congruent with current palliative care objectives to enhance quality of life and reduce pain.

Both VR and AI interventions have the potential to manage cancer-related pain and this can be evidenced by clinical studies. Initial trials have shown that the intensity of perceived pain reduces, the range of motion is improved and patient satisfaction increases with the addition of these technologies to physiotherapy programs. Even though more extensive, extended research is needed to verify these results, the initial evidence is optimistic and has preoccupied the pursuit of common digital therapeutics in the management of cancer (Shin et al. 2024).

The remaining issues are the development of user-friendly interfaces, the need to provide equal access to technology, data privacy and the training of healthcare providers to use VR–AI. These barriers are gradually being overcome by constant innovation and interdisciplinary cooperation. Since VR headsets are becoming more affordable and AI algorithms becoming more trustworthy, incorporating these technologies into a clinician's daily routine will become more viable.

6.2. Cancer pain: scope and challenges

6.2.1. *Prevalence and impact*

Pain is one of the most disturbing and frequent symptoms experienced by cancer patients. Research indicates that approximately 30–50% of patients have high levels of pain during diagnosis and this level increases to more than 70% in patients with advanced or terminal conditions.

Types of pain are different: nociceptive pain is caused by tissue damage, and neuropathic pain is caused by nerve damage by a chemotherapy or tumor compression; in many cases, both nociceptive and neuropathic pain occur, making treatment difficult. Cancer-related pain, in addition to the physical effect, takes a heavy toll on a psychological level. Constant pain is closely correlated with emotional issues, such as anxiety, depression, irritability and sleeping disorders, which reduces quality of life and restricts the possibilities of patients participating in everyday activities or rehabilitation programs. Pain that is poorly treated (or not treated) negatively affects the process of treatment compliance, recovery and eventually patient survival (Friedrichsdorf et al. 2016).

6.2.2. *Traditional physiotherapy weakness*

Traditional physiotherapy is still one of the major nonpharmacological options in cancer pain management and rehabilitation. Some common methods are manual therapy, joint mobilizations, exercises, stretching, electrotherapy as well as heat and cold.

Although these methods have benefits, they also have obvious limitations. Firstly, the sessions are usually awkward, exhausting or monotonous to the patients and this interferes with prolonged participation. Secondly, outcomes are different as they are affected by the skill, experience and compliance of the therapist. Thirdly, these methods tend to be characterized by the lack of immediate feedback, constant control and individual patient flexibility, which reduces the effectiveness in the long term. The following deficiencies justify the use of VR and AI to improve physiotherapy procedures (Perret et al. 2006).

6.3. VR in pain management

6.3.1. *What VR?*

VR is an interactive, immersive and computer-generated 3D environment, which allows the user to interact with digital content in real time. It is more commonly experienced with the help of head-mounted displays (HMDs), sensor-impregnated gloves or motion-tracking systems, which replicate real or imagined environments that may be navigated and interacted with. VR is gaining progressively more popularity in the sphere of healthcare, particularly in physiotherapy, as it facilitates the ability to change sensory perception and offers engagement of the mind, which helps to cope with pain. VR is fully immersive and absorbs the visual, auditory and proprioceptive data, which means that the discomfort and physical restrictions are not focused on anymore. In the case of cancer patients whose pain can be caused by tumors, surgery, chemotherapy and radiation therapy, VR represents an interesting, noninvasive form of pain management that enhances the physiotherapeutic experience (Bhowmik 2024).

6.3.2. *Mechanisms of VR for pain relief*

A number of neurophysiological and psychological processes are involved in the analgesic effect of VR. All of these mechanisms are synergistic, and they decrease the subjective perception of pain and increase engagement in rehabilitation activities:

– *Distraction hypothesis*: the distraction hypothesis is one of the most researched and basic processes of pain relief induced by VR. The process of perception of pain depends not only on the perceptions we receive but also on the thinking. VR provides an immersive, multisensory experience that takes up all of the attention of a patient, leaving them with less cognitive capacity to interpret nociceptive (pain-associated) information. This attentional shift interferes with the pain pathways, particularly in the anterior cingulate cortex, insula and somatosensory cortices, leading to reduction of pain sensation (Torta et al. 2017). As an example, when subjected to painful processes, patients who are in an environment of a soothing virtual forest have a much lower discomfort level compared to those who are not using VR, which demonstrates the strong effect of cognitive distraction.

– *Body ownership and neuroplasticity*: neuroplastic reprogramming with the body ownership illusion is another critical mechanism. VR worlds may mimic usual movements of limbs or body processes, despite the fact that the physical body of the patient is not functioning properly. This is analogous to mirror therapy, which is applicable in phantom limb pain, involving providing visual feedback of an intact body part in motion that can fool the brain into remapping the body part representation of the missing or painful one. The long-term effect of VR simulation in pain-free movement is to retrain the brain record centers involved in pain processing, which promotes neuroplasticity, i.e. the brain capacity to rearrange itself by creating new neuronal connections and, eventually, alleviating chronic pain.

– *Endorphin*: this primary neurotoxin triggers dopamine and serotonin release, which causes chronic pain and neuropathy.

– *Endorphin release*: this is the main neurotoxin. It stimulates the dopamine and serotonin release and leads to chronic pain and neuropathy. Immersive and engaging VR triggers the reward system in the brain, resulting in the release of endogenous opioid such as endorphins and dopamine. These neurochemicals generate analgesic effects and wellness. Dopaminergic is stimulated to stimulate motivation and reduce emotional distress caused by chronic pain (Bonanno et al. 2024). The mechanism is particularly useful in long-term cancer rehabilitation, when the morale and engagement of patients is as important as the reduction of physical symptoms.

6.3.3. *Clinical applications of VR in cancer physiotherapy*

VR is not a universal solution since it can be adapted to different clinical situations in oncology treatment by its flexibility. Its uses cut across acute and chronic pain management, rehabilitation and psychological support. The uses of VR in cancer pain management and physiotherapy are given in Table 6.1:

– *Burns and surgical recovery*: patients with tumor surgeries are usually in excruciating pain after surgery. VR is able to engage patients in wound cleaning and dressing changes and prevent acute pain and anxiety, which will help make physiotherapy more bearable.

– *Chemotherapy-induced peripheral neuropathy (CIPN)*: numerous chemotherapy medicines damage nerves, resulting in pain and tingling, as well as balancing skills. VR activities, including virtual balance boards or walking simulations, entail training the sensory reeducation, the recovery of proprioceptors and gait stability.

– *Pediatric oncology*: children undergoing chemotherapy, radiation or painful diagnostic procedures often experience high levels of distress. VR distraction therapy – such as playing immersive games or exploring virtual underwater worlds – has been shown to significantly reduce pain and anxiety in pediatric patients, facilitating better cooperation during physiotherapy sessions (Addab 2020).

– *Palliative care and end-of-life support*: VR provides terminal patients with immersive experiences. This includes travel through the virtual world, landscapes or music, which may improve mood and reduce perceived pain without medication.

Clinical application area	Type of VR intervention	Target outcome/mechanism	Observed or reported benefits
Post-surgical rehabilitation	Immersive nature scenes; guided movement VR	Distraction analgesia and motor retraining	Reduced post-operative pain; increased tolerance during physiotherapy sessions
Chemotherapy-induced peripheral neuropathy (CIPN)	Sensory reeducation and gait simulations	Reestablish proprioceptive feedback and reduce neuropathic pain	Improved balance; reduced tingling and discomfort
Pediatric oncology	Game-based interactive VR environments	Cognitive distraction; anxiety reduction	Lower procedural pain; improved therapy cooperation and adherence
Palliative and end-of-life care	Relaxation and immersive travel VR experiences	Emotional and sensory modulation	Improved mood; reduced perception of pain without pharmacologic aid
Post-mastectomy physiotherapy	Virtual mirror therapy (VMT)	Neuroplastic reprogramming and visual illusion of restored limb function	Reduced phantom pain; enhanced upper limb mobility
Long-term cancer survivorship	Home-based or tele-VR sessions	Remote rehabilitation and engagement	Enhanced therapy continuity and patient satisfaction

Table 6.1. *Applications of VR in cancer pain management and physiotherapy*

6.3.4. *Evidence-based observations*

The idea to use VR in the cancer physiotherapy is supported by progressively more clinical studies, as well as by patients. The subjects complain of reduced pain, increased motivation and compliance, compared to when in traditional therapies alone.

It has been demonstrated that even short-term exposure to VR results in strong analgesia. The reduction in pain scores up to 40% reported by cancer patients who used VR in a rehab setting was quite impressive in randomized trials. They also reported VR exercises to be more enjoyable and less tiring and achieved higher rates of completion and improved functional outcomes in the long term.

In addition to patient reports, neuroimaging reveals that physiological consequences can be exhibited when patients are subjected to immersive VR activities. fMRI reveals that the pain-related parts of the brain become less active.

6.4. AI in pain management

6.4.1. *Role of AI in healthcare*

Advanced technologies in the current healthcare industry are turning into AI-based data analysis, clinical decision support and predictive analytics. AI cuts across several fields including machine learning (ML), deep learning (DL), natural language processing (NLP) and computer vision, all of which are able to process complex biomedical data. AI enhances accuracy in the diagnosis, predictability of disease evolution, individual therapeutic plans and efficient allocation of clinical funds in pain management (Kuo 2023). AI devices recognize trends within patient data, for example, the risk of pain flare-ups or adverse drug reactions, which a clinician could have missed. Personalized and predictive care methods are becoming increasingly relevant in valuing AI in the field of pain rehabilitation and oncology.

AI enhances the quality of diagnostic tests and allows continuous management in the form of self-education. AI technologies can integrate and analyze large volumes of multi-modal data, including images, EHRs and data from personal monitoring devices, making them ideal to address the complex, multi-dimensional nature of cancer pain, which includes nociceptive, neuropathic and psychological components.

6.4.2. *AI in physiotherapy*

The use of AI in the area of remote digital health has taken great strides. For cancer patients receiving physiotherapy, the use of AI increases both safety and effectiveness with the provision of accurate and real-time motion tracking in combination with predictive analytics and personalized remote real-time guidance. The following sections describe some of the prominent ways in which AI is being used in physiotherapeutic practice.

6.4.2.1. *Motion capture and biomechanics*

AI technology in motion capture has advanced to the point where sensors, cameras and wearables, and other AI computer vision and DL resources are used to track a patient's movements in real time. AI systems assess a patient's posture, angle of joints and gait and movement symmetry, providing valuable and impartial assessment feedback on critical biomechanical parameters necessary in functional recovery and compensatory movement avoidance to a patient – and in a manner that promotes injury or pain (Rebelo et al. 2023).

6.4.2.2. *Predictive pain models*

Pain is a subjective phenomenon and is one of the most complex experiences, composed of many interconnected contributing factors, and AI attempts to solve this problem through predictive modeling. AI predictive pain modeling uses a combination of clinical records, patient-reported outcomes and multiple other parameters to predict possible pain levels and identify patients who may be at risk of chronic pain.

6.4.2.3. *Chatbots and conversational agents*

AI-enabled chatbots and conversational agents function as virtual physiotherapy assistants and are especially useful for home-based rehabilitation. They communicate with patients to assess pain levels, remind patients of exercises and provide psychological encouragement. They also offer educational materials on coping with pain, stretching and ergonomic practices in real time.

6.4.3. *AI-driven personalization*

The fact that AI can dynamically personalize the treatment in physiotherapy is transforming the practice. Unlike the traditional rehabilitation frameworks, which tend to be inflexible, AI facilitates pinpoint accuracy in physiotherapy by adapting to real-time patient performance, physiological metrics and patient feedback.

For instance, rehabilitation exercise AI task adjusters would be able to change the exercise task, complexity and duration based on the patient's performance and pain feedback. If patients report pain and easily complete the task, the system will immodestly increase the exercise challenge. If discomfort is detected, whether through pain scores, or metrics of painful and poorly executed movements, the task will be made easier. This type of responsiveness keeps rehabilitation within the therapeutic window.

For patient rehabilitation to be therapeutic, it must be challenging enough to encourage improvement but not so difficult as to aggravate or demoralize the patient (Winstein et al. 2016).

AI can use multi-dimensional data such as heart rate variability in the form of fatigue, data-collection electromyograph muscle activation and subjective data reports of emotions to adjust exercise. This thoughtful construct pleases the biopsychosocial pain model, which states pain should also be alleviated through the emotional, cognitive and psychological frameworks.

During periods of treatment, patients may experience high energy levels followed by extreme fatigue when oncology physiotherapy is rendered. AI can schedule and modify goals, as well as suggest alternative exercises, such as guided breathing and mindfulness to disengage patients while still promoting mindfulness and breathing exercises hypothetically.

6.5. Integrating VR and AI: a synergistic approach

6.5.1. *Real-time adaptive VR systems*

The most significant innovation in the integration of VR in physiotherapy for managing cancer pain is the use of adaptive VR systems in conjunction with AI. These systems incorporate real-time physiologic data such as heart rate variability, skin response, fMsc, facial expressions, muscular tension and eye movements to assess the patient's contextual factors and emotional state in a VR physiotherapy system, as well as pain and distress levels. AI adjusts the virtual scene to achieve the most effective therapeutic result. For example, the system will recognize a patient's augmented distress and fatigue, and will soften the chosen therapeutic physical tasks, framing the session with relaxing music and soothing visualizations, or even provide a pause for relaxation (Pizzoli 2021). In contrast, periods of comfort and low pain allow for greater challenges to be added in order to promote more mobility and endurance. This system's VR physiotherapy integration offers a system of rehabilitation under the most flexible and patient-centered conditions possible, with minimal need for therapist manual adjustment to accommodate pain and psychological distress.

6.5.2. *Closed-loop feedback systems*

AI-driven, closed-loop feedback systems integrated within VR monitor patients and adjust therapies in real time. These systems monitor a patient's advanced biosensors such as electromyography (EMG) and electroencephalography (EEG) recordings, as well as behavior and thermal imaging during therapeutic sessions. AI analyzes real-time data streams to detect pain and fatigue, then classifies fatigue and pain level to detect improvement. Then, it modulates therapeutic parameters in a real-time exercise pace, task difficulty and type of immersive environment. The system replicates the behavior of a therapist through adjusting the treatment according to the observed outcomes. The urgency of the automated adjustments ensures gradual progressiveness that ensures that the therapy does not exceed the pain thresholds of the patients (El-Tallawy et al. 2024). Such systems allow individualized remote rehabilitation, a critical feature for in-person rehabilitation programs for under-provided cancer survivors.

6.5.3. *Gamification and behavioral reinforcement*

AI-guided VR technology inventions incorporate Serlin's research, analyzing gamification effects on therapeutic practices, specifically on motivational factors during VR therapy sessions. This research is pivotal in reengineering custom VR gamification for cancer patient physiotherapy. AI real-time adaption VR cancer physiotherapy games include emotional and behavioral analytics, played during therapy sessions. AI–VR cancer physiotherapy gamification uses therapeutic principles of game design. VR's therapeutic game narrative incorporates story-driven elements, deep engagement, distraction and potential pain mitigation. Distraction involves story-driven elements that engage patients to participate. Deep engagement includes missions to fulfill progressively deep therapy sessions to increase mastery of pain control. Tasks include turning achievements and pain control into sustaining game rewards.

AI-described adaptive processes reinforce motivational principles of mastery, challenge and control. These elements contribute to increased pain tolerance, sense of achievement and encourage participation during therapy. AI game adaptive processes facilitate control increase, optimize distraction during pain therapy sessions and elevate mood.

6.6. Case studies and clinical implementations

6.6.1. *Use case I*

AI–VR rehabilitation system designed for patients recovering from head and neck cancer due to surgeries that affect their swallowing functions was the starting point for therapists' introduction to AI rehabilitation systems. In these therapies, patients used the system to perform rehabilitation swallowing exercises in VR. They worked in VR on the motor functions that needed to be mastered for swallowing rehabilitation. What made the system especially effective was AI–VR's ability to tailor the rehab focus (Khanal et al. 2022). With real-time tracking of muscle activity and advanced computer vision technology, the system uses adaptive algorithms to adjust exercises in real-time to patients' capabilities, actively challenging them while balancing session goal attainment with therapeutic expectations. This AI focus on real-time balancing goal attainment improved the therapy quotient, thereby decreasing the discomfort expectation. For the patients, gamification of VR made the therapy motivative and VR gamification of therapy made it rehabilitation for session adherence. In clinical evidence, the system increased functional rehabilitation of swallowing and session attendance, with many patients voicing preference for the system, while the clinical expectation and relationship.

6.6.2. *Use case II*

Virtual mirror therapy (VMT) has been successfully used to treat post-mastectomy pain and functional limitations for breast cancer survivors. Through this method, patients used VR helmets and engaged with AI-powered avatars that copied their limb movements, including the impacted arm and shoulder. These avatars served as digital mirrors, replicating movements to produce a visual illusion of restored motor function. Many patients reported significant relief of phantom limb pain, neuropathic pain, improved muscle coordination and range of motion after several weeks of ongoing therapy (Zaheer et al. 2021). The AI component was vital as it perceived each patient's range of motion and customized the visual feedback and movements. The combination of visual relearning and adaptive AI made VMT a highly effective option for complex post-surgical pain and functional recovery in breast cancer physiotherapy.

6.6.3. *Pediatric cancer and VR distraction therapy*

The use of AI and VR technologies in hematology/oncology is a new field of research. VR technologies have been used in the past for immersion distraction therapy. These include environments where children go on underwater adventures, magical forests or interactive storytelling games. Such VR therapies utilize AI technologies to analyze and adapt to a patient in real time by detecting behavioral and physiological indicators, such as heart rate, facial expressions and voice, to adjust the simulations to a child's needs and desires. The system, for example, can pivot from energetic and active scenes or storylines to tranquil ones for rest. This active control over environment manipulation was a primary reason for the ease in engagement and a calm state. Clinical outcomes and the changing nature of this therapy have been documented in the clinical setting, particularly in the alleviation of anxiety in medical procedures and perception of pain in the treatment process (Girishan Prabhu et al. 2024). Parents and staff reported that the use of chemical restraints on children during procedures was less and the reduction was noticeable.

6.7. Technical and ethical considerations

6.7.1. *Data privacy and consent*

The integration of wearable sensors and cloud-based AI platforms in cancer physiotherapy enables the continuous collection of real-time physiological and behavioral data. This data, which includes heart rate variability, facial expressions, muscle activity and pain responses, is highly sensitive and must be handled with utmost care. Without robust data governance frameworks, there is a risk of unauthorized access, misuse or unintended sharing of personal health information. Patients must be clearly informed about what data is being collected, how it will be used and who will have access to it (Alhasan 2025). Consent procedures must be dynamic and ongoing, allowing patients to modify or withdraw consent as technologies evolve. Furthermore, AI systems should offer explainability: patients and clinicians alike must understand how decisions or recommendations are generated.

6.7.2. *Bias in AI algorithms*

AI models are only as good as the data on which they are trained. If datasets lack representation from diverse populations, the resulting algorithms may be biased, potentially leading to misdiagnoses or inappropriate treatment recommendations. In the context of cancer pain, variables such as ethnicity, socioeconomic status, gender and age influence pain expression and tolerance. Algorithms must be validated across multiple cohorts to ensure fairness and accuracy. Researchers and developers should engage in inclusive data collection practices and conduct impact assessments to identify and mitigate potential algorithmic biases (Metcalf et al. 2021).

6.7.3. *Technology access and digital divide*

While AI and VR promise significant advancements, their benefits may remain inaccessible to underserved populations due to cost, connectivity or a lack of digital literacy. Expensive hardware, limited Internet access and a shortage of trained personnel in rural or low-income regions can exacerbate existing healthcare disparities (Afni 2023). To bridge this divide, stakeholders must invest in affordable devices, develop low-bandwidth solutions and introduce community-based programs such as mobile clinics or remote physiotherapy platforms tailored to local needs.

6.8. Future directions

6.8.1. *Multimodal integration*

Multimodal integration of a variety of sophisticated technologies will be the key to the future of cancer physiotherapy. VR and AI may be combined with haptic feedback, EMG, augmented reality (AR) and even olfactory stimulation to make the process of rehabilitation more positive. Haptic devices provide a patient with a sense of touch as they undergo virtual exercises. EMG sensors (real time) are used to measure muscle movements (Kumar et al. 2026a). AR is a combination of real-life images and new computer-generated imagery that enhances the effectiveness of exercises. The inclusion of olfactory stimuli may be used to deal with pain and anxiety, evoke memory and elevate emotional condition.

6.8.2. *Remote and home-based rehabilitation*

Lightweight VR headsets and mobile AI tools provide an opportunity for home-based cancer rehabilitation. Physiotherapy can be done by the patient at home, which will decrease the dependence of the patient on the infrastructure of the hospital. The patients are actively monitored remotely, their progress is registered and AI is used to analyze therapy intensity to comply. In this way, adherence to therapy, maintenance of care and travel burden to low mobility or immunity patients are increased (Kumar et al. 2026b).

6.8.3. *Prediction of chronic pain*

Advanced AI systems are under development that would utilize clinical history, behavioral trends and physiological analyses to identify the level of pain and forecast the future path of pain (Harati et al. 2025). These tools allow clinicians to predict pain with sufficient accuracy to tailor physiotherapy before the symptoms can become worse, which will eliminate escalation, complications and exacerbations. Such an evidence-based approach would help sustain wellness and enable a more effective distribution of resources (Rejula et al. 2021).

6.8.4. *AI sources and virtual therapists*

AI therapeutic virtual therapists have the ability to detect posture, adjust the shape of an individual and give them a form of motivation, which constantly supports a user. These AI coaches can provide emotional support during a session with speech recognition and affective computing, which allows for soothing of the psychologically caused pain. This combination of machine smart and nurture is astonishing (Gudur et al. 2024).

6.9. Conclusion

VR and AI technologies are transforming pain management for cancer physiotherapy. These technologies provide novel therapeutic alternatives that are more engaging and patient-centered. As a form of pain management and anxiety relief, VR technologies allow patients to focus on an environment that provides guided visualization, sensory engagement and distraction from pain. AI improves these technologies. For pain management therapy, AI analyzes and adjusts the therapy in real time, according to the patient's physiological and psychological responses. These technologies are repositioning emotional recovery and psychological therapies as vital components of the rehabilitation framework. Psychological therapy and emotional support are vital because cancer patients often experience pain, fatigue and depression. Many cancer patients also find traditional physiotherapy uninspiring and demotivating. The VR and AI integration provides therapy that is more engaging and responsive to patient needs. These technologies are also the foundation of remote therapy and home-based rehabilitation, which helps patients who are more isolated and perhaps more in need of support. The increase in the affordability and accessibility of AI and VR technologies will provide more value in a wider range of practice environments. Ongoing research will be needed to ensure merger efficacy, followed by comprehensive clinical trials and crafting of ethical frameworks concerning privacy, informed consent and access equity as ethical imperatives. If done right, VR and AI will be essential and significant components of oncology rehabilitation, improving preserved functional abilities, quality of life across rehabilitation phases and overall rehabilitation outcomes.

6.10. References

Addab, S. (2020). The feasibility, clinical utility, tolerability and initial clinical efficacy of virtual reality distraction with children undergoing medical procedures at a specialized pediatric orthopedic hospital. Thesis, McGill University.

Afni, N. (2023). Disparities in healthcare access: Addressing systemic barriers in urban and rural communities. *Journal of Health Literacy and Qualitative Research*, 3(2), 112–122.

Alhasan, T.K. (2025). Managing legal risks in health information exchanges: A comprehensive approach to privacy, consent, and liability. *Journal of Healthcare Risk Management*, 44(4), 12–24.

Alzeer, J. (2023). Integrating medicine with lifestyle for personalized and holistic healthcare. *Journal of Public Health and Emergency*, 7. doi: 10.21037/jphe-23-71.

Bhowmik, A.K. (2024). Virtual and augmented reality: Human sensory-perceptual requirements and trends for immersive spatial computing experiences. *Journal of the Society for Information Display*, 32(8), 605–646.

Bonanno, M., Papa, D., Cerasa, A., Maggio, M.G., Calabrò, R.S. (2024). Psycho-neuroendocrinology in the rehabilitation field: Focus on the complex interplay between stress and pain. *Medicina*, 60(2), 285.

Cascella, M., Cascella, A., Monaco, F., Shariff, M.N. (2023). Envisioning gamification in anesthesia, pain management, and critical care: Basic principles, integration of artificial intelligence, and simulation strategies. *Journal of Anesthesia, Analgesia and Critical Care*, 3(1), 33.

El-Tallawy, S.N., Pergolizzi, J.V., Vasiliu-Feltes, I., Ahmed, R.S., LeQuang, J.K., El-Tallawy, H.N., Varrassi, G., Nagiub, M.S. (2024). Incorporation of "artificial intelligence" for objective pain assessment: A comprehensive review. *Pain and Therapy*, 13(3), 293–317.

Friedrichsdorf, S.J., Giordano, J., Desai Dakoji, K., Warmuth, A., Daughtry, C., Schulz, C.A. (2016). Chronic pain in children and adolescents: Diagnosis and treatment of primary pain disorders in head, abdomen, muscles and joints. *Children*, 3(4), 42.

Girishan Prabhu, V., Stanley, L., Morgan, R., Shirley, B. (2024). Designing and developing a nature-based virtual reality with heart rate variability biofeedback for surgical anxiety and pain management: Evidence from total knee arthroplasty patients. *Aging & Mental Health*, 28(5), 738–753.

Gudur, A., Patil, V., Jain, A., Garg, N. (2024). Hybrid genetic algorithm and deep learning approach for lung nodule detection and classification in chest X-rays. *International Journal of Intelligent Systems Applications in Engineering*, 12(3s), 577–587.

Harati, K., Mosaddeghi-Heris, R., Kiani, K., Saligheh Rad, M., Morovatshoar, R., Kamali, M., Ghasemi, M., Faizabadi, S., Madady, A., Sina, N. et al. (2025). *The AI Revolution: Predicting and Managing the Next Global Health Challenges and Emerging Disease Outbreaks*, PreferPub, San Jose, CA.

House, G., Burdea, G., Grampurohit, N., Polistico, K., Roll, D., Damiani, F., Hundal, J., Demesmin, D. (2016). A feasibility study to determine the benefits of upper extremity virtual rehabilitation therapy for coping with chronic pain post-cancer surgery. *British Journal of Pain*, 10(4), 186–197.

Kalusivalingam, A.K., Sharma, A., Patel, N., Singh, V. (2021). Enhancing patient care through IoT-enabled remote monitoring and AI-driven virtual health assistants: Implementing machine learning algorithms and natural language processing. *International Journal of AI and ML*, 2(3) [Online]. Available at: https://www.cognitivecomputingjournal.com/index.php/IJAIML-V1/article/view/75.

Khanal, S.R., Paulino, D., Sampaio, J., Barroso, J., Reis, A., Filipe, V. (2022). A review on computer vision technology for physical exercise monitoring. *Algorithms*, 15(12), 444.

Kumar, A., Batta, P., Ahuja, S., Rathore, P.S. (eds) (2026a). *AI-driven Innovations in Physiotherapy and Oncology 1*. ISTE Ltd, London and Wiley, Hoboken, NJ.

Kumar, A., Batta, P., Ahuja, S., Rathore, P.S. (eds) (2026b). *AI-driven Innovations in Physiotherapy and Oncology 2*. ISTE Ltd, London and Wiley, Hoboken, NJ.

Kuo, C.L. (2023). Revolutionizing healthcare paradigms: The integral role of artificial intelligence in advancing diagnostic and treatment modalities. *International Microsurgery Journal (IMJ)*, 7(1). doi: 10.24983/scitemed.imj.2023.00177.

Meier, T.A., Refahi, M.S., Hearne, G., Restifo, D.S., Munoz-Acuna, R., Rosen, G.L., Woloszynek, S. (2024). The role and applications of artificial intelligence in the treatment of chronic pain. *Current Pain and Headache Reports*, 28(8), 769–784.

Metcalf, J., Moss, E., Watkins, E.A., Singh, R., Elish, M.C. (2021). Algorithmic impact assessments and accountability: The co-construction of impacts. In *Proceedings of the 2021 ACM Conference on Fairness, Accountability, and Transparency*, 735–746.

Perret, D.M., Rim, J., Cristian, A. (2006). A geriatrician's guide to the use of the physical modalities in the treatment of pain and dysfunction. *Clinics in Geriatric Medicine*, 22(2), 331–354.

Pizzoli, S.F. (2021). User-centered digital intervention: Tailoring relaxing places. PhD Thesis, IEO, Milan.

Prabhu, V.G., Stanley, L., Morgan, R. (2020). A biofeedback enhanced adaptive virtual reality environment for managing surgical pain and anxiety. *International Journal of Semantic Computing*, 14(3), 375–393.

Rasa, A.R. (2024). Artificial intelligence and its revolutionary role in physical and mental rehabilitation: A review of recent advancements. *BioMed Research International*, 2024(1), 9554590.

Rebelo, A., Martinho, D.V., Valente-dos-Santos, J., Coelho-e-Silva, M.J., Teixeira, D.S. (2023). From data to action: A scoping review of wearable technologies and biomechanical assessments informing injury prevention strategies in sport. *BMC Sports Science, Medicine and Rehabilitation*, 15(1), 169.

Rejula, V., Anitha, J., Belfin, R.V., Peter, J.D. (2021). Chronic pain treatment and digital health era-an opinion. *Frontiers in Public Health*, 9, 779328.

Shi, Y. and Wu, W. (2023). Multimodal non-invasive non-pharmacological therapies for chronic pain: Mechanisms and progress. *BMC Medicine*, 21(1), 372.

Shin, H.J., Cho, I.T., Choi, W.S., Kim, H.R., Kang, M.B., Yang, W.J. (2024). Digital therapeutics in Korea: Current status, challenges, and future directions – A narrative review. *Journal of Yeungnam Medical Science*, 42, 8.

Sun, Y. (2022). Virtual and augmented reality-based assistive interfaces for upper-limb prosthesis control and rehabilitation. *arXiv preprint*. doi: 10.48550/arXiv.2205.02227.

Torta, D.M., Legrain, V., Mouraux, A., Valentini, E. (2017). Attention to pain! A neurocognitive perspective on attentional modulation of pain in neuroimaging studies. *Cortex*, 89, 120–134.

Winstein, C.J., Stein, J., Arena, R., Bates, B., Cherney, L.R., Cramer, S.C., Deruyter, F., Eng, J.J., Fisher, B., Harvey, R.L. et al. (2016). Guidelines for adult stroke rehabilitation and recovery: A guideline for healthcare professionals from the American Heart Association/American Stroke Association. *Stroke*, 47(6), e98–e169.

Zaheer, A., Malik, A.N., Masood, T., Fatima, S. (2021). Effects of phantom exercises on pain, mobility, and quality of life among lower limb amputees; a randomized controlled trial. *BMC Neurology*, 21(1), 416.

7

Machine Learning for Optimizing Exercise Intensity in Oncology Rehabilitation

In the context of holistic cancer care, oncology rehabilitation aims to restore function, minimize fatigue and enhance quality of life during survivorship. Even with this goal, one of the longest-standing problems remains finding the right intensity of exercise for the individual. The variations in physical capacity, the stage of cancer, the current and previous therapies and recovery gaps all demand individual and personal level rehabilitation. This is not what mainstream approaches take into account. This issue can be potentially addressed by machine learning (ML): it can make exercise adaptation highly personalized, responding to individual objectives and numerous, complicated interventions in real time. This chapter demonstrates the possibility of ML to transform data-driven rehabilitation in oncology by defining particular kinds of ML methods, including supervised learning to make predictions, unsupervised learning to work with patient groups and reinforcement learning to perform adaptive exercise. An example of application given in the chapter is biofeedback control, where exercise control systems vary based on wearable streams of data that check the active and passive biomechanical and physiological measures. These systems reduce the risks and ineffectiveness of rehabilitation. The chapter lists such challenges as data silos and heterogeneity, algorithm biases, privacy and lack of effective clinical workflows to facilitate the use of ML. Lastly, it explains the future of transparent AI systems, interdisciplinary cooperation and necessity to introduce ethical implications to ensure that the application of ML in cancer rehabilitation will be effective and representative to everyone. ML has the ability to revolutionize the production and provision of customized exercise oncology programs with such advances.

7.1. Introduction

Cancer rehabilitation is one of the dimensions of survivorship care. The patients, who have undergone interventions such as chemotherapy, radiotherapy or surgery, tend to lose significant physical functioning, stamina and muscle mass. These side

---

Chapter written by Anand GUDUR and Himanshu WAGH.

effects may greatly undermine the quality of life, enhance the comorbidity risk and decrease long-term survival. There is a rising amount of evidence that favors the use of organized physical activity as a means of preventing these negative consequences, enhancing recovery and enhancing physical and mental health (Vella et al. 2023). Nevertheless, the traditional model of rehabilitation, which is one-size-fits-all, is not a solution to the needs of oncology patients, which are dynamic and more individual.

The cancer survivors represent a varied group, as they vary in terms of age, type of cancer, stage, comorbidity, treatment, functional limitation and psychosocial status. An exercise program created as one-size-fits-all will either under- or over-train, restrict the benefits or become draining, and cause injury or flare-up of symptoms. Rehabilitation thus needs to be designed to be personalized, based on the specific baseline of a patient as well as their physiological responses and progress toward recovery (Stinear et al. 2020).

Machine learning (ML) provides a solution of transformation. The convolutions in complex, multidimensional data are caused by the ML algorithms that reveal patterns that were not taught directly. ML is also used in the rehab of oncology where it is used to combine real-time clinical data (heart rate, respiratory rate, muscle activity), performance (speed and range of movement) and outcomes (fatigue, pain, motivation) to generate dynamic and personalized exercise prescriptions. It is important to note that real-time adaptive control of intensity is essential in the attainment of the best results.

With ML systems, the link between the response and exercise intensity depends on constant data monitoring and analysis to estimate a response of the patient. It is particularly crucial in cancer patients whose level of fatigue and stress can significantly change. Wearables and motion-capture devices input their information into ML models that identify the indicators of overworking, including elevated heart rate variability (HRV), less step symmetry or slower muscle recovery, which provokes an automatic decrease in intensity to a safer point (Singh et al. 2018).

Thousands of exercise sessions with varying patient demographics can be found in historical datasets that allow ML models to learn the patterns of associations between baseline attributes and exercise responses. These forecasting insights can enable clinicians to have a confident starting intensity and monitor gradual changes in an objectively driven way, contributing to making the rehabilitation personal.

Aside from immediate response predictions, ML can assist in longitudinally tracking and evaluating outcomes. ML algorithms can quantify a patient's progress by tracking, over weeks or months, the plateau periods and possible setbacks in their variable physiologic and functional data (Srichawla 2024). These observations enable the clinician to make much more impactful interventions by refocusing rehabilitation efforts.

Adding to the traditional supervised learning methods of regression and classification, new ML techniques, particularly reinforcement learning (RL), can be highly beneficial for real-time exercise optimization. RL uses trial and error to enable algorithms to make a series of decisions to achieve the best possible outcome for a given criterion. For cancer rehabilitation, an RL-based system could learn which exercise intensities work best functionally for a patient over time, and adapt accordingly.

Most importantly, the use of ML in oncology rehabilitation can uplift patient engagement and adherence. Many cancer survivors experience motivational issues, especially in the emotionally and physically draining periods of treatment and recovery. ML systems having personalized real-time feedback, gamified progress and exercise goals make workouts seem more achievable (Vadia et al. 2025).

While the use of ML in clinical rehabilitation has a lot of potential, there are still barriers to its use, such as the need for diverse, high-quality datasets, issues surrounding the privacy and security of data, the interpretability of complicated models and efforts to accommodate ML systems into the current clinical workflow. Nonetheless, given the rapid pace of developments in patient monitoring, cloud computing and wearable technologies, ML-based rehabilitation has a higher chance of practical use.

7.2. Exercise intensity in oncology rehabilitation

7.2.1. *Defining exercise intensity*

The physiological stress imposed on the body during an activity is termed 'exercise intensity'. It is one of the main constituents of exercise prescription. It will determine the safety and effectiveness of any rehabilitation program, as well as the outcomes of the rehabilitation program. Particularly, the health status of clinical oncology patients makes defining and controlling the intensity of rehabilitation exercise even more critical. Exercise intensity is conventionally measured using clinical approaches, the most common being heart rate, VO_2 max and the lactate threshold (Jamnick et al. 2020). Such data-based approaches are able to quantitatively estimate cardiovascular and metabolic exertion.

Subjective measures are still necessary. The Barg Rating of Perceived Exertion (RPE) scale can enable the patient to objectively rate their effort which is very important since objective measurements can be misleading through pain, fatigue and emotional stress. The muscular fatigue is also checked using strength and resistance protocols, which are characterized by changes in maximal voluntary contraction or changes in EMG signals – or both.

The described tools remain valid, though they need to be modified within the oncology setting. Chemotherapy, radiation and surgery change stamina, cardiovascular functionality, autonomic control and locomotion. Therefore, the best intensity should be obtained through clinical judgment and not formulae or general population averages. To be safe and successful in rehabilitation, a carefully designed program incorporating physiological data, functional evaluation and patient-reported outcomes (PRO) is needed (Patel et al. 2012).

7.2.2. Challenges in modulating intensity

Regulating exercise levels in oncology rehabilitation creates many clinical and logistical issues. The overall level of variability in the physical functioning, the levels of fatigue and the overall health status of cancer patients are very high, particularly in patients receiving active treatment. Their physiological reactions to exertion are not as predictable as in healthy persons, and it is likely that a combination of factors governs them:

– *Treatment-related fatigue*: cancer-related fatigue (CRF) is one of the most common and crippling side effects of chemotherapy and radiation therapy. This unremitting excessive fatigue is disproportionate to its degree of activity and is not alleviated by rest (Strasser 2021). Consequently, low-intensity exercises may be problematic to the patients, and their tolerance may be highly fluctuating on a daily basis.

– *Immunosuppression and risks of infection*: a large number of cancer treatments impair immune functions, which makes them more susceptible to infections. High-intensity exercise can have an acute suppressive effect on immune parameters or provide a context (e.g. overcrowded gyms) that increases exposure risk. Thus, care should be taken with intensity to safeguard the immunocompromised.

– *Comorbidities*: a big percentage of cancer patients are comorbid, for example, with cardiovascular disease, diabetes or lung problems. These comorbidities limit the intensity of the exercise and require constant attention to eliminate negative outcomes such as arrhythmias, ischemia or respiratory distress.

– *Age-related deterioration*: a lot of cancer patients are very old, and they may experience age-related sarcopenia, low aerobic potential and degeneration of the joints.

– *Tolerance variability*: the most complicated issue, perhaps, is inter- and intra-individual variabilities in patient tolerance to exercise. Others might improve rapidly with progressively intensifying improvement, or others might revert or have symptom exacerbation using the same regimen. Variations in the exercise tolerance in one session as compared to another may be caused by fatigue, nausea, neuropathy or psychological distress.

An exercise protocol where the intensity of the exercise is pre-established and not altered is not only ineffective, but even counterproductive. Excessive exercise may result in fatigue and other medical issues, whereas under-exercise may result in ineffective, sub-therapeutic exercising, which causes a decrease in gains in strength, endurance and functional capacity. We require dynamic, adjustable and real-time changes of intensity to solve this.

Wearable sensors and other monitoring gadgets (e.g. heart rate monitors, accelerators and EMG devices) give real-time information about exertion and safety levels, which allows monitoring risks in real time (Jafleh et al. 2024). Once this data is integrated into ML and other AI solutions, clinicians will have the chance to practice prescription more precisely with the help of the current and past physiological data and predictive analytics. This will make sure that the data is not challenged too much or too little and that the rehabilitation will take place in a safe manner and achieve the goals of the patient.

7.3. Machine learning in healthcare and rehabilitation

7.3.1. *Machine learning overview*

Machine learning (ML) is one of the most important branches of artificial intelligence (AI) and refers to the ability of a computer to learn patterns from data without being programmed to do so. ML can be divided into three main categories: supervised learning, unsupervised learning and RL. Each of these categories has their benefits and approaches to address specific needs within the healthcare domain.

In supervised learning, the algorithm is provided data and the corresponding answer is given. This type of learning is most appropriate for tasks that involve forecasting, such as determining the risk level of patients, estimating recovery periods or predicting therapy protocol adherence. In oncology rehabilitation, supervised models can predict fatigue, track the cardiovascular response or strain musculoskeletal systems by forecasting levels based on real-time and historical data (Chen et al. 2025).

In rehabilitation, unsupervised learning can be used to group silent patients based on their response on a continuum to exercise, symptoms or biomechanical characteristics without prior definition. This is critical for heterogeneous cancer patients, where symptomatology and recovery pathways diverge widely.

In RL, a system is designed where an agent learns by interacting with an environment and receives a reward or punishment system. In rehabilitation, RL can be used to design flexible training programs that adapt in response to patient performance, assessing the best intensity and type of exercise to maximize results.

Together, these ML paradigms offer the dynamic modeling of patient trajectories, facilitating the personalization of rehabilitation strategies that adapt to the evolving state of the patient. For instance, a supervised model may flag a patient as high risk in terms of treatment-induced fatigue, while other RL algorithms adjust exercises on the fly to help avoid exacerbation (Rudroff 2025). Unsupervised clustering can simultaneously expose hidden patient subtypes that respond more positively to certain treatments, thus helping with more precise targeting.

7.3.2. *Machine learning (ML) in broader rehabilitation contexts*

The prospects for incorporating ML into oncology care are promising. It has shown significant progress in several rehabilitation areas, especially stroke rehabilitation. Here, ML models monitor limb functions, gauge neural plasticity, and recover and predict ability. Motion capture technology works with supervised learning to assess motor skills so that therapists can track changes in motor rehabilitation on some objective quantitative metrics and not on subjective visual estimation.

In cardiac rehabilitation, ML predicts adverse outcomes and personalizes exercise prescription and detection of early cardiac distress during exercise. For example, some steady-state ECG algorithms detect anomalies in real time and ask clinicians to intervene. Moreover, during a training session, adaptive algorithms adjust exercise to provide a specific type of exercise that can be set using a variable heart rate, and oxygen consumption is a preset target (Gajda et al. 2024).

In physiotherapy for the musculoskeletal system, ML identifies overuse musculoskeletal injuries through movement pattern analysis using real-time joint and muscle movement sensors and computer vision. Algorithms can also predict which patients are at risk of delayed recovery or non-compliance, allowing for early intervention.

The applications highlight some of the most important strengths of ML in rehabilitation:

– monitoring movement quality using inertial measurement units (IMUs), accelerometers or video analysis to quantify motion patterns with high precision;

– predicting fall risk by analyzing gait stability, stride asymmetry and postural sway, especially in elderly or neurologically impaired populations;

– adjusting exercise workloads dynamically in real time, using biofeedback loops that interpret physiological and biomechanical cues;

– tracking compliance and performance trends, offering personalized feedback and motivational cues through digital platforms or virtual coaches.

Smart systems allow accurate, data-driven and very personalized rehabilitation. Oncology is a complex area with side effects of treatment, comorbidity and immune status. However, the track record of success of ML in other areas proves its worth (Pandya et al. 2021). Further development of such ML methods with oncology-specific adjustments is projected to become the key to providing more efficient, scalable and responsive rehabilitation plans to cancer patients.

7.4. ML techniques for exercise intensity optimization

ML offers a potent technology to customize and maximize exercise intensity during the rehabilitation of cancer patients. Among the most impactful ML paradigms – supervised learning, unsupervised learning and RL – each provides unique advantages for handling complex, high-dimensional clinical and physiological data.

7.4.1. *Supervised learning for exercise intensity prediction*

Supervised learning models work with labeled datasets where some features are tied to the outputs and labeled datasets are used. For instance, some features could be the characteristics of the patient and the outputs could be the levels of exercise intensity that are optimal. Support vector machines, random forests and gradient boosted trees are some of the most common supervised algorithms. By taking the age, type of cancer, type of treatment, stage of recovery, physical performance and other similar attributes into account, the models can analyze and determine the exercise intensity that will be most appropriate for the patient.

A supervised model could take a dataset of breast cancer survivors who have been through varying rehabilitation protocols as a training example. The model can be trained to determine the exercise zone of a 30-min cycling session at a target heart rate of a target aerobic exercise using real-time inputs such as HRV, oxygen saturation and patient-reported fatigue levels (Boutros et al. 2025). This prediction can be important clinically for adjusting the resistance or duration to be patient-centered on a particular day.

There are also deep learning models, for example, artificial neural networks that are designed to figuring out complex nonlinear relationships of various inputs to determine the exercise intensity. Although these models are complex, working neural networks can facilitate real-time decisions and are designed to be scalable and adaptive, making them a great fit for smart rehabilitation systems.

7.4.2. *Unsupervised learning and clustering for patient stratification*

Compared to other learning techniques, unsupervised learning explores hidden structures in unexplored unlabeled datasets. K-means, DBSCAN and hierarchical clustering can identify patient groups based on physiological, behavioral and biochemical characteristics (Aljohani 2024). Such clusters can help in creating bespoke protocols for exercise intensity where labels do not exist.

Take, for example, unsupervised clustering on longitudinal datasets from wearable sensors incorporating accelerometry, HRV and galvanic skin response sensors. Such clusters can expose underlying patterns pertaining to patient fatigue and recovery. Patients can cluster naturally into groups and exercise regimens may also be tolerated at higher intensities for the "high resilience" patients, while "low tolerance" patients may do better with low-intensity, high-frequency protocols. Diagnostic and prognostic performance outcomes may not exist, but the underlying unsupervised learning can still help clinicians adjust targeted personalized rehabilitation based on the performance outcome.

Cardiovascular and neurostimuli response patterns will dynamically define patient clusters. Methods such as principal component analysis (PCA) and t-SNE help in complex dimensionality reduction (Devadas 2024). Such techniques streamline complex structures for more targeted identification and retrieval of unique patient groups from datasets and enriched clinical data. These techniques can help data-driven personalized care in clinical oncology.

7.4.3. *Reinforcement learning (RL) for adaptive exercise prescription*

RL works well with self-paced rehabilitation exercises since it consists of dynamic and time-dependent decisions. Here, an agent learns through direct interaction with the environment observing a state (the patient's physiologic data), taking an action (modifying the exercise) and then receiving a response (new state of fatigue, pain or performance) over some time. The agent learns to maximize cumulative rewards (increased endurance, less fatigue), while avoiding conditions that may lead to unsafe situations.

Let us consider an example of an RL-based agent working on a treadmill. Over some time, the agent evaluates and learns the importance of real-time data such as oxygen saturation and fatigue levels, lactate concentration (lactate threshold), EMG signals and perceived exertion scoring. If fatigue is less and the score is greater than oxygen saturation of 95%, the agent learns to increase the treadmill speed or incline in small increments (Yang et al. 2024). When fatigue is excessive, and the score describes an active cardiovascular state, the agent learns to reduce exercise intensity of the treadmill or incorporate active rest.

The safe progression through an exercise with this form of RL maintains participant engagement. Real-world rehabilitation robotics and virtual coaching platforms in rehabilitation exercises can be optimized with RL strategies such as Deep Q-Networks (DQN) or proximal policy optimization (PPO). For optimized exercise intensity in oncology rehabilitation, see ML approaches in Table 7.1.

ML technique	Core function	Input data sources	Application in exercise optimization	Advantages/outcomes
Supervised learning (e.g. random forest, SVM and ANN)	Predicts optimal exercise intensity based on labeled datasets	Wearable sensor data (HR, HRV and EMG), HER and fatigue scores	Determines safe and effective intensity zones for aerobic/strength training	Improved accuracy in intensity prescription; reduced fatigue
Unsupervised learning (e.g. K-Means, DBSCAN and PCA)	Identifies hidden patterns or clusters in patient data	Physiological and behavioral parameters	Stratifies patients into tolerance groups for tailored regimens	Enables subgroup-based customization of rehab protocols
Reinforcement learning (RL)	Learns optimal intensity through feedback and rewards	Real-time biofeedback, heart rate and EMG	Continuously adjusts intensity during training based on patient response	Dynamic adaptation; ensures safety and sustained engagement
Deep learning (DL)	Captures complex nonlinear relationships	Multimodal sensor data, motion capture and video	Predicts movement efficiency, fatigue levels and recovery trends	High precision personalization and pattern detection

Federated learning	Trains shared models without sharing sensitive data	Multi-center clinical datasets	Enables large-scale model generalization while preserving privacy	Enhanced model robustness and patient data protection
Explainable AI (XAI) techniques	Improves model interpretability	Model outputs and decision rules	Provides rationale for intensity recommendations	Enhances clinician trust and clinical integration

Table 7.1. *Machine learning approaches for optimizing exercise intensity in oncology rehabilitation*

7.5. Data sources for ML modeling

7.5.1. *Wearable devices and biosensors*

Wearable technology is rapidly changing the way we monitor patients in different fields, including rehabilitation oncology. Smartwatches, fitness bands, chest straps and other multifunctional wearables emit, transmit and store a variety of physiological and biomechanical data, a collection that is nearly constant and of very high resolution. Devices of this kind monitor:

– *Heart rate and heart rate variability (HRV)*: fluctuations in HRV show the functioning of the autonomic nervous system, revealing the changes in fatigue, stress and recovery over a period. HRV changes can also indicate overtraining and exhaustion (Stephenson et al. 2021).

– *Accelerometry*: wearable sensors record and classify movements and give measures of intensity, the number of steps and gait and sedentary behavior measurements.

– *Skinner temperature and sweat rate*: the parameters measure the level of hydration and, when combined with other physiological parameters, can be used to determine the level of exertion during an exercise session.

– *Blood oxygen saturation (SpO_2)*: oxygen saturation during exercise helps to avoid hypoxia, which is particularly important with patients who have lung cancer or other diseases that affect lung function.

– *Electromyography (EMG)*: EMG sensors monitor fatigue, and muscle activity, and thus muscular performance relative to specified targets.

Live stream data is extremely valuable to an ML model that requires vast amounts of granular data to capture the less obvious patterns and trends. Ingesting wearable-collected data into the ML pipelines, clinicians can anticipate how patients will react to the exercise, modify the rehabilitation regimen and avoid negative outcomes, including overexertion or falls (Adetunji 2024). Scalability, as well as allowing home-based or tele-rehabilitation programs, is also supported by remote monitoring and enhances accessibility and compliance.

7.5.2. *Patient-reported outcomes (PROs)*

Although sensor data records physiological and mechanical data, it is not a complete picture of a patient subjectively experiencing such symptoms as fatigue, pain, mood or perceived exertion. The gap is addressed by PROs which present direct patient-reported data information that is often collected using validated questionnaires and scales.

Some of the most used PRO instruments in oncology rehabilitation are:

– *FACIT-F (functional assessment of chronic illness therapy-fatigue)*: evaluates the severity and impact of fatigue, a prevalent and debilitating symptom among cancer patients.

– *EORTC QLQ-C30 (European organization for research and treatment of cancer quality of life questionnaire)*: assesses a wide array of domains including physical, emotional and social functioning, as well as symptoms such as nausea, pain and sleep disturbances.

The information of sensors can be combined with subjective inputs of machine-learned models to create panoramic images of patient health. As an illustration, when wearable data indicate an average heart rate recovery post exercise, but the patient expresses an elevated level of fatigue, the system alarms about a discrepancy. This could lead to an assessment or an intervention (MacKnight 2022) and is a sophisticated method of personalizing exercises. Besides the information, the voice of the patient is identified. The longitudinal observation of the PROs is also used to determine the progress and to predict the tendencies of the decline or enhancement of the clinical picture.

7.5.3. *Electronic health records (EHRs)*

Electronic health records (EHRs) are rich in structured and unstructured clinical data, but they are poorly exploited in the rehabilitation. EHRs contain medication history, laboratory results, imaging studies and clinical notes in addition to the necessary background information:

– *Cancer type, treatment regimen and adverse effects*: understanding whether a patient underwent chemotherapy, radiation or immunotherapy allows the system to account for therapy-induced fatigue or cardiotoxicity.

– *Comorbid conditions and medication profiles*: conditions such as diabetes, neuropathy or cardiovascular disease significantly impact exercise tolerance and safety. Concurrent medications may influence heart rate, balance or endurance (Zaki et al. 2024).

– *Laboratory and imaging results*: biomarkers (e.g. hemoglobin levels, inflammatory markers) and imaging findings can indicate physical readiness and help avoid contraindications to exercise.

The context-aware nature of the system due to the implementation of EHR information into machine learning models would guarantee safer and tailored exercise prescriptions.

As an example, the system automatically varies the intensity of exercise in case hemoglobin decreases or bone metastases is identified and changes in the protocols (Engle et al. 2024). Additionally, EHR data, coupled with real-time streams of wearables, PROs, etc., will provide a complete data ecosystem that will support the process of risk stratification and rehabilitation scheduling in real time.

7.6. Challenges and limitations

7.6.1. *Data quality and heterogeneity*

One of the major problems in ML in the field of oncology rehabilitation is the accessibility of quality and reliable data. Wearable and biosensors are related to noise, artifacts, and uneven and unreliable data gathering. As an illustration, accelerometers provide erratic results in case of loose straps or misplacement. Motion artifacts or poor skin contact can act to distort HRV.

Data heterogeneity is also created by the fact that the devices are manufactured by different producers, sampling rates and sensor calibration. Lacking data, which is due to battery malfunctions, loss of connection or non-compliance on the part of the user, also worsens the model accuracy (Kumar et al. 2026a). This variation can be mitigated through standardizing data-collection procedures and preprocessing interventions such as imputation, filtering, normalization, etc., which can enhance the strength of models.

7.6.2. *Model interpretability*

Even AI methods such as deep neural networks and ensemble algorithms remain black boxes, which is an obstacle to clinical implementation. Clinicians should be given clear clarifications why a certain exercise intensity or a regimen is the one to be used. They might not have confidence or the actions on the outputs of the model without interpretable reasons, particularly regarding the delicate field of cancer rehabilitation.

Explainable AI (XAI) methods (SHAP values or decision rules) are becoming commonly combined to enhance transparency, yet it is difficult to find the right balance between accuracy and interpretability (Kumar et al. 2026b).

7.6.3. *Ethical and legal issues*

Implementation of ML in healthcare must be carried out with a close consideration of ethical and law-related challenges:

– *Privacy*: wearables are generating granular sensitive health data, which is susceptible to attack, as wearables track us at all times.

– *Consent*: the patients are required to provide informed consent and have a full understanding of the type of data that is gathered, its usage and the individuals that may access the data.

– *Bias*: the models that are trained on smaller or homogeneous cohorts might not apply to larger groups of the population, which results in unfair results. The problem of algorithmic bias requires a variety of datasets and ML methods that are conscious of fairness (Singh et al. 2022).

7.6.4. *Clinical validation*

Most of the tools in the field of exercise personalization based on ML, although promising in pilot studies, are still in scholarly or research environments. They should be subjected to a lot of clinical validation before adoption into real-world clinical work habits. Safety, effectiveness and clinical utility have to be proven by randomized controlled trials, prospective cohort trials and real-world evidence gathering (Sherman et al. 2016).

These tools can only be trusted by clinicians, patients and regulatory bodies through a strong validation process.

7.7. Future directions

7.7.1. *Multimodal data fusion*

Multimodal data fusion refers to the use of multiple data sources, including EMG, EEG, gait analysis and voice biomarkers, to give a more detailed image of a patient in terms of both their physiological and cognitive conditions. EMG can be used to detect the fatigue and activation patterns of muscles, whereas EEG can be used to assess the levels of cognitive load and stress. Motion tracking records the quality of movement, volume, extent of motion and balance. Biomarkers that a voice can detect include pitch, speech rate and tremor, which can be used to detect fatigue, emotional stress and respiratory strain. A combination of these modalities allows ML systems to constantly evaluate the exercise tolerance and modify the recommendations dynamically (Paniagua-Gómez and Fernandez-Carmona 2025). This method is particularly useful in the field of oncology, where physical and cognitive exhaustion may be confused.

7.7.2. *Customizable digital twin*

The patient digital twins are personalized virtual models demonstrating how the body of a person will react to various interventions through exercises. When combined with wearables, clinically integrated digital history, imaging and treatment documents, models can assess different exercise scenarios and predict patient responses (Gudur et al. 2024b). A digital twin of a breast cancer survivor can assess the effects of moderate-intensity cycling over a six-week period while adjusting for fatigue from chemotherapy, as well as cardiotoxicity. This gives clinicians greater confidence when personalizing prescriptions, as the outcomes can be optimized while risks are minimized (Chen et al. 2023).

7.7.3. *Federated learning*

Federated learning supports collaborative model building across institutions while keeping sensitive patient information at its originating site. Unlike most other areas of healthcare, rehabilitation oncology is characterized by siloed, limited datasets. For instance, hospitals and research entities are able to build collaborative ML models while training on local datasets, keeping the datasets on-site (Anuyah et al. 2024). This provides a model that is aggregated from a variety of diverse datasets which enhances its generalizability across various patient populations. This is privacy-preserving and compliant with most privacy regulations, which makes it suitable for the large-scale development of precise rehabilitation strategies (Marques et al. 2024).

7.7.4. *Integration with genomic data*

Future exercise optimization driven by ML models may also leverage genomic and epigenomic information. There are genetic variations that determine recovery and fatigue as well as cardiovascular responses to and after exercise (Gudur et al. 2024a). Combining both genetic information with the wearable sensor readings and PRO, MC models can produce extremely personalized exercise programs. Indicatively, patients with polymorphisms that enhance performance of the mitochondria or lower inflammation might have to be put on altered regimens to avoid overworking. The next stage of data-driven oncology rehabilitation is biological personalization.

7.8. Conclusion

Self-administered rehab with the help of ML could revolutionize oncology. ML can predict the best intensity to apply to a patient by perfecting exercises in real time, which creates a predictive model. This real-time modification also allows clinicians to use information on numerous sources to adjust physical activity. Recent rehabilitation regimens tend to use the one-size-fits-all regimens, disregarding the differences in the side effects and the recovery patterns of individuals. Individualized oncologic rehab should be based on the lesions and clinical profile of the individual patient in order to maximize functional recovery, minimize risk of over-exertion and uphold compliance. An ML model could identify the early evidence of fatigue based on the minor physiological indicators and reduce the intensity of a session. These smart systems might bring autonomous care of patients and present decision assisting over to clinicians. Issues related to quality of data, equitable access, privacy and the seamless integration of the ML-guided rehab into clinical processes need to be considered in order to make the practice of the ML-guided rehab a viable part of clinical practice.

7.9. References

Adetunji, O. (2024). Advanced digital-twin modeling for predictive monitoring of postoperative cardiac patients using wearables and EHR data. *GSC Biological and Pharmaceutical Sciences*, 27(1), 295–314.

Aljohani, A. (2024). Optimizing patient stratification in healthcare: A comparative analysis of clustering algorithms for ehr data. *International Journal of Computational Intelligence Systems*, 17(1), 173.

Anuyah, S., Singh, M.K., Nyavor, H. (2024). Advancing clinical trial outcomes using deep learning and predictive modelling: Bridging precision medicine and patient-centered care. *arXiv preprint*. doi: 10.30574/wjarr.2024.24.3.3671.

Boutros, P., Kassem, N., Barteit, S. (2025). Advancements in wearable biosensors: Transforming cardiovascular health monitoring and disease management. *Progress in Molecular Biology and Translational Science*, 215, 279–310.

Chen, X., Shi, X., Yu, Z., Ma, X. (2023). High-intensity interval training in breast cancer patients: A systematic review and meta-analysis. *Cancer Medicine*, 12(17), 17692–17705.

Chen, Y., Wu, G., Wang, Q., Ding, Y., Wu, L., Shi, H., Sun, Z., Ou, Z., Miao, Y., Ji, X. et al. (2025). Applications of robotic exoskeletons as motion assistive systems in cancer rehabilitation. *Research*, 8, 0855. doi: 10.34133/research.0855.

Devadas, R.M. (2024). 3D visualization using t-SNE dimensionality reduction technique. In *Interactive and Dynamic Dashboard: Design Principles*, Vadivel, A., Meena, K., Sumathy, P., Selvaraj, H., Shanmugavadivu, P., Shaila, S.G. (eds). CRC Press, Boca Raton.

Engle, J., Marshall, G., Lefkowitz, T., Maltser, S. (2024). Fractured knowledge: Making sense of exercise in patients with bone metastases. *American Journal of Physical Medicine & Rehabilitation*, 103(3S), S58–S61.

Gajda, R., Gajda, J., Czuba, M., Knechtle, B., Drygas, W. (2024). Sports heart monitors as reliable diagnostic tools for training control and detecting arrhythmias in professional and leisure-time endurance athletes: An expert consensus statement. *Sports Medicine*, 54(1), 1–21.

Gudur, R.A., Bhosale, S.J., Gudur, A.K., Datkhile, K.D. (2024a). Effect of cyp1a1, cyp1b1 and cyp2c gene polymorphisms on doxorubicin and paclitaxel. *Bioinformation*, 20(10), 1244.

Gudur, R.A., Bhosale, S.J., Gudur, A.K., Datkhile, K.D. (2024b). Genetic polymorphisms in glutathione S-transferase (GST) gene and their correlation with toxicity of chemotherapy in breast cancer patients. *Asian Pacific Journal of Cancer Prevention: APJCP*, 25(7), 2271.

Jafleh, E.A., Alnaqbi, F.A., Almaeeni, H.A., Faqeeh, S., Alzaabi, M.A., Al Zaman, K., Alnaqbi, F., Almaeeni, H., Alzaabi, M. (2024). The role of wearable devices in chronic disease monitoring and patient care: A comprehensive review. *Cureus*, 16(9), e68921. doi: 10.7759/cureus.68921.

Jamnick, N.A., Pettitt, R.W., Granata, C., Pyne, D.B., Bishop, D.J. (2020). An examination and critique of current methods to determine exercise intensity. *Sports Medicine*, 50(10), 1729–1756.

Kumar, A., Batta, P., Ahuja, S., Rathore, P.S. (eds) (2026a). *AI-driven Innovations in Physiotherapy and Oncology 1*. ISTE Ltd, London and Wiley, Hoboken, NJ.

Kumar, A., Batta, P., Ahuja, S., Rathore, P.S. (eds) (2026b). *AI-driven Innovations in Physiotherapy and Oncology 2*. ISTE Ltd, London and Wiley, Hoboken, NJ.

MacKnight, J.M. (2022). Non-cardiac conditions that mimic cardiac symptoms. In *Sports Cardiology, An Issue of Clinics in Sports Medicine*, Dean, P. (ed.). Elsevier, Philadelphia, PA.

Marques, L., Costa, B., Pereira, M., Silva, A., Santos, J., Saldanha, L., Silva, I., Magalhães, P., Schmidt, S., Vale, N. (2024). Advancing precision medicine: A review of innovative in silico approaches for drug development, clinical pharmacology and personalized healthcare. *Pharmaceutics*, 16(3), 332.

Pandya, S., Thakur, A., Saxena, S., Jassal, N., Patel, C., Modi, K., Shah, P., Joshi, R., Gonge, S., Kadam, K. et al. (2021). A study of the recent trends of immunology: Key challenges, domains, applications, datasets, and future directions. *Sensors*, 21(23), 7786.

Paniagua-Gómez, M. and Fernandez-Carmona, M. (2025). Trends and challenges in real-time stress detection and modulation: The role of the IoT and artificial intelligence. *Electronics*, 14(13), 2581.

Patel, S., Park, H., Bonato, P., Chan, L., Rodgers, M. (2012). A review of wearable sensors and systems with application in rehabilitation. *Journal of Neuroengineering and Rehabilitation*, 9(1), 21.

Rudroff, T. (2025). Digital biomarkers and AI for remote monitoring of fatigue progression in neurological disorders: Bridging mechanisms to clinical applications. *Brain Sciences*, 15(5), 533.

Sherman, R.E., Anderson, S.A., Dal Pan, G.J., Gray, G.W., Gross, T., Hunter, N.L., LaVange, L., Marinac-Dabic, D., Marks, P.W., Robb, M.A. et al. (2016). Real-world evidence: What is it and what can it tell us. *The New England Journal of Medicine*, 375(23), 2293–2297.

Singh, N., Moneghetti, K.J., Christle, J.W., Hadley, D., Froelicher, V., Plews, D. (2018). Heart rate variability: An old metric with new meaning in the era of using mhealth technologies for health and exercise training guidance. Part two: Prognosis and training. *Arrhythmia & Electrophysiology Review*, 7(4), 247.

Singh, H., Mhasawade, V., Chunara, R. (2022). Generalizability challenges of mortality risk prediction models: A retrospective analysis on a multi-center database. *PLOS Digital Health*, 1(4), e0000023.

Srichawla, B.S. (2024). Future of neurocritical care: Integrating neurophysics, multimodal monitoring, and machine learning. *World Journal of Critical Care Medicine*, 13(2), 91397.

Stephenson, M.D., Thompson, A.G., Merrigan, J.J., Stone, J.D., Hagen, J.A. (2021). Applying heart rate variability to monitor health and performance in tactical personnel: A narrative review. *International Journal of Environmental Research and Public Health*, 18(15), 8143.

Stinear, C.M., Lang, C.E., Zeiler, S., Byblow, W.D. (2020). Advances and challenges in stroke rehabilitation. *The Lancet Neurology*, 19(4), 348–360.

Strasser, F. (2021). Management of cancer-related fatigue. In *Survivorship Care for Cancer Patients: A Clinician's Handbook*, Rauh, S. (ed.). Springer, Cham, 203–231.

Vadia, N., Patel, P., Sutariya, V. (2025). Advances and opportunities in digital diabetic healthcare systems. In *Decision Support System for Diabetes Healthcare: Advancements and Applications*, Desai, U., Acharya, B., Shukla, M., Gurupur, V. (eds). River Publishers, Aalborg, 15–52.

Vella, S.A., Aidman, E., Teychenne, M., Smith, J.J., Swann, C., Rosenbaum, S., White, R.L., Lubans, D.R. (2023). Optimising the effects of physical activity on mental health and wellbeing: A joint consensus statement from Sports Medicine Australia and the Australian Psychological Society. *Journal of Science and Medicine in Sport*, 26(2), 132–139.

Yang, L., Amin, O., Shihada, B. (2024). Intelligent wearable systems: Opportunities and challenges in health and sports. *ACM Computing Surveys*, 56(7), 1–42.

Zaki, S., Alam, M.F., Sharma, S., El-Ashker, S., Ahsan, M., Nuhmani, S. (2024). Impact of concurrent exercise training on cardiac autonomic modulation, metabolic profile, body composition, cardiorespiratory fitness, and quality of life in type 2 diabetes with cardiac autonomic neuropathy: A randomized controlled trial. *Journal of Clinical Medicine*, 13(13), 3910.

8

AI-Driven Digital Twins for Simulating Physiotherapy Outcomes in Cancer Care

There has been an increased focus on the quality of life after treatment and cancer survivorship. As part of this focus, physiotherapy has been incorporated to help resolve function loss, pain and fatigue after chemotherapy, radiotherapy or surgery. Nevertheless, traditional rehabilitation practices still use generalized frameworks and fail to account for individual differences. Implementing AI-supported digital twins (DTs) technology could help transform the personalization potential for physiotherapy practices in cancer care. DTs are virtual models of an individual's biomechanical, physiological and clinical characteristics that are continuously updated and integrated with imaging, wearable sensors, electronic health record data and self-reported questionnaires. When integrated with AI technology, DTs provide the potential for the simulation, prediction and real-time optimization of outcomes of physiotherapy plans. This chapter attempts to synthesize the state of the art in AI-driven DTs in the rehabilitation of cancer patients. We analyze the fundamentals of DT physiologic modeling that spans multiple scales, the contribution of AI as an autonomous agent and the infrastructure for movement simulation and therapy personalization. We detail specific use cases, such as shoulder rehabilitation after mastectomy, fatigue in cancer rehabilitation, gait training and pain management to demonstrate the wide range of uses. We discuss the potential for the rehabilitation of cancer patients through evidence from similar fields such as AR-based telerehabilitation, AI virtual clinics and musculoskeletal physiotherapy. Data siloing and integration, explainability of models, ethics, and software and clinical regulation are some of the challenges analyzed.

8.1. Introduction

Surgery, chemotherapy and radiations are some of the common treatments of cancer and cause physical impairment that severely influences the mobility, strength and general functionality of a patient. The general issues that can be experienced after treatment are the restricted range of motions, muscle weakness, persistent pain,

---

Chapter written by Rashmi GUDUR and Mrudula NIMBARTE.

fatigue, neuropathy and lymphedema. These problems may harshly restrict the everyday process and decrease the quality of life; that is why physiotherapy is so important in the recovery process. Physiotherapy aims to restore function, improve mobility, alleviate pain and enhance overall well-being, making it a cornerstone in the continuum of cancer care (Baladaniya and Baldania 2025).

Nevertheless, conventional practices in physiotherapy tend to follow set rehabilitation protocols which do not attend to the individual patient's needs. This type of practice often does not incorporate the factors of age, type of cancer, comorbidities and psychological, physical and emotional fitness, which may cause responses to therapy to differ. Some patients may obtain fewer benefits or suffer the consequences of under- or over-treatment (George et al. 2021). In the absence of personalized rehabilitation protocols, optimizing rehabilitation outcomes for cancer survivors becomes challenging.

Conversely, increasingly sophisticated technologies such as artificial intelligence (AI) and digital twin (DT) technology are entering the rehabilitation sectors of healthcare. DT technologies in rehabilitation healthcare are defined as the virtual and constantly updated model created in real time, which represents the patient's biomarkers and physical and behavioral characteristics as captured by sensors and other clinical devices (e.g. EHRs, wearables, imaging technology and other monitors) (Abd Elaziz et al. 2024).

DTs in cancer rehab can mimic certain movements of the joints, the activation of muscles, the shifting of the body, balance and coordination. They can also track the physiological aspects of the body, such as heart rate, saturation of oxygen in the blood and fatigue level. Coupled with some AI, these types of DTs can become predictive systems, showing the possible effects of certain physiotherapy approaches on a patient's recovery in the future. For the clinician, this means the rehabilitation approaches can be tested on the DT before applying them in real life, minimizing the risk and improving the result.

DTs rely on AI, specifically on machine learning (ML) and deep learning (DL). These systems detect patterns in the vast amounts of treatments given to a patient, as well as the overall population in order to predict possible outcomes of the given treatment. For example, an AI DT for a patient with breast cancer and limited shoulder mobility after surgery can predict the outcomes of certain approaches such as resistance training versus manual therapy versus neuromuscular re-education, and predict which one will work best. This is the primary benefit of AI DTs in therapy because of their constant flexibility.

These systems are real-time learners, unlike traditional rehabilitation in which the suggestions are prepared beforehand and are only changed when the patient improves. As an example, in case of stalling or regression of a patient, the AI component corrects the therapy and proposes new or alternative exercises. It is dynamic feedback that generates an individualized, flexible response that improves patient engagement and adherence (Vallée 2024).

Remote rehabilitation is another application of AI DTs, which has gained more significance due to the development of telehealth and home care. Motion-linked sensors are put in patients and smartphones are used to collect the data, which can then be tracked and assessed in real time by DTs through monitoring and evaluating the home exercises. Providers can also access this data to make timely adjustments and ensure active care (Batool and Lopez 2023). This real-time approach enhances access to physiotherapy and eases pressure on clinical facilities. AI DTs in cancer rehabilitation enhance integrated rehabilitation approaches. Oncologists, physiotherapists, data specialists and biomedical engineers can all engage in co-analyses of the DTs to collaboratively calibrate rehabilitation plans to clinical pathways and the patient's lived experience.

8.2. Background and theoretical framework

8.2.1. *Digital twins in healthcare*

Initially, the idea of DTs began in engineering and manufacturing, where virtual representations of systems were built to test and optimize various aspects without the expense and danger of real-world trial and error. With time, this idea found its way into healthcare with the human body being considered for virtual modeling to anticipate various physiological responses, disease progressions and responses to therapy. In contrast to the DTs in other fields, healthcare DTs are dynamic, real-time evolving virtual representations of individual patients, made personalized through health records, personal biological health data and real-time environmental data.

In oncology, DTs are changing the way cancer diagnosis, treatment and patient care are managed. They allow for advanced modeling of tumor growth and responses to various treatments, including radiotherapy and chemotherapy, and the pharmacokinetics and pharmacodynamics of several anticancer drugs (Ioele et al. 2022). They further aid in modeling the treatment of pain, cancer-related pain and the progression of cancer and adverse reactions to drugs. Such comprehensive simulations assist in personalizing therapy while reducing the likelihood of adverse reactions.

Moreover, decision trees give clinicians and researchers the opportunity to ask "what-if" questions and fine-tune treatment strategies to accommodate biological and behavioral differences for each patient.

8.2.2. *Musculoskeletal and physiotherapy digital twins*

The use of DTs in musculoskeletal (MSK) rehabilitation and physiotherapy is an emerging field of rehab medicine and even more so for cancer care. Cancer patients suffer from disease and treatment-related MSK impairments and can greatly benefit from rehabilitation during this stage of care. Patients in this population typically present with several MSK-related impairments that, if left unaddressed, may affect rehabilitation, and more importantly, the patient's overall quality of life, such as weaknesses and stiffness in joints and muscles, peripheral neuropath and overall fatigue, and restricted mobility. As such, effective physiotherapy becomes even more indispensable, and DTs have the potential to improve this.

In MSK rehabilitation, DT systems use biomechanical modeling and wearable sensors – IMUs, EMGs and force sensors – to monitor gait, muscle activity, joint angle and external physical load in real time. Patient-oriented AI analyzes the data streams and produces the personalized avatars that imitate the MSK dynamics of the patient. The DT can then predict biomechanical responses to simulated physiotherapy and adjust the intervention using predicted outcomes (Mikołajewska et al. 2024). Real-time adaptiveness is added with the introduction of AI into these DT systems. As an example, a DT conducted by a cancer survivor with lymphedema would be able to simulate several different exercises to the shoulder to determine the most effective protocols to recover muscles.

The system considers the adjustments required in the intensity of movements, frequency or type of movement when calculating the movement of the rehab of a person based on progress. This ability goes beyond the rehab system merely responding to a user's needs. It truly makes the rehab process individualized and makes rehab more effective.

8.2.3. *Virtual physiological human (VPH) and multiscale modeling*

The virtual physiological human (VPH) initiative seeks to build a multiscale model of the human body to comprehend various biological scales, studying the body from the molecules and the cells to the tissues and the organs, and the whole organism. VPH can be best understood not as a single system, but as a constellation of different systems that integrate various sources of data, models and simulations.

The aim is to create a digital, anatomically and physiologically accurate representation of a human being, both healthy and diseased. Looking at the VPH in the context of DT technology for cancer patient, physiotherapy provides invaluable insight into how systemic MSK inflammation and cancer cachexia impact the local function of the MSK system. For example, VPH-informed DTs can predict the impact of alterations to a muscle fiber's structure, mitochondrial activity and nerve conduction on the gait and range of motion of the whole limb. They also model the higher levels of neuromuscular control, motor unit recruitment and reflex activity, functions that are often lost in patients after chemotherapy and radiation. Multiscale modeling also captures macroscopic and microscopic movements, enabling clinicians to discern the functional impairments on their root and focus their intervention.

As an example, if a VPH-based DT establishes that reduced knee extension during gait is caused by neuromotor inhibition as opposed to joint restriction, the focus of therapy will be neuromuscular reeducation, as opposed to just passive stretching (Huynh et al. 2020).

8.2.4. *The role of artificial intelligence*

The use of AI in developing and performing DTs in physiotherapy and oncology practices is invaluable. ML, DL, reinforcement learning, systems pharmacology and generative models in AI provide the ability to individualize real-time simulation and predictive analytics, and provide decision support to clinicians.

DL and ML methods for predictive analytics and real-time assessment of the DT, for example, use and integrate everything from EHRs, wearable sensors, imaging and genomic data, and continually refine the accuracy of the DT. Complications such as treatment-related neuropathy, impaired wound healing and functional decline may be predicted much earlier as advanced neural networks identify patterns and correlations beyond the scope of human perception (Gutman et al. 2024).

Personalized rehabilitation protocol design greatly benefits from reinforcement learning. AI can design a strategy to "rehabilitate" virtual twins, for example, to find the best recovery exercises for a post-operative shoulder dysfunction patient after a mastectomy, to improve the outpatient rehabilitation outcomes.

Generative models, and particularly generative adversarial networks (GANs), provide tools to simulate and synthesize data and help fill in the gaps of models defined by rare and underrepresented patient types by realistic simulation of patient responses to various therapies. Additionally, AI-powered system pharmacology can simulate how an individual's unique genetic, metabolic and physiological traits affect drug responses – enabling integration of pharmacotherapy planning with physiotherapy interventions in a truly holistic care model (Ahmad 2025).

Combined, these AIs will turn the DT into a dynamic system that is intelligent and capable of learning. It also directly informs real-time clinical decision-making and enhances patient outcomes in a way that allows the transition to more responsive, predictive and preventive care, which varies with the dynamic recovery trajectory of the individual patient.

8.3. Current research landscape

8.3.1. *Bibliometric and thematic mapping*

Bibliometric analysis of recent publications indicates that in recent years, there has been a steep increase in DT studies in the rehabilitation field which signifies the increase in scholarly and clinical interest. The analysis of keywords and co-citation clusters shows that there are a few notable themes, namely, movement monitoring, adaptive and customized therapy protocols, telerehabilitation and sensor-integrated assessment tools. These themes highlight a shift in practice toward the old and the new data-driven and patient-focused interventions. Specifically, motion tracking can be typically associated with the use of wearable devices and motion capture systems that convey real-time biomechanical information into DT models. Adaptive treatment schemes make use of AI algorithms to adjust the intensity, duration and type of exercise, in response to the ongoing feedback of the physiological and performance information of the patient (Alzahrani and Ullah 2024). Telerehabilitation is another type of integration of DT, which is strengthened by the pandemic-induced transition to remote care, as it allows a person to get a set of interventions at home that can be personal. The key to all of these efforts lies in the use of sensors, inertial measurement units (IMUs), electromyography (EMG) and force sensors, which aid in the development of dynamic, patient-specific models, which can be used to aid the accurate diagnosis and treatment planning.

8.3.2. *Oncology-focused DT research*

Oncology DT studies have been majorly devoted to the process of tumor progression modeling, response to different treatment modality modeling and clinical decision-making via virtual trials and pharmacokinetics/pharmacodynamics modeling. These DT models have the ability to combine genomics, histopathology, imaging and treatment history to generate detailed simulations of the disease path of individual patients. Nevertheless, these capabilities have been relatively rarely applied to post-treatment rehabilitation and there is a void in the literature. Current studies reveal that DTs may be applied in the management of pain, particularly in those patients who have surgery or are under chemotherapy, as it allows the predicting of pain flares and the optimal timing of medication. In surgery oncology,

DTs have been used to model procedures and design interventions at high levels of anatomical accuracy, which may decrease post-operative complications that may impede the rehabilitation process. Nonetheless, there are very limited direct applications of DTs to model MSK or functional recovery patterns in cancer survivors. This disparity may be bridged, thus allowing a more consistent continuity of care, where the rehabilitation plans are informed by the same digital infrastructure that influenced diagnosis and treatment (Afridi and Khan Shah 2024).

8.3.3. *Technology prototypes in rehabilitation*

The viability and possibility of implementing the DT methods can be demonstrated through emerging technology prototypes in the field of rehabilitation. DT control strategies installed on the exoskeleton robots have shown a high level of precision with an error in real-time tracking of less than 0.05 m and diagnostic errors of over 80%. Such systems monitor the intent of patients with the help of DTs and adjust the motion assistance to simplify the therapy process and improve recovery outcomes (Mikołajewska et al. 2024). Telerehabilitation based on the augmented reality (AR) has demonstrated specific potential in post-mastectomy patients by facilitating the shoulder mobility, pain level and general quality of life through simulation of movement exercises and full-immersion feedback.

8.4. Framework for AI-driven DT in cancer physiotherapy

8.4.1. *Data modalities and integration*

Physiotherapy DT models can only be effective when they combine multimodal data that incorporate anatomy, physiology, function, as well as lived experience of the patient. The most important data streams are imaging CT, MRI and ultrasound which describe the anatomy and the alterations after the operation. These images provide the space of the structure which is required to make precise anatomical models. IMUs, EMGs, pressure and skin sensors are biomechanical sensors that measure the joint angles, muscle activation and gait patterns. They create biomechanical models of movement and physical ability of a patient together and particularly in the post-oncologic surgery setting.

Physiological data on heart rate, pain, fatigue and other progressive symptoms are offered by real-time and dynamic information. They measure functional response, forecast overuse and determine systemic rehabilitation outcome (Malesevic et al. 2023). EHRs provide context (diagnoses, surgical history, chemo/radiation, comorbidities, medications and systemic conditions), which are medically accurate and personalized DT simulations. Pain, fatigue and validated quality-of-life scores which are patient-reported outcomes are also necessary.

The inclusion of subjective information can be used to ensure that DT models are adaptable and sensitive to the lived experience of the patient and self-reported barriers that can be missed by sensors. The data inputs are pooled into a DT framework which is oriented toward the VPH paradigm, a more detailed framework that simulates and predicts the human organism on biological scales to intervene.

Data source/modality	Data type	Example tools/devices	Purpose in digital twin physiotherapy simulation
Medical imaging	Structural and anatomical data	MRI, CT and ultrasound	Provides accurate anatomical reconstruction and detects post-surgical alterations
Biomechanical sensors	Kinematic and kinetic data	Inertial measurement units (IMUs), electromyography (EMG) and pressure sensors	Captures real-time motion, joint angles and muscle activity for biomechanical modeling
Physiological sensors	Functional and systemic data	Heart rate monitors, pulse oximeters and fatigue trackers	Monitors cardiovascular response, fatigue thresholds and physiological load
Electronic health records (EHRs)	Clinical and treatment history	Hospital databases and cloud EHR systems	Supplies clinical context including cancer type, therapy regimen and comorbidities
Patient-reported outcomes (PROs)	Subjective functional data	Fatigue, pain and QoL questionnaires	Adds personalized insights into perceived pain, fatigue and quality of life
Genomic and biomarker data *(future integration)*	Molecular and predictive data	Genomic sequencing and proteomic assays	Enhances precision by predicting therapy responsiveness and fatigue susceptibility

Table 8.1. *Integration of data modalities for AI-driven digital twin-based cancer physiotherapy*

8.4.2. *Modeling and simulation architecture*

The architecture of DTs which are designed for simulating outcomes in digital physiotherapy are created with a multi-layered and modular framework. There is a base layer that develops anatomical models. MSK system digital reconstructions are based on imaging data. This layer ensures that simulation of therapies is consistent with the anatomy of the exact patient such as post-operative tissues, asymmetries in limbs and other variations.

The second layer is concentrated at biomechanical modeling. It also replicates movement dynamics, muscle behavior and joint dynamics in therapy. It approximates intra- and post-operative mechanical stress and strain in exercises, which are powered by the information of wearable sensors (Alzahrani and Ullah 2024).

Functional and physiological modeling is covered by the third layer. It combines the work of the cardiovascular system, metabolic and nervous system. This layer forecasts the fatigue limits, work–rest proportions and warns about possible physiological abnormalities.

All of the data is optimized by the AI layer, and the most effective and safest physiotherapy protocols are identified. ML algorithms adjust exercise plans dynamically according to the projected results and patient feedback.

The last layer is the feedback system which provides the DT with real-time updates and closes a loop. It enables the virtual world to change and react in real time to the current treatment of the patient.

8.4.3. *Simulation of physiotherapy interventions*

DTs can be used to simulate different methods of physiotherapy. Alternative protocols can be pre-tested by the therapists. As an illustration, one DT could compare the stretching protocols with strengthening protocols and predict the results over time and rehabilitation at various stages.

Simulation protocols are able to test the intensity of exercise, duration of exercise, injury prevention and fatigue. Besides, DTs may provide personalized training through AR or VR and guide individuals interactively and in real time (Kostadimas et al. 2025). This interaction will maintain patient attention and keep the therapy within the biomechanical and physiological boundaries.

8.5. Use cases and scenario examples

8.5.1. *Post-mastectomy shoulder rehabilitation*

DTs can be designed with the help of imaging data, such as MRIs, ultrasounds, motion sensors to model post-mastectomy shoulder biomechanics. These virtual models approximate the range of motion, stiffness and compensatory muscle patterns of a patient, hence enhancing individualized rehabilitation plans (Killen et al. 2020). The DTs track the movements and provide real-time feedback when augmented with AR technology. This improves patient compliance, clinician supervision and general compliance to home exercises.

8.5.2. *Cancer-related fatigue and gait training*

DTs simulate the way cancer-related fatigue deteriorates muscle functioning, gait dynamics and energy consumption. Simulating fatigue accumulation, clinicians will be able to develop adaptive strength and endurance programs to decrease overexertion and risk of falls.

Individualized gait training has the potential to improve the balance, mobility and functional independence of elderly cancer survivors, especially in elderly individuals (Schwenk et al. 2016).

8.5.3. *Movement optimization and pain management*

DTs can predict pain exacerbations and propose pacing strategies by incorporating computational pain models, which use nociceptive and movement responses along with patient self-reporting through movement patterns in addition to patient self-reporting with the combination of nociceptive and movement responses (Kumar et al. 2026a). These understandings maximize treatment and reduce the possibility of pain-related failures.

8.5.4. *Telerehabilitation scaling*

DT simulation telerehabilitation is personalized by continuous analysis of the wearable sensor data with DT-based platforms. RMS can scale and adapt interventions in real time to provide high-quality care in resource-restricted or rural areas, which can be adopted at scale and is inexpensive (Kumar et al. 2026b).

8.6. Evidence of effectiveness

8.6.1. *Outcome results*

The DTs simulate MSK mechanics, therapy responses and recovery patterns correctly in non-oncology physiotherapy. They make changes to treatment plans dynamically according to patient-specific biomechanical and physiological functionality, which improves compliance and functional outcomes. As an example, patients under post-injury or post-operative programs healed quicker and were more satisfied with AI-based DT systems (Liang and Liew 2025). In terms of oncology, there is scanty evidence regarding the application of AR-based DT platform to enhance the upper limb functionality following the mastectomy. Patient-reported QoL is also improved with these interventions by preventing fatigue and pain during guided and immersive sessions of physiotherapy.

8.6.2. *Oncology DT impact*

DTs are primarily applied in oncology in the context of pharmacological modeling. Simulations are used to achieve the optimal dose of a drug, anticipate side effects and determine their survival (Wu et al. 2022). DT-controlled pain management anticipates exacerbations and assists in advance dosage, prolonging pain-free activity and reducing opioid consumption. However, there is still limited research in the area of oncology physiotherapy of DTs. There is no evidence on the large-scale clinical trials connecting DT-based rehab and functional recovery or QoL of cancer survivors.

8.6.3. *Analogous systems in MSK physiotherapy*

AI-based systems in MSK physiotherapy, for example, post-orthopedic surgery rehabilitation, have reduced wait time, individualized care and increased patient satisfaction. These achievements confirm that there is a proof of concept in the application of DT oncology rehabilitation. They advocate the inclusion of AI-based DTs in cancer rehab in order to provide more effective, accurate and patient-oriented care (Chaparro-Cárdenas et al. 2025).

8.7. Ethical, practical and regulatory challenges

8.7.1. *Data privacy and sharing*

DT systems are AI-based, multimodal systems that combine the most sensitive data, including data of medical imaging, wearable sensors, electronic health records (EHRs) and patient-reported outcomes. This data is crucial to protect by means of safe collection, storage and transmission. The interoperability standards allow the data to be shared among systems, and privacy is also assured by encryption, anonymity and high-level restrictions (Gujar 2025). Effective data governance principles are needed as one of the foundations of regulatory compliance and patient trust.

8.7.2. *Model explainability and clinician trust*

The adoption of DT systems in medicine will require trust. It is a result of openness in decision-making in the system. When a DT makes such predictions without explanations, the clinicians might be reluctant to be involved in shared decision-making. Thus, the approaches to explainable AI (XAI) are urgent. XAI assists clinicians in visualizing the relationships between individual inputs and outputs so that they can also support refute, or override predictions, when necessary, in the clinical context.

8.7.3. *Equity and accessibility*

DTs should not exacerbate prevailing inequities in healthcare. Populations that are deprived of digital technologies or have low health literacy may be unwittingly left out in innovations. Thus, DTs are to be as accessible as possible, inexpensive, easy and adjustable to address the individual requirements of the least affluent, with vulnerable groups being among them (Istambul and Abdulghani 2024).

8.7.4. *The computational and resource challenges represent*

The last area of discussion is the problem of computational and resource challenges. DTs require a lot of computational resources and real-time data processing. Health systems with low resources have certain challenges and might need enormous resources and funding to implement and support such technologies.

8.7.5. *Legal liability*

The legal liability of the defendant and the evidence and decision-making criteria used to conclude the case are considered during the clinical validation. To prove the effectiveness and safety of DT interventions, there is a need to conduct clinical validation of these interventions (Gudur et al. 2023), which would involve randomized controlled trials (RCTs) of high quality. It should also be established by the law how responsibility and culpability are defined in case of negative consequences of AI-based decisions to maintain ethical and professional responsibility.

8.8. Future directions and research agenda

8.8.1. *Clinical trials*

DTs are found to have potential in simulation of physiotherapy outcomes in cancer treatment, but require solid clinical validation. The most important pilot trials should be on breast and prostate cancer rehabilitation because these cancers are oftentimes known to cause permanent functional deficits, which could be cured through structured physiotherapy. Preliminarily, a trial can be used to assess the feasibility, safety and accuracy of prediction of outcomes of DT-guided interventions, which in turn facilitates the RCTs.

8.8.2. *Multi-mode data integration*

The researcher will utilize multi-modes of data integration by combining the information using the various modes of research. Future DT models ought to integrate a wider variety of data modalities to enhance the level of fidelity and personalization. Genomic information is capable of forecasting personal therapy effects and fatigue levels.

Wearables, blood biomarkers or patient-reported outcomes can be used to determine systemic fatigue markers that can be used to implement dynamic changes in exercise intensity and pacing (Harrington et al. 2020). By incorporating these data, more accuracy will be achieved in planning physiotherapy and predicting outcomes.

8.8.3. *Advanced AI methods*

The modern AI approaches, including reinforcement learning and large language models (LLMs), have the potential to enhance the decision support in the DT platforms. Reinforcement learning develops adaptive therapy mechanisms that are subject to feedback and performance. The use of LLMs facilitates the natural-language interface, which can be used by clinicians and patients to engage with the DT system in an intuitive manner (Guttha 2025).

8.8.4. *AR/VR and gamification*

DT-based rehabilitation programs could be enhanced with the use of AR and VR and gamification strategies in order to improve patient engagement, adherence and motivation more, especially in home-based or telerehabilitation environments.

8.8.5. *Regulatory frameworks*

There should be regulatory frameworks to ensure the safe usage of DTs in the healthcare industry. Such frameworks are expected to address clinical validation, cybersecurity, data governance and explicit ways of clinician oversight and accountability (Gudur et al. 2024).

8.8.6. *Interdisciplinary teamwork*

This is the primary component of the creation and implementation of DTs in cancer rehabilitation. It should bring in AI researchers, physiotherapists, oncologists, ethicists, regulatory bodies and most importantly patients – ensuring that solutions that are clinically relevant, ethical and user-centered are in place (Rani and Hasanpuri 2025).

8.9. Conclusion

Cancer rehabilitation becomes more open with the use of an AI-based DT. DTs aid in the preparation of data-driven and customized physiotherapy sessions. DTs make the virtual replicas realistic by integrating the imaging, motion capture, clinical records and patient reports. The treatment can be simulated and modified in real time by the clinicians, enhancing safety, effectiveness and personalization. The majority of the existing oncology DT research is concerned with predictive and modeling of treatment responses.

Application of DTs in physiotherapy among survivors of cancer is still premature. The MSK DT prototypes and virtual rehab platforms have emerged and demonstrated possibilities of being adopted in clinics. To illustrate, systems with AI enhance compliance, forecast results and enhance interaction. Application of these to oncology rehabilitation would address cancer fatigue, post-operative shortages and long-term disability. In order to exploit the full potential of DT-guided physiotherapy, their efficacy and safety should be established in large clinical trials. This would be possible through a collaboration between oncology, physiotherapy, bioengineering and data science. The ethics should safeguard privacy, guarantee the transparency of algorithms and have equal access.

The responsible application of AI-based DTs could revolutionize the field of oncology rehabilitation and improve its functionality, assist in developing a specific recovery strategy and the overall survivorship care.

8.10. References

Abd Elaziz, M., Al-qaness, M.A., Dahou, A., Al-Betar, M.A., Mohamed, M.M., El-Shinawi, M., Ali, A., Ewees, A.A. (2024). Digital twins in healthcare: Applications, technologies, simulations, and future trends. *Wiley Interdisciplinary Reviews: Data Mining and Knowledge Discovery*, 14(6), e1559.

Afridi, A. and Khan Shah, N. (2024). Digital transformation in healthcare rehabilitation: A narrative review. *Journal of Process Management and New Technologies*, 12(1–2), 16–30.

Ahmad, F. (2025). Optimizing treatment: The role of pharmacology, genomics, and AI in improving patient outcomes. *Drug Development Research*, 86(3), e70093.

Akkaya, N. (2022). The impact of injury on depression in athletes: A systematic review. Thesis, Pepperdine University, Malibu.

Alzahrani, A. and Ullah, A. (2024). Advanced biomechanical analytics: Wearable technologies for precision health monitoring in sports performance. *Digital Health*, 10, 20552076241256745.

Baladaniya, M. and Baldania, S. (2025). Exploring the impact of physical therapy on patient outcomes across the cancer care continuum: A narrative review. *Cureus*, 17(7), e87144. doi: 10.7759/cureus.87144.

Batool, A. and Lopez, A. (2023). Healthcare access and regional connectivity: Bridging the gap. *Journal of Regional Connectivity and Development*, 2(2), 260–271.

Chaparro-Cárdenas, S.L., Ramirez-Bautista, J.A., Terven, J., Córdova-Esparza, D.M., Romero-Gonzalez, J.A., Ramírez-Pedraza, A., Chavez-Urbiola, E.A. (2025). A technological review of digital twins and artificial intelligence for personalized and predictive healthcare. *Healthcare*, 13(14), 1763.

George, M., Smith, A., Sabesan, S., Ranmuthugala, G. (2021). Physical comorbidities and their relationship with cancer treatment and its outcomes in older adult populations: Systematic review. *JMIR Cancer*, 7(4), e26425.

Gudur, A., Patil, V., Gopal, L., Thapliya, S. (2023). Multi-objective optimization for breast cancer risk prediction models with particle swarm optimization. *International Journal of Intelligent Systems and Applications in Engineering*, 12, 531–554.

Gudur, R.A., Bhosale, S.J., Gudur, A.K., Kale, S.R., More, A.L., Datkhile, K.D. (2024). Influence of (C1236T and C3435T) polymorphisms of ABCB1 gene on chemotherapy treatment outcome and toxicity in breast cancer patients. *Asian Pacific Journal of Cancer Prevention: APJCP*, 25(5), 1567.

Gujar, P. (ed.) (2025). Data standardization and interoperability. In *Data Usability in the Enterprise: How Usability Leads to Optimal Digital Experiences*. Apress, Berkeley, CA, 89–110.

Gutman, B., Shmilovitch, A.H., Aran, D., Shelly, S. (2024). Twenty-five years of AI in neurology: The journey of predictive medicine and biological breakthroughs. *JMIR Neurotechnology*, 3(1), e59556.

Guttha, P.R. (2025). Transforming healthcare and life sciences: Applications of large language models and artificial intelligence. *Innovations*, 7(7) [Online]. Available at: https://scholarlyarticle.vncinstitute.com/index.php/IJMI/article/view/40.

Harrington, S.E., Stout, N.L., Hile, E., Fisher, M.I., Eden, M., Marchese, V., Pfalzer, L.A. (2020). Cancer rehabilitation publications (2008–2018) with a focus on physical function: A scoping review. *Physical Therapy*, 100(3), 363–415.

Huynh, W., Lee, M., Kiernan, M.C. (2020). Neurorehabilitation approaches for disorders of the peripheral nervous system. In *Oxford Textbook of Neurorehabilitation*, Dietz, V. and Ward, N.S. (eds). Oxford University Press, Oxford.

Ioele, G., Chieffallo, M., Occhiuzzi, M.A., De Luca, M., Garofalo, A., Ragno, G., Grande, F. (2022). Anticancer drugs: Recent strategies to improve stability profile, pharmacokinetic and pharmacodynamic properties. *Molecules*, 27(17), 5436.

Istambul, M.R. and Abdulghani, T. (2024). Designing inclusive digital platforms for marginalized communities: A design science research approach. In *Widyatama International Conference on Engineering 2024 (WICOENG 2024)*. Atlantis Press, Paris, 104–113.

Killen, B.A., Falisse, A., De Groote, F., Jonkers, I. (2020). In silico-enhanced treatment and rehabilitation planning for patients with musculoskeletal disorders: Can musculoskeletal modelling and dynamic simulations really impact current clinical practice? *Applied Sciences*, 10(20), 7255.

Kostadimas, D., Kasapakis, V., Kotis, K. (2025). A systematic review on the combination of VR, IoT and AI technologies, and their integration in applications. *Future Internet*, 17(4), 163.

Kumar, A., Batta, P., Ahuja, S., Rathore, P.S. (eds) (2026a). *AI-driven Innovations in Physiotherapy and Oncology 1*. ISTE Ltd, London and Wiley, Hoboken, NJ.

Kumar, A., Batta, P., Ahuja, S., Rathore, P.S. (eds) (2026b). *AI-driven Innovations in Physiotherapy and Oncology 2*. ISTE Ltd, London and Wiley, Hoboken, NJ.

Liang, J. and Liew, F.F. (2025). The clinical significance of femoral and tibial anatomy for anterior cruciate ligament injury and reconstruction. *Current Medical Imaging*, 21(1), E15734056361050.

Malesevic, N., Svensson, I., Hägglund, G., Antfolk, C. (2023). An integrated approach for real-time monitoring of knee dynamics with IMUs and multichannel EMG. *Sensors*, 23(21), 8955.

Mikołajewska, E., Masiak, J., Mikołajewski, D. (2024). Applications of artificial intelligence-based patient digital twins in decision support in rehabilitation and physical therapy. *Electronics*, 13(24), 4994.

Páez, A., Rovers, M., Hutchison, K., Rogers, W., Vasey, B., McCulloch, P. (2022). Beyond the RCT: When are randomized trials unnecessary for new therapeutic devices, and what should we do instead? *Annals of Surgery*, 275(2), 324–331.

Rani, S. and Hasanpuri, V. (2025). Data security and ethical considerations in healthcare digital twins. In *Digital Twin Technology for Better Health: A Healthcare Odyssey*, Dixit, M., Bhatele, K.R., Tiwari, D. (eds). CRC Press, Boca Raton.

Schwenk, M., Grewal, G.S., Holloway, D., Muchna, A., Garland, L., Najafi, B. (2016). Interactive sensor-based balance training in older cancer patients with chemotherapy-induced peripheral neuropathy: A randomized controlled trial. *Gerontology*, 62(5), 553–563.

Vallée, A. (2024). Envisioning the future of personalized medicine: Role and realities of digital twins. *Journal of Medical Internet Research*, 26, e50204.

Wu, C., Lorenzo, G., Hormuth, D.A., Lima, E.A., Slavkova, K.P., DiCarlo, J.C., Virostko, J., Phillips, C.M., Patt, D., Chung, C. et al. (2022). Integrating mechanism-based modeling with biomedical imaging to build practical digital twins for clinical oncology. *Biophysics Reviews*, 3(2), 021304. doi: 10.1063/5.0086789.

9

Natural Language Processing of Patient Feedback to Personalize Oncology Physiotherapy

In the context of cancer care, personalized oncology physiotherapy has to take into account the complex interrelations of the physical, emotional and psychological aspects of the individual patient. The refinement of physiotherapy has predominantly focused on structured data sources such as clinician evaluations and standardized patient-reported outcome measures. Much of the valuable data, however, lies in the unstructured patient data: notes from consultations, online reviews, satisfaction surveys, free-text responses in digital health systems and other unstructured narratives. Artificial intelligence, especially the natural language processing (NLP) branch, offers data extraction and text interpretation tools on unstructured data, which can convert qualitative data into quantitative value assessments to create personalized physiotherapy plans. This chapter presents the innovative and potential uses of NLP in patient feedback for oncology physiotherapy. It defines patient-generated content types and feedback systems, the NLP analytic methods and data content analyses for individualizing care plans, monitoring patient mood and sentiment, adverse event detection, engagement optimization and the tailored personalization of care plans on integrating feedback systems. It offers a critical review of sentiment analysis and other NLP methods such as topic modeling, named entity recognition, and the newer transformer models such as BERT. It explains ethical issues as well as the natural difficulties of text interpretation, especially in complicated fields such as oncology. We discuss the inequity of cross-lingual systems, the limited NLP models specific to the domain of oncology physiotherapy narratives, and several other gaps. Incorporating NLP into oncology rehabilitation programs can provide real-time decision support for clinicians, streamline communication between patients and providers, and promote dynamic care approaches. This chapter aims to assist researchers, clinicians and digital health developers in using NLP to assist in transforming personalization.

Chapter written by Trupti YADAV and Faisal Hussain HUSSAIN.

9.1. Introduction

The last 10 years has seen remarkable advancements in cancer physiotherapy. It has moved past the generalized rehabilitation approach and has become focused and specialized to cater to individual cancer survivors in terms of their physical, psychological and emotional needs (Lown et al. n.d.). Survivors of cancer treatment face complex challenges, including fatigue, pain, mobility loss, lymphedema, neuropathy, cognitive dysfunction and emotional issues (Burgos-Bragado et al. 2025). These challenges require a deeper level of complex personalization in rehabilitation, which cannot be addressed by the traditional generic programs. Unfortunately, it seems that personalization seems to be lacking even with the improvements in clinical methods. This has a lot to do with clinical approaches that strategically ignore the patient in therapeutic interactions (Tyagi et al. 2024).

As the world of healthcare continues to evolve in patient-centered care, the clinical approach that has the most potential and needs the most attention is the voice of the patient. Patient voice, which has been accrued through means such as interviews, surveys, narrative logs and online feedback is most valuable in terms of the symptoms, treatment effectiveness, therapy and the needs that remain unmet (Fabian et al. 2022). The patient voice has largely been ignored due to the challenges that qualitative data presents in the traditional framework of evidence-based care.

This is where the implementation of natural language processing (NLP), a branch of artificial intelligence, becomes a game changer (Vaswani et al. 2017). NLP provides the ability to comprehend, interpret and meaningfully produce human language. The use of NLP entails the ability to understand and analyze large quantities of free-text patient data and derive useful conclusions. In the case of oncology physiotherapy, this entails the recognition of patient symptomatology, therapy emotional distress, exercise preferences and even predicting therapy drop-out. Real-time improvements to the physiotherapy care and techniques can be made based on these insights.

Incorporating NLP into cancer physiotherapy beautician's workflow offers the ability to transform from generalized care models to individualized tailored rehabilitative programs. NLP-powered systems analyze and interpret clinical notes, PROMs, therapy session transcripts, digital surveys and online therapy support groups. For the NLP systems, these documents contain useful and important information, and the hidden and implicit narratives reflect the patient's condition and motivation, and the efficacy of the therapy. We can imagine a physiotherapy clinic, where every week, a patient narrative is collected, describing their experience with a prescribed exercise regimen.

A patient may observe that certain movements are more challenging or that tiredness has increased after chemotherapy. It is almost impossible to manually examine hundreds of entries particularly in a number of patients. A physiotherapist can benefit from an NLP algorithm quickly reading these notes, identifying similar phrases or emotions, and underlining them. This not only saves time it also allows physiotherapists to respond to rehabilitation plans quicker.

The NLP methodology of qualitative information is very effective. Tone can be evaluated in order to identify comments related to dissatisfaction, anxiety or frustration. This feedback can be further grouped in terms of such themes as pain, exercise tolerance or treatment side effects. Recognition names entity recognition identifies important words such as medical concepts, drugs or body parts. The combination of these tools enables the therapists to directly involve the experiences of patients and their needs into the treatment planning.

9.2. Sources of patient feedback in oncology physiotherapy

NLP applications require quality data that is contextual. The access to rich textual feedback is necessary in oncology physiotherapy when the treatment is highly individualized, patients have diverse experiences. The ability to derive insights on the basis of unstructured data on the language facilitates the formation of responsive and adaptive strategies in care. Critical sources of patient-generated or clinician-recorded feedback that are appropriate to undergo NLP analysis include the following.

9.2.1. *Clinical notes and consult transcripts*

One of the most reliable and simplest sources of textual data are clinical documentation. The physiotherapists document the information of their consultation, sessions of therapy and follow-up visits, which provides abundant information on the symptoms, emotional responses, compliance to treatment and in-session behaviors of a patient (Fernandes et al. 2023).

This data can be used with NLP to detect patterns such as fatigue after-exercise, changes in mobility and psychosocial distress. By clustering and assessing notes over several patients, the most frequent issues – such as discomfort during the range of motion, dysphoria in group therapy or session anxiety – can be identified and resolved to enhance programmatic effectiveness. Transcripts from multidisciplinary consultations may also refer specifically to challenges in physiotherapy or modifications to be made, which makes these discussions important materials for comprehensive NLP analysis.

9.2.2. *Patient-reported outcome measures (PROMs) with free-text fields*

Most patient-reported outcome measures (PROMs) tend to follow a structured scale to assess a patient's physical functioning, pain, fatigue or quality of life. However, many still have optional qualitative free-text fields as well. These fields give patients the chance to explain their scores in greater detail as well as provide rationale or other issues that lie outside the predefined parameters (Pham et al. 2024).

For instance, a patient might rate fatigue as a "3" but in the "comments" box would explain "I feel exhausted only after the stair-climbing exercises, not the cycling". This kind of detail is important because it helps in the prescription of more personalized therapy. Techniques of narrative analysis such as keyword identification, sentiment analysis and thematic analysis can address more complex details in qualitative responses tied to specific scoring patterns or treatment results. This type of analysis, in aggregate across patients, helps in the fine-tuning of therapeutic strategies and the highlighting of problematic interventions.

9.2.3. *Online health forums and support groups*

Patients gain experience and share experience and advice and support on sites such as Reddit (e.g. r/cancer), Patients such as Me and the Cancer Survivors Network, where forums are filled with discussions on cancer treatment, some even talking about physical rehabilitation (Katz et al. 2020).

These discussions are more informative than what is shared in the doctor's office because patients share without the fear of being judged, and without time limitations. In these discussions, patients share coping strategies, and some frustrations with rehab exercises including explaining reasons for non-compliance. NLP tools can be used to extract themes, detect emerging issues (such as dissatisfaction with tele-rehab) and monitor evolving patient concerns in real-time. Ethical data sourcing, consent and anonymization are crucial when using such platforms for research or clinical feedback loops.

9.2.4. *Satisfaction surveys and feedback forms*

Hospitals and rehabilitation centers frequently solicit feedback through open-ended surveys following physiotherapy sessions or program completion. Such records are useful not only in quality assurance but in making individual changes in care.

NLP can be used to analyze the answers of a survey to enable institutions to perform a large-scale sentiment analysis, predictors of patient satisfaction or dissatisfaction, and gaps in service delivery. Clustering and topic modeling may be used to reveal common themes, for example, the lack of personalization, excellent communication with staff or the inability to schedule. The collected insights are used to inform the staff training, service redesign and investing in patient engagement technologies.

9.2.5. *Social media and digital health apps*

As mobile health (mHealth) technologies and cancer rehabilitation apps are gaining popularity, patients are now becoming more digital when reporting on their daily symptoms or activity logs or therapy compliance. They feed this data in the form of text into either symptom trackers, pain journals or chat-based virtual assistants within the apps.

This patient-generated, real-time content can be analyzed in NLP to identify a change in the predicted trajectories of progress, a precursor of complications or changes in the tone of emotion. For example, consistent entries such as *"struggling more than usual today"* or *"skipped again, too tired"* may suggest therapy fatigue or psychological distress.

9.3. NLP techniques applied to patient feedback

9.3.1. *Preprocessing of unstructured feedback*

Apart from collecting patient feedback, we must also prepare and analyze it with various NLP tools and techniques. Most texts from interviews, surveys and digital logs ignore structures and noise. To address this, several techniques must first be executed:

– Tokenization involves breaking the text into smaller chunks, which may include words or sentences. Understanding a respondent's lengthy, complex feedback may be difficult and thus, breaking it into computationally analyzable elements helps.

– Lemmatization and stemming deal with reducing words to their base forms, which may help standardize variations and thus increase model performance. "Running," "runs" and "ran", for example, all reduce to "run".

– Stop-word removal techniques focus on words that appear frequently but contribute little to the value of content. "The," "is" and "and" are common examples. Most of the time, these words prevent algorithms from determining the true meaning of patient insights.

These techniques assist in transforming a patient's feedback from simple narratives and structured text, allowing for the subsequent NLP models to analyze effectively without being burdened with noise and redundancy.

9.3.2. *Sentiment analysis*

It is true that sentiment analysis is a frequently utilized NLP method for determining and gauging the emotional tone present within patient stories. This method distinguishes between positive, negative, or neutral and sentiments through contextual and lexical analysis. Utilizing sentiment analysis within patient narratives in physiotherapy or physiotherapy for oncology offers insight into emotion sentiment (Amiri et al. 2023).

For example, if several patients expressing sentiments of frustration, disappointment or discomfort describing their experiences using phrases such as "painful and pointless exercises" or "I dread each session" sentiment indicators such as these suggest disengagement and/or dissatisfaction. In contrast, sentiment indicators such as "I feel stronger every day" and "the therapist really listens" are positive. These emotions can be utilized in constructing motivators, exploring other treatments and, most importantly, adjusting therapeutic goals to improve compliance and strengthen the therapeutic alliance.

Because emotionally-focused care plans developed sentiment analysis, sentiment models can better align their care plans emotionally than the sentiment models which were developed. More sophisticated models focus on emotionally charged sentiments and are better at identifying and deconstructing clinically pertinent and clinically irrelevant variables and statements, enhancing actionable and interpretable sentiment analysis outputs.

9.3.3. *Topic modeling*

Unsupervised learning, especially latent Dirichlet allocation (LDA), helps recognize themes and trends hidden within patient feedback (Robert et al. 2021). Such algorithms identify and cluster related words and associate them with specific themes, thus revealing dominant issues, one recurring concerns or similar experiences patients share.

As in the case of feedback analytics, LDA helps identify the themes in feedback such as “exercise-induced fatigue”, “access issues with tele-rehabilitation” or “group therapy sessions” and “satisfaction”. Identifying themes helps digital literacy, exercise intensity and fatigue and the rate at which feedbacks are priorities the healthcare provider improves helps healthcare priorities and reforms resource allocation for intervention.

Healthcare resource allocation improves with focus on the feedback fatigue, exercise intensity and digital literacy. Topic modeling also facilitates contrastive analysis across groups, such as early- and late-stage cancer patients, to identify dominant concerns, thus allowing for focused adjustments to the physiotherapy plan in the context of each cohort.

9.3.4. *Named entity recognition (NER)*

Named entity recognition (NER) involves recognizing and parsing specific clinical texts. In relation to oncology physiotherapy, it can include drugs, symptoms, anatomical reference, therapeutic exercises, names of the healthcare providers and even places and dates.

Entity extraction of “cisplatin”, “lymphedema”, “resistance band exercises” or “home-based rehab” helps NER to convert qualitative data to structured data that can be used for electronic medical records (EMRs) and decision support systems (Ohno et al. 2025). This conversion allows clinical workflows to be used for real-time insights and adaptive rehabilitation.

Compared to generalized NER systems, NER systems based on medical literature such as BioBERT enhance precision in identifying and extracting pertinent physiotherapy oncology terminologies.

9.3.5. *Text classification*

In text classification, feedback from patients is classified in supervised machine learning into different categories. This is done using classification algorithms, which can be support vector machines (SVM), logistic regression or neural networks of deep learning, to categorize feedback into labels such as:

– Side effects: for example, “I feel dizzy after every session”.

– Therapy adherence: for example, “I skipped the last two days”.

– Emotional distress: for example, “I am desperate regarding my recovery”.

These categories assist in the prioritization of clinical care and triage of patients, as well as in the process of escalation. The system will be able to automatically flag and alert relevant care teams on the need to intervene in time and improve patient safety as well as reduce the response time through the training of models on labeled data. Table 9.1 shows NLP techniques and their application in oncological physiotherapy.

NLP technique	Core function	Example of patient feedback insight	Clinical application
Sentiment analysis	Detects emotional tone (positive/negative/neutral) in patient feedback	"I'm too tired after each session" → negative sentiment	Identifying therapy fatigue, emotional distress or disengagement
Topic modeling (e.g. LDA)	Discovers common themes in large feedback datasets	Themes: pain management; exercise difficulty; tele-rehab satisfaction	Helps prioritize interventions and allocate physiotherapy resources
Named entity recognition (NER)	Extracts entities such as symptoms, drugs or exercises	"Swelling after resistance band exercises" → *symptom* + *exercise*	Enables structured data integration with EHRs for real-time monitoring
Text classification	Categorizes text into predefined classes (e.g. side effects, adherence, etc.)	"Skipped exercises this week" → *nonadherence*	Triggers clinician alerts for early intervention
Transformer models (BERT and BioBERT)	Understands context and nuance in clinical narratives	"I felt weak but motivated after therapy" → *positive recovery signal*	Enhances personalized decision-making and trend analysis

Table 9.1. *NLP techniques and their applications in oncology physiotherapy*

9.4. Personalizing oncology physiotherapy using NLP insights

NLP can revolutionize oncology physiotherapy. NLP can uncover customized actionable information by extracting unorganized patient data: session notes, mobile app diaries, surveys and interview transcripts. The ensuing systems facilitate dynamic, personalized care that is real-time responsive to the changing needs and experiences of a patient (Ango et al. 2025).

9.4.1. *Flexible therapy scheduling*

During and after cancer treatment, fatigue and changes in energy levels are the norm and are likely to test physiotherapy. A hint of fatigue can be found through narrative reporting. The expression of feeling drained or being unable to get through the session is indicative of treatment fatigue. NLP tools determine the frequency of these phrases as well as their context and allow the physiotherapy team to modify the care plan to reduce the frequency, duration and intensity of sessions (Shahzadi et al. 2024). Together, these methods can assist the teams in avoiding fatigue and providing more patient-centered, caring care.

9.4.2. *Flexible exercise program design*

The inconsistency in patient responses, preferences and tolerance to therapy in the context of oncology rehabilitation requires individual exercise programs. NLP systems are used to examine feedback to identify preferences, the milestones attained, setbacks and the level of tolerance, thus maximizing the rehabilitation. Pain or stiffness can show what elements require modification by monitoring changes, for example, less pain or greater range of movement (Ortega et al. 2020). In the case with NLP, therapists pluck out free-text comments, and this helps them to concentrate on the most useful exercises for every client.

9.4.3. *Monitoring adherence and engagement*

The constant practice of physiotherapy is necessary, but most patients lack motivation, are confused or demotivated. Such emotional conditions are expressed in indirect language, including, but not limited to, phrases of doubt (I am not sure that this is helping me) and disengagement (I did not bother doing the exercises this week). NLP algorithms will identify these indicators in the early stage, so that therapists can interfere before dropout or non-adherence. Clinicians may identify the disengagement signs and determine categories of support conversations, clarify patients or make new objectives to re-motivate the patient (Williams et al. n.d.).

9.4.4. *Psychosocial support integration*

The patients of cancer rehabilitation are addressing the physical and emotional sides of the recovery process. Sentiments of solitude, and feelings of anxiousness, frustration and hopelessness, are often part of the emotional distress, as a consequence of changes during their rehabilitation. When multidisciplinary teams receive flow triggers, they can make appropriate referrals to counselors, psychologists and/or social workers (Chang et al. 2011). This purposeful step ensures that all aspects of care are integrated, addressing physical functioning and the equally important psychosocial component during cancer survivorship.

9.4.5. *Decision support capabilities*

When integrated into EHRs or physiotherapy workflows, NLP-derived insights can help streamline and automate several aspects of real-time clinical decision support. Rather than manually reviewing extensive logs and patient journals, therapists are able to access auto-generated summaries that describe the patient's major themes such as progress, barriers and emotional status. Automated insights that are based on alerts, sentiment scores and engagement scores are likely to improve the speed and quality of the response and/or intervention. As NLP improves, so will the ability to deliver subtle, context-aware summaries, and this will, in turn, improve the efficiency and customization of the services delivered in physiotherapy.

9.5. Case studies and systems in practice

9.5.1. *IBM Watson Oncology*

IBM Watson Oncology was one of the first ambitious projects on integrating NLP into clinical decision-making. It primarily "tries" to provide support to oncologists when recommending personalized treatment plans, but more importantly, it shows how robust NLP can help understand the scalability of NLP technology in challenging healthcare settings. Watson processes enormous amounts of unstructured medical literature, clinical notes and patient records, transforming them into structured information that decision support systems can use. While it does not directly focus on physiotherapy, the architecture illustrates the possible use of NLP in rehabilitation from oncology (Faheem and Dutta 2023). The ability to understand context-specific language, interpret ambiguous clinical narratives and integrate with electronic medical records (EMRs) makes Watson a foundational model for future applications in personalized oncology physiotherapy.

9.5.2. *Patient voice*

Patient voice is an NLP-based platform that processes unstructured patient feedback qualitatively to transform it into structured data to evaluate and improve the service. It analyzes unstructured responses to surveys, identifies sentiments, issues and themes to inform its recommendations for patient-centered care (Vashishtha and Kapoor 2023). Although it is nowadays utilized in general healthcare, as well as general mental health services, the methodology can be transferred to oncology physiotherapy with high transferability. Focusing on subtle manifestations of suffering, exhaustion or emotional load, patient voice will assist clinicians to better comprehend the live experiences of patients and change the approach to rehabilitation. It is designed in a modular fashion, enabling the adaptation of the pipelines to the requirements of the domain, for example, to identify cancer-rehabilitation words.

9.5.3. *CAIRN (Computer-Aided Instruction in Rehabilitation Needs) prototype*

CAIRN system is a conceptual prototype which unites NLP-derived information based on patient feedback with EMRs and patient-reported outcomes measures (PROMs). CAIRN uses unsupervised clustering algorithms to classify patients based on their requirement of rehabilitation services, level of pain and exercise capacity (Baker and Xiang 2023). This is used to develop individualized exercise prescriptions and monitoring plans and provides a completely adaptive physiotherapy model that is based on patient language.

9.6. Challenges and limitations

9.6.1. *Uncertainty and draftsmanship in words*

The inherent ambiguity and richness of natural language is one of the main issues of applying NLP to the analysis of patient feedback. Patients frequently tend to use metaphorical, emotional or colloquial terms to explain what is happening to them "like I was hit by a truck" or "my legs turn to jelly". Standard NLP models, particularly those not trained in the medical or rehabilitative context, may misinterpret or overlook the true clinical significance of such statements (Jiang et al. 2025).

9.6.2. *Lack of data and labeling*

High-performing NLP models need huge amounts of annotated datasets and present diverse uses. In the NLP area of oncology physiotherapy, specifically annotated datasets are not present. Manual annotation is expensive, time-consuming and requires an expert's annotated dataset for accuracy (Oliveira et al. 2025). Most resources for NLP systems remain underdeveloped which limits their robustness.

9.6.3. *Lack of generalizability*

Domain-specific NLP models might struggle with generalizability and applicability to other settings and populations. Variations and disparities in language, communication styles, dialects and even treatment plans can lead to an unreliable model (Gudur et al. 2024a).

9.6.4. *Workflow integration*

NLP promises the most utility when it complements existing systems and documentation processes. Systems in place for physiotherapy documentation workflows need to be made more efficient. NLP tools can provide valuable insights, but there should be no additional workflows to manage for the clinicians.

9.6.5. *Ethical and regulatory issues*

NLP systems on patient-level data must consider patient data and digital feedback under privacy, legal and ethical frameworks. Trust and transparency are essential for NLP systems, which depend on provisions for data privacy and informed consent, consented data anonymization, HIPAA and GDPR compliance.

9.7. Future directions

9.7.1. *Development of domain-specific NLP models*

Achieving the most precise and clinically relevant outcomes of NLP systems within the scope of oncology physiotherapy requires training them on language from the specific field. Feedback from patients during the process of cancer rehabilitation involves distinctive language, atypical emotional responses and descriptions of physical symptoms that general models are likely to overlook or misinterpret

(Matsuzaka and Yashiro 2022). Custom-trained NLP models that utilize annotated physiotherapy notes, session transcripts and patient outcome descriptions would understand the details better and improve the sensitivity and specificity of the insights extracted from the text.

9.7.2. *Multilingual and cross-cultural adaptation*

Due to the international nature of cancer care, LNP tools should be flexible to language and cultural diversity. The pain, fatigue, motivation and satisfaction are manifested differently in patients according to language and sociocultural backgrounds. NLP systems therefore have to have the ability to be multilingual and have the ability to be receptive to culturally entrenched expressions. This accommodation promotes fair access and proper translation to non-English speaking or underserved groups of people (Doke-Bagade and Bagade 2024).

9.7.3. *Integration with multimodal data*

The multimodal data integration: it is divided into two steps: the inclusion of the multimodal data and its integration with the multimodal data. The analysis of patient feedbacks should not be conducted separately. The combination of NLP knowledge and wearables, motion sensors and physiological sensor information would allow researchers to have a better picture of the health condition and rehabilitation of a patient. This type of multimodal fusion can be used to make individualized, evidence-based changes to care plans (Kaverinsky and Malakhov 2023).

9.7.4. *Explainable NLP models*

To achieve clinical credibility, NLP needs to provide understandable outputs. Clinicians must know how one of these models came up with its results, be it finding distress, or forecasting risk of dropping out. The explainable AI methods are needed to guarantee safe and ethical implementation in the healthcare environment.

9.7.5. *Real-time and longitudinal analysis*

A longitudinal and quasi-experimental analysis will be performed in the case with descriptive statistics. The evaluation of feedbacks by the patient in the long-run can show the trend of recovery, interest or the decline. Longitudinal and real-time NLP analyses allow timely intervention, assist in planning adaptive therapy and finding patients with non-compliance or treatment fatigue risks (Kumar et al. 2026a).

9.8. Ethical considerations

With NLP finding its way to the field of physiotherapy related to oncology, it is crucial to discuss the issues of ethics and operations that are bound to appear with this technology. The nature of any NLP application in patient care has to be based on consent and transparency. Patients must also be made well aware of how their text-based feedback, obtained through surveys or interviews, or digital means, will be gathered, managed and used to make treatment decisions. Transparency fosters trust, ensures autonomy and enables patients to make informed choices about participating in data-driven care models (Kumar et al. 2026b).

The issue of bias and fairness cannot be overlooked. NLP models developed without considering an imbalance of datasets run the risk of undermining the inequities that exist in the healthcare system. In the case of these models, examination for bias should also cover the disparity of age, gender, ethnicity, language and economic status. In any case, inclusive and equitable disregarding bias enables the value of patient-centered care in the formulation of personalized recommendations for inclusive inequity. Last of all, the ethical value of care must be maintained during the NLP deployment cycle.

Even though machine learning models can easily spot a set of patterns and develop certain predictive analytics, the control of any changes of treatment should rest with an accountable healthcare professional (Gudur et al. 2024b). They should be the one to make the final call. Clinical relationships thrive on the personal connection involving a patient and assessments that precisely reflect the individual's conditions. Primary data without the humane value introduces the risk of the patient being merely a point on a graph. NLP must be a source of analytical support, not an emotional substitute.

9.9. Conclusion

Using NLP for integrated digital health technology in physiotherapy within oncology can enable practice shifts towards personalization, flexibility and patient centricity. Unstructured patient commentary, note progress and conversations transcribed into text or other digital formats contain pointers towards unrecognized patterns of physical pain, emotional suffering and burnout. Real-time adaptation of physiotherapy is possible because NLP identifies critical missing vectors of patient experience that would otherwise lead to a stagnation of care development. NLP augments digital health technologies and allows clinicians to act on patient-initiated signals concerning pain, motivation to engage with therapy and adherence to prescribed therapy bypassing traditional screening, which improves therapy outcomes, and patient satisfaction. Clinicians can use NLP technology to appreciably understand

and communicate with their patients in trusting, collaborative and engaging ways. Progressing NLP is a technology and promotes practice in patient-initiated dialogue; however, it can present several challenges, including privacy issues, algorithm accuracy concerning bias, voluntary and informed participation, and fairness. NLP technology must involve all relevant members to promote and distribute technologies equitably. The next decade could be transformative in NLP integration with digital health technologies for rehabilitation and cancer care. Successful integration will involve the collaboration of oncologists, physiotherapists, data scientists and linguists, among others.

9.10. References

Amiri, Z., Heidari, A., Darbandi, M., Yazdani, Y., Jafari Navimipour, N., Esmaeilpour, M., Sheykhi, F., Unal, M. (2023). The personal health applications of machine learning techniques in the internet of behaviors. *Sustainability*, 15(16), 12406.

Ango, R., Singh, T.M., Dadireddy, M.K. (2025). Machine learning for healthcare applications: Transforming patient care, diagnosis, and treatment. In *Next-Generation Therapeutics Using Internet of Things and Machine Learning*, Shuaib, M. and Alam, S. (eds). IGI Global, Hershey, PA, 49–72.

Baker, S. and Xiang, W. (2023). Artificial intelligence of things for smarter healthcare: A survey of advancements, challenges, and opportunities. *IEEE Communications Surveys & Tutorials*, 25(2), 1261–1293.

Burgos-Bragado, J.M., Jiménez-Sánchez, C., Brandín-de la Cruz, N., Carpallo-Porcar, B., Le-Roux, L., Blas-Laina, J.L., Alamillo-Salas, J., Gracia-Gimeno, P., Calvo, S. (2025). Integrating telerehabilitation into the prehabilitation and rehabilitation pathway in colorectal cancer: A protocol for a randomized controlled trial. *medRxiv*.

Chang, P., Szymanski, K.M., Dunn, R.L., Chipman, J.J., Litwin, M.S., Nguyen, P.L., Sweeney, C.J., Cook, R., Wagner, A.A., DeWolf, W.C. et al. (2011). Expanded prostate cancer index composite for clinical practice: development and validation of a practical health related quality of life instrument for use in the routine clinical care of patients with prostate cancer. *The Journal of Urology*, 186(3), 865–872.

Doke-Bagade, P.E. and Bagade, O. (2024). Applications of AI-based models in the field of biomedicine. In *Biosystems, Biomedical & Drug Delivery Systems: Characterization, Restoration and Optimization*, Kulkarni, S., Haghi, A.K., Manwatkar, S. (eds). Springer, Singapore, 259–286.

Fabian, K.P., Kowalczyk, J.T., Reynolds, S.T., Hodge, J.W. (2022). Dying of stress: Chemotherapy, radiotherapy, and small-molecule inhibitors in immunogenic cell death and immunogenic modulation. *Cells*, 11(23), 3826.

Faheem, H. and Dutta, S. (2023). Artificial intelligence failure at IBM 'Watson for Oncology'. *IUP Journal of Knowledge Management*, 21(3), 47–75.

Fernandes, M.B., Valizadeh, N., Alabsi, H.S., Quadri, S.A., Tesh, R.A., Bucklin, A.A., Sun, H., Jain, A., Brenner, L.N., Ye, E. et al. (2023). Classification of neurologic outcomes from medical notes using natural language processing. *Expert Systems with Applications*, 214, 119171.

Gudur, R.A., Bhosale, S.J., Gudur, A.K., Kale, S.R., More, A.L., Datkhile, K.D. (2024a). The effect of CYP2C19* 2 (rs4244285) and CYP17 (rs743572) SNPs on adriamycin and Paclitaxel based chemotherapy outcomes in breast cancer patients. *Asian Pacific Journal of Cancer Prevention: APJCP*, 25(6), 1977.

Gudur, R., Patel, D.J., Amin, P., Singh, J. (2024b). An investigation of the differences in long-term patient survival rates between robotic and thoracoscopic lobectomy. *Onkologia i Radioterapia*, 18(8), 001–006.

Jeong, I., Chung, W.G., Kim, E., Park, W., Song, H., Lee, J., Oh, M., Kim, E., Paek, J., Lee, T. et al. (2025). Machine learning in biosignal analysis from wearable devices. *Materials Horizons*, 17. doi: 10.1039/D5MH00451A.

Jiang, Y., Zhao, J., Yuan, Y., Zhang, T., Huang, Y., Zhang, Y., Wang, Y., Li, Y., Guo, X., Zhao, Y. et al. (2025). Never compromise to vulnerabilities: A comprehensive survey on AI governance. arXiv preprint, arXiv:2508.08789.

Katz, M.S., Staley, A.C., Attai, D.J. (2020). A history of #BCSM and insights for patient-centered online interaction and engagement. *Journal of Patient-Centered Research and Reviews*, 7(4), 304.

Kaverinsky, V. and Malakhov, K. (2023). Natural language-driven dialogue systems for support in physical medicine and rehabilitation. *South African Computer Journal*, 35(2), 119–126.

Korhonen, O., Väyrynen, K., Krautwald, T., Bilby, G., Broers, H.A., Giunti, G., Isomursu, M. (2020). Data-driven personalization of a physiotherapy care pathway: Case study of posture scanning. *JMIR Rehabilitation and Assistive Technologies*, 7(2), e18508.

Kumar, A., Batta, P., Ahuja, S., Rathore, P.S. (eds) (2026a). *AI-driven Innovations in Physiotherapy and Oncology 1*. ISTE Ltd, London and Wiley, Hoboken.

Kumar, A., Batta, P., Ahuja, S., Rathore, P.S. (eds) (2026b). *AI-driven Innovations in Physiotherapy and Oncology 2*. ISTE Ltd, London and Wiley, Hoboken.

Lown, E.A., Banerjee, A., Dvorak, C., Hartogensis, W., Hecht, F.M., Mangurian, C., Melton, A., Vittinghoff, E. (n.d.). Comparing the effects of acupressure on cancer treatment side effects in children and youth—The ACT-CC study. Report, Patient-Centered Outcomes Research Institute (PCORI), Washington, D.C.

Matsuzaka, Y. and Yashiro, R. (2022). Applications of deep learning for drug discovery systems with bigdata. *BioMedInformatics*, 2(4), 603–624.

Ohno, Y., Aomori, T., Nishiyama, T., Kato, R., Fujiki, R., Ishikawa, H., Kiyomiya, K., Isawa, M., Mochizuki, M., Aramaki, E. et al. (2025). Performance improvement of a natural language processing tool for extracting patient narratives related to medical states from Japanese Pharmaceutical Care Records by increasing the amount of training data: Natural language processing analysis and validation study. *JMIR Medical Informatics*, 13(1), e68863.

Oliveira, M.B., Mendes, F., Martins, M., Cardoso, P., Fonseca, J., Mascarenhas, T., Saraiva, M.M. (2025). The role of artificial intelligence in urogynecology: Current applications and future prospects. *Diagnostics*, 15(3), 274.

Ortega, M.A., Fraile-Martínez, O., García-Montero, C., Pekarek, L., Guijarro, L.G., Castellanos, A.J., Sanchez-Trujillo, L., García-Honduvilla, N., Álvarez-Mon, M., Buján, J. et al. (2020). Physical activity as an imperative support in breast cancer management. *Cancers*, 13(1), 55.

Pawar, D., Phansalkar, S., Sharma, A., Sahu, G.K., Ang, C.K., Lim, W.H. (2023). Survey on the biomedical text summarization techniques with an emphasis on databases, techniques, semantic approaches, classification techniques, and similarity measures. *Sustainability*, 15(5), 4216.

Pham, T.D., Teh, M.T., Chatzopoulou, D., Holmes, S., Coulthard, P. (2024). Artificial intelligence in head and neck cancer: Innovations, applications, and future directions. *Current Oncology*, 31(9), 5255–5290.

Robert, Delir Haghighi, P., Burstein, F., Urquhart, D., Cicuttini, F. (2021). Investigating individuals' perceptions regarding the context around the low back pain experience: Topic modeling analysis of twitter data. *Journal of Medical Internet Research*, 23(12), e26093.

Shah, A.M., Muhammad, W., Lee, K., Naqvi, R.A. (2021). Examining different factors in web-based patients' decision-making process: systematic review on digital platforms for clinical decision support system. *International Journal of Environmental Research and Public Health*, 18(21), 11226.

Shahzadi, M., Rafique, H., Waheed, A., Naz, H., Waheed, A., Zokirova, F.R., Khan, H. (2024). Artificial intelligence for chimeric antigen receptor-based therapies: A comprehensive review of current applications and future perspectives. *Therapeutic Advances in Vaccines and Immunotherapy*, 12, 25151355241305856.

Tyagi, A.K., Kumari, S., Tiwari, S., Subbiah, P. (2024). Industry 5.0 at the Core of Society 5.0: A revolution in the modern era of the health care sector. In *Sustainable Innovation for Industry 6.0*, Sharma, A., Moses, O., Sharma, R.B., Gupta, S. (eds). IGI Global, Hershey, PA, 341–363.

Vashishtha, E. and Kapoor, H. (2023). Enhancing patient experience by automating and transforming free text into actionable consumer insights: A natural language processing (NLP) approach. *International Journal of Health Sciences and Research*, 13(10), 275–288.

Vaswani, A., Shazeer, N., Parmar, N., Uszkoreit, J., Jones, L., Gomez, A.N., Kaiser, Ł., Polosukhin, I. (2017). Attention is all you need. In *31st Conference on Neural Information Processing Systems (NIPS 2017)*, Long Beach, CA.

Willingham, T.B., Stowell, J., Collier, G., Backus, D. (2024). Leveraging emerging technologies to expand accessibility and improve precision in rehabilitation and exercise for people with disabilities. *International Journal of Environmental Research and Public Health*, 21(1), 79.

10

AI-Enhanced Biomechanical Feedback Systems for Radiation Therapy Recovery

The typical effects of radiation therapy (RT) are soft tissue fibrosis, musculoskeletal injuries and fatigue. Even though RT is necessary in most cancers, it may drive patients towards functional disability. Rehabilitation must then not be just an exercise in functional restoration; this needs to be done in both physical and psychological dimensions. Prejudiced mechanical feedback systems are now becoming available in order to satisfy these requirements. Such mobile sensors track the movement of a person, the angles of their joints, muscular activity and posture. This has been enhanced by the use of artificial intelligence (AI). Machine learning and deep learning algorithms are used to process data collected by many sensors over time, identifying specific patterns and tailoring the rehabilitation process to an individual patient. AI can be improved in healthcare by increasing the accuracy of therapies and providing the patient with instant feedback. This enhances the motivation and compliance, which results in overall performance improvement. The next section is a review of the application of biomechanical feedback systems, which are currently AI-enhanced, in post-radiation therapy. It talks about the important systems, clinical applications and their effects on patient recovery. It mentions the benefits of accuracy, flexibility and continuous monitoring, stating the variability of the data, data privacy issues and the necessity of additional clinical trials.

10.1. Introduction

Radiation therapy has remained one of the most effective and extensively used therapies for treating breast, lung, prostate, cervical and head–neck tumors. Although it is effective on malignant cells, it usually affects the healthy tissues around it, leading to long-term musculoskeletal and neuromuscular issues, including fibrosis, pain, and stiffness, lymphedema and motion loss. Such issues may significantly decrease their capacity of a patient to accomplish daily activities and decrease their quality of life.

Chapter written by Dhairyasheel PATIL and Himanshu WAGH.

Rehabilitation is important after treatment in order to recover lost functions. The usual traditional therapy takes a certain duration of guided exercises that aim at alleviating pain, stiffness, movement as well as muscle strength. Nevertheless, it does not take advantage of real-time evaluation, and that can result in an imbalance in progress, particularly in the post-radiation treatment (Grosu-Bularda et al. 2024), which may be complex. This deficiency has spurred the emergence of technology-based rehabilitation, in particular, the biomechanical feedback system.

These systems use a collection of sensors for monitoring and real-time assessing rehabilitation exercises. These sensors include inertial measurement units (IMUs), electromyography (EMG) and pressure sensors, along with force platforms in order to gather and process kinematic and kinetic information for monitoring and assessing rehabilitation exercises, which include movement of treated muscles and joints. These systems improve rehabilitation and reduce the risk of injury by providing movement suggestions to patients using visual, auditory and haptic feedback (Li et al. 2025).

The relative newness of AI (Artificial Intelligence) technology helps with developing a more advanced biomechanical feedback. Most AI systems can do more than track data; they can interpret data in real-time, find anomalies, make personal suggestions and perform predictive analysis (Shah et al. 2024). Machine learning models can be trained with data containing a population's different patterns of movement to identify outlier movement patterns that may represent an impairment or risk. AI can identify compensatory movements that may be slight and hidden, signifying a dysfunction or an incorrect rehab technique.

The most significant benefit of the AI-facilitated biomechanical feedback systems is their ability to customize rehabilitation therapy. The recovery journey of cancer survivors varies greatly, and depends on factors such as the location of the tumor, the radiation dosage, the presence of comorbid diseases, the person's age and their baseline fitness. AI can use a range of biomechanical data to assess the level of the various movement patterns of a patient over time, identify a trend and revise therapy goals. This ensures that the level of therapy is challenging in a relevant safe manner maximizing the benefits of therapy. The most significant benefit is remote monitoring and telerehabilitation.

Patients currently or recently undergoing radiation therapy often experience fatigue and may not be able to travel often, and may live far away, or be in an immunocompromised state. AI-enabled wearables can facilitate rehabilitation therapy at home, and the relevant data is continuously sent to the rehab therapist (Kaul and Maurya 2025). Therapists no longer need to be in the same location to evaluate the progress of patients with the help of cloud-based platforms. Therapists can make the necessary changes to protocols and converse with patients without being physically present. Such accessibility not only increases continuity of care with rehabilitation plans but also improves compliance and connects underserved and rural populations to care.

These systems can be designed to bolster the psychological dimensions of rehabilitation. Patients can use cloud-connected systems as AI analytics in combination with interactive systems, progress dashboards and rehabilitation exercises provide patients with insights that motivate engagement and the psychological dimensions of self-rehabilitation.

There are, however, important challenges as well as considerations to take into account when deploying AI-enhanced biomechanical feedback systems. In the non-laboratory world, these systems need to be robust, reliable and technologically advanced. Accuracy of sensors and real-time data processing need to be maintained. AI models require vast and representative training data to approach becoming unbiased and fair. The collaboration between clinicians of various fields is necessary to incorporate these systems into practice, so practitioners should be trained to handle and analyze the outputs of AI.

There should also be validation studies and standardized protocols for widespread clinical adoption. Concerns about ethics and privacy need to be mitigated. As for the biomechanical and physiological data, there are privacy, informed consent and autonomy issues that need to be resolved (Singh 2023). Data security, especially for cloud computing, is critical around the HIPAA and GDPR. Regarding these systems, patients should receive balanced information, including the benefits and the limitations, to empower them to make informed decisions.

10.2. Radiation therapy sequelae and rehabilitation needs

Radiation therapy remains one of the foundations for treating certain cancers, particularly solid tumors, including those of the breast, head and neck, cervix and prostate. As much as the aim of radiation therapy is to destroy cancerous cells, the therapy also unwittingly impacts nearby healthy tissues. These side effects of radiation therapy may be quite severe and long-lasting as they can negatively impact the physical functioning of the patient and thus the patients' overall functional ability in the everyday activities of their life (Singh et al. 2024).

The most common post-radiation therapy complications include:

– radiation-induced fibrosis, which results in stiffening of tissues and can cause chronic pain and reduced mobility;

– muscle weakness and fatigue, limiting endurance and physical performance;

– joint immobility, particularly around the area of treatment, which restricts the range of motion;

– alignment being interrupted by postural dysfunction, which is due to muscular imbalances and joint restrictions;

– gait abnormalities, resulting from compensatory movement patterns due to pain or stiffness;

– *lymphedema and swelling*, particularly in cases involving lymph node irradiation, leading to fluid retention and discomfort in the limbs.

These problems may develop over time, some seven weeks or months following radiation. Otherwise, they can cause progressive functional deterioration, chronic disability and low quality of life. Accordingly, it is necessary to implement effective post-radiation rehabilitation, which will be an important part of survivorship care, with the goal of restoring musculoskeletal functioning, mobility and overall well-being.

Conventionally, therapy is based on physical examinations and programmed exercise. Therapists note the gait, measure the range of joint motion and assess strength as a way of identifying deficits and consequently tailor general exercise programs that address the affected areas and stimulate recovery (Kumar et al. 2026a).

Although regular physiotherapy might be useful, it does not always provide the personalized approach and objective data necessary to make a treatment decision. Individual patients vary widely in their biomechanical patterns, healing timelines and tolerance to therapy. Because of this, there is a need to try different types of approaches. Additionally, manual evaluations can be inconsistent and become time consuming and frustrating, making it impossible to monitor small changes over long periods of time, and offer a standardized treatment (Kumar et al. 2026b).

These issues have led to a demand for more objective and evidence-based approaches to rehabilitation, more flexible systems to accommodate therapy as clientele needs and demands progress or change within a session. Biomechanical feedback is one of the more advanced and innovative systems. Such systems can offer real-time movement assessments and evaluations by using software algorithms, cameras and other systems within the rehabilitation center to perform advanced real-time assessments of the patients. Clinicians will be able to understand the mechanics of the patients' movements, and offer more advanced and constructive approaches to problems. These systems can be adjusted to offer real-time feedback both visually and by audible commands. Commands can stimulate and direct the patients' movements to achieve different goals during the rehabilitation session

(Wang et al. 2022). AI is the next approach to increasing the adaptability of biomechanical feedback.

AI algorithms, especially those using machine learning and deep learning techniques, can study large volumes of patient movement data, anticipate recovery paths and tailor personalized interventions relevant to individual needs. Such systems with the help of AI would be able to change the intensity of the therapy and prescriptions of exercises in real-time, as well as learn to identify potentially dangerous complications, such as lymphedema or muscle imbalance, at the initial stage of development. This can be used to make timely interventions. Evidence-based practices can be developed to help treat impaired patients using AI systems developed to assist radiation survivors with similar impairments and trained on progression-oriented exercises based on the patients' needs and requirements (Fatunmbi n.d.).

These systems facilitate remote rehabilitation as well. They facilitate cloud-based patient–clinician exercise monitoring, which allows patients to do prescribed exercises at home. Moreover, they use sensor analysis and biomechanical analysis technologies to encourage patients who require rehabilitation. Engaging therapies can be structured based on therapeutic goals, and gamification and progress-visualization techniques can be used to make patients feel in control of their achievements. The evidence-based practices that are created based on the needs of the patients can be done through AI systems that are customized to meet the needs of radiation survivors with similar impairments and trained on progression-oriented exercises.

The systems also facilitate remote rehabilitation through patient–clinician exercise tracking on the cloud and allow the patient to perform prescribed exercises at home. They also apply sensor and biomechanical analysis technology to offer motivational support to patients that require rehabilitation. The process of gamifying treatment therapy and visualizing progress can be structured in the form of therapeutic goals to make patients feel their agency and achievement (Mallawatantri and Fernando 2024). AI systems that consider the requirements of radiation victims with similar limitations and are trained on the basis of the exercises aiding their progress, can develop treatment courses based on the prescription of evidence-based practices sensitive to the needs of each person.

The systems also facilitate remote rehabilitation through patient–clinician exercise tracking on the cloud and allow the patient to perform prescribed exercises at home. These systems also use sensor and biomechanical analysis technology to offer motivational support for patients in need of rehabilitation. Engaging treatment therapy can be designed around therapeutic goals using gamification and progress visualization techniques to enhance patients' sense of agency and accomplishment.

Integration of the technology into clinical workflows, the initial outlay of funding, the privacy of client data and the necessity of training clinicians are all issues requiring a solution. Fortunately, many of these challenges are likely to lessen as the technology becomes more sophisticated and affordable.

10.3. Biomechanical feedback technologies in RT recovery

10.3.1. *Wearable sensors and electromyography (EMG)*

Wearable biomechanical feedback systems use sensors such as accelerometers, gyroscopes and magnetometers. These small, light sensors can be placed on various body segments, including wrists, ankles, trunks and thighs, and report multiple parameters on body motion continuously. Accelerometers, gyroscopes and magnetometers work in conjunction to provide information on the movement of body parts, posture and orientation. These sensors provide substantial information on a person's posture and body movements during rehabilitation as they monitor limb movement, gait cycle and trunk stability during radiation therapy, without any invasive procedures.

These sensors can be combined to monitor and assess the gait of the patient, looking for uneven symmetry in limb use, identifying compensatory movements and postural adjustments or deviations, all of which can be very subtle but are critical for identifying a functional deficit and possible recovery delays (Nascimento et al. 2020). Other biomechanical sensors, especially EMG sensors, measure the electrical activity associated with muscle activity and work very well with the other systems.

For practical and non-invasive measures, surface EMG (sEMG) systems are used, which provide important data on the neuromuscular activations and work coordination, muscle fatigue and recovery, timing and movement sequencing. For survivors of radiation therapy, muscle strength and activation may be compromised due to fibrosis and nerve damage. In these cases, therapy can be guided by EMG. AI-enabled rehabilitation platforms use EMG to capture muscle activity and combine it with kinematic data from motion sensors to create personalized exercise prescriptions. These platforms can monitor the patient's movements and provide timely corrective feedback to encourage safe and effective exercise and minimize the risk of injury or regression (Jubair et al. 2025). In addition, these systems automate progress quantification over time, providing valuable data to patients and clinicians regarding recovery progress.

10.3.2. *Surface imaging and optical tracking*

In clinical and laboratory work, surface imaging coupled with optical tracking systems functions as biomechanical assessment tools. These systems can produce intricate 3D reconstructions of movement patterns by using high-resolution cameras and reflective or active markers positioned temporarily (for the duration of the assessment session) on the patient's body. They accurately record the kinematics of the body as a whole, limbs, joints and even the more intricate parts of the body.

These systems produce advanced analyses of complex, integrative movement such as gait, balance and physical activity in compensatory movement strategies (Díaz et al. 2019). These tools identify functional movement pattern (dyskinesia) and its compensatory patterns in movement systems damaged by radiation.

Integrated with AI, predictive models of recovery based on surface imaging data can be developed. For example, a patient's identification of altered gait patterns (abnormal gait) during the expected stage of rehabilitation can be tracked by AI (predictive modeling) and recommend rehabilitation strategies. Predictive modeling based on surface imaging data holds great promise for rehabilitation in radiation therapy where such predictive modeling, along with functional coupling and the performance of the active domain predictive modeling, can minimize disability.

The optical tracking system has also proven beneficial because it can be performed easily without any discomfort or risk to the patient and global clinical use on an ongoing basis (Alsunaidi et al. 2021). The use of optical tracking combined with wearable sensors enables multi-modal data fusion by integrating real-time internal muscle activation data, with data on body external movement patterns providing a complete overview of the functional performance of the individual on a real-time basis.

10.3.3. *Biomechanical modeling and finite element analysis*

Finite element analysis (FEA) and biomechanics focus on testing and modifying advanced computer algorithms for abstract simulation of global tissues and joints for various and complex use scenarios and case studies. Bone, muscles, ligaments and soft tissues are described with various fabricated and simulated physical properties and various volumetric and custom anatomical data are used; primarily, MRI and CT are used for data preparation.

FEA, during the rehabilitation of patients recovering from radiotherapy, of how patients respond to exogenous and endogenous stretches during exercise or movement is pivotal for understanding how treatment has altered elasticity as the tissues become and remain irradiated or are burned (Emerzian et al. 2023). FEA assists in estimation or prediction of certain outcomes in therapy like in the case of FEA on the shoulder of a head-and-neck radiation patient when fibrotic stretching or exercises are determined as initial activities for therapy and regression analysis are used to estimate the rate of regression, which is critical to prevent tissue damage or excessive fibrotic. In the design of assistive devices, particularly prosthetics and exoskeletons, biomechanical modeling aids in tailoring and optimizing their use. Understanding tissue loading helps to reduce tissue over- and under-stress and improve patient mobility to become more independent. Computed estimation aids are transformative for unique patient rehabilitation.

Technology type	Core components	AI integration	Clinical applications	Expected outcomes
Wearable sensors (IMU, EMG and pressure sensors)	Accelerometers, gyroscopes and EMG electrodes	Machine learning models for motion analysis and fatigue detection	Tracking limb movement, posture correction and muscle activity in radiation survivors	Enhanced accuracy in movement monitoring, prevention of improper motion and improved rehabilitation adherence
Surface imaging and optical tracking	3D motion cameras; reflective markers	Deep learning for pattern recognition and motion deviation analysis	Gait and posture assessment and detection of compensatory movement in post-radiation cases	Objective movement evaluation and improved therapy personalization
Biomechanical modeling and finite element analysis (FEA)	Computational anatomical models; simulation software	Predictive AI for tissue stress modeling and therapy optimization	Personalized therapy design and load distribution analysis in affected tissues	Reduction in therapy-related strain and optimized recovery protocols
Robotic and exoskeleton systems	Actuated joints, force sensors and AI controllers	Reinforcement learning for adaptive motor support	Assisted walking and limb mobilization for patients with neuromuscular impairment	Increased muscle engagement and reduced therapist workload
AR/VR-based feedback systems	Head-mounted displays; motion capture systems	Adaptive AI feedback for task difficulty adjustment	Virtual-guided therapy for postural correction and neck mobility	Improved patient motivation and real-time performance enhancement

Table 10.1. *Key biomechanical feedback technologies and their AI applications in radiation therapy recovery*

Rather than depending on one-size-fits-all therapy schemes, practitioners are able to create individualized therapy programs based on the clients' anatomy, injuries and restrictions using biomechanical simulations (Tanaka and Yamamoto 2020). As AI advances, the combination with biomechanical modeling will continue to refine automation and bespoke rehabilitation, further improving the recovery process.

10.4. Role of AI in biomechanical feedback systems

10.4.1. *AI in motion prediction and tracking*

AI technology has entered rehabilitation and made it possible to track and predict motion with remarkable accuracy. Machine learning and deep learning techniques can identify and streamline the vast and complicated pattern recognition tasks associated with the analysis of human movement (Amiri et al. 2024). In clinical rehabilitation, AI assesses quality of movement in real time, identifies deviations from normal motion patterns and predicts possible joint trajectories.

Potential real-time applications of movement analysis in rehabilitation include the use of convolutional neural networks and recurrent neural networks (long short-term memory models included) that analyze time series data of wearable sensors (e.g. IMUs). These AI models predict movement phases which help clinicians identify gait abnormalities and compensatory strategies. These systems can provide real-time corrective prompts, enhancing motor learning and therapeutic outcomes. AI-based motion tracking systems are particularly effective in home-based rehabilitation where clinical supervision is limited and clinicians need to provide guidance asynchronously.

10.4.2. *Data fusion and adaptive feedback*

AI has the ability to integrate and coherently summarize the vast information coming from different sensors to understand a person's biomechanical and physiological state. Data fusion techniques can combine information from IMUs, EMG, optical motion capture and pressure sensors to describe in detail the mechanics and breadth of a person's body performance to capture the dominant movements, the muscles that are active and the overall performance.

Having alternative ways of determining the motor functioning of a person is invaluable, particularly in demanding rehabilitation cases. The adaptive feedback is one of the key benefits of AI (Grosu-Bularda et al. 2024). Depending on the performance of the patient during a session, the software is capable of making dynamic changes to the activities needed and the difficulty of the activities. As an example, in case the patient is performing well, the system can increase the patient's challenges. On the contrary, if the patient makes errors or displays signs of weariness, the system will be able to simplify the task, guarantee safety, give motivating instructions, as well as promote active engagement. Having this flexibility improves and focuses the rehabilitation process, making it a more personalized experience for the patient. It also helps in maintaining the patient's engagement and motivation and prevents rehabilitation burnout.

10.4.3. *AI-driven robotic and wearable exoskeleton systems*

AI-driven robotic systems and wearable exoskeletons are helping patients with neuromuscular conditions and movement restrictions, including those recovering from post-radiation sequelae. These systems help patients with complex movements, for example, walking or reaching, by guiding the patients' limbs through paths and trajectories. AI integrated into these robotic systems allows therapy to shift from a more robotic, rigid therapy system to a more responsive one.

AI systems learn from the patients' involuntary actions, such as muscle contractions, joint resistance or asymmetrical movements during gait, to adjust support during movement, helping to prevent a robotic system from over- or under-assisting. Over-assisting results in loss of engagement, under-assisting may lead to frustration or injury and passivity may be associated with injury rehabilitation (Dianti n.d.). There is a clear benefit in using AI-integrated robotic systems during rehabilitation post-radiation therapy as it involves alleviating muscle stiffness and fatigue, thus enhancing functional recovery. This results in therapy burden being safely, more efficiently and ergonomically tailored to individual patients' needs, harmonizing the burden with the work of the physiotherapist.

10.4.4. *Augmented and virtual reality feedback*

Immersive and interactive therapy environments that promote active participation are being made possible by augmented reality (AR) and virtual reality (VR) technologies revolutionizing rehabilitation (Patil et al. 2022). These technologies are tethered with AI, and they provide unprecedented individualization and feedback and monitoring in real-time.

AR systems will give the patient real-time visual holographical indicators to steer their motion, place their limbs correctly and adjust their stance. VR systems enable patients and therapists to assess functional capacities in controlled settings, which replicate real-life conditions, including climbing the stairs or picking up objects. In addition, the technologies apply AI to detect and react to user behavior, as well as to adjust activities in the virtual world. For example, if a patient is performing a movement with difficulty, the system provides task simplification in a controlled manner with guided repetitive task practice.

10.5. Examples and case studies of integrated systems

As rehabilitation, particularly for oncology patients, has benefited from advanced systems that incorporate AI along with biomechanical feedback systems, there are also several rehabilitation systems that have pioneered augmented rehabilitation technologies. One important category is upper limb rehabilitation gloves that have multi-modal sensor technologies incorporated within them, including flex sensors, IMUs and force sensors. These systems gather highly detailed information including the angle of each joint, movements of the fingers and the amount of force that is applied, all of which are processed by advanced machine learning systems, which classify movements, identify irregularities and propose corrective actions (Wei and Wu 2023). Such systems markedly enhance the motor learning experience by providing real-time feedback that is responsive to the needs of the individual, which is a first of its kind truly responsive rehabilitation technology.

Real-time EMG-driven robotic gait trainers with the ability to dynamically change key parameters including stride length, cadence and resistance, are also advanced systems that seamlessly incorporate AI technologies within rehabilitation. These systems help patients regain functional mobility following pelvic and lower limb radiation. These systems advance the rehabilitation process by modifying their assistance in ways that foster and support the learning of new skills needed for functional mobility.

Integration of AI systems with motion capture systems has also fostered new ways to monitor and assess rehabilitation. For example, AI systems monitor and assess shoulder mobility of post-radiation therapy breast cancer survivors and assess compensatory movements. The AI systems provide detailed feedback in advanced rehabilitation systems which is critical for both therapists and patients.

To address the rehabilitation needs of patients with head and neck cancer, particularly for neck stiffness and posture-related problems, AR rehabilitation systems are used (Erdoğan et al. 2023). These systems provide interactive exercise feedback and performance scoring in real-time.

As a result of the use of AI in these rehabilitation systems, there have been remarkable benefits in cancer care such as increased patient adherence and accuracy in tracking rehabilitation results, decreases in musculoskeletal pain, improvements in range of motion and joint control, and increased precision in outcome measurements. This reflects a paradigm shift in cancer rehabilitation physiotherapy.

10.6. Benefits of AI-enhanced biomechanical feedback for RT recovery

AI-powered rehabilitation systems offer innovative new technologies to recovery therapy. These new systems are individualized and personalized. Due to the specificity and personalization of the AI feedback systems and exercises, each patient's unique and individualized physiological and biomechanical qualities, rehabilitation therapy can be much more effective and can be done more safely to reduce therapy-related injuries (Fang et al. 2024). An important characteristic of these systems is the therapeutic feedback systems done in AI rehabilitation therapy in real-time.

Patients receive feedback when AI and sensor systems detect deficiencies in the patient's performance or when the patient's biomechanics deviate from the prescribed rehabilitation exercises. These feedback systems and techniques become much more effective to rehabilitation than simply relying on the patient's memory for the exercises and preventing injuries.

Moreover, AI enables objective monitoring through continuous collection and analysis of movement metrics, range of motion and performance trends. Clinicians can review this information for more accurate evaluations for multiple dimensions and modify the rehabilitation plan more effectively to assist the patient in achieving their rehabilitation recovery goals.

Lastly, AI can be integrated as an adjunct telehealth component with home care through mobile and wearable technology to help patients continue their rehabilitation therapy seamlessly incorporated into everyday activities. Patients can travel and care for loved ones and support long-distance rehabilitation goals.

To conclude, motivational engagement features, such as gamification, progress tracking systems and visual feedback, are integrated into these systems. Such features increase adherence and create a sufficiently rewarding experience, making therapy fun, goal-centered and enjoyable (Petkevičius 2023).

10.7. Challenges and limitations

When considering the implementation of AI-trained biomechanical feedback systems for the recovery of radiation therapy patients, there are still a few issues that need to be resolved for effective use and acceptance by the medical community. One of the issues that needs to be resolved is that of the availability of the necessary data. AI systems require sufficient amounts of data and for that data to be high quality and varied to provide personalized, accurate feedback GPT-3 (Liang et al. 2022). However, data of this sort is particularly scarce with regard to the recovery of post-radiation therapy patients. This is because the data is niche in nature, patients out vary, and there is no standardized approach to data collection.

Complexity of the systems is another critical issue to be resolved. These systems include biomechanical sensors, real-time processing units, AI algorithms and feedback components. Achieving seamless integration and maintaining reliability, real-time responsiveness and system security all require intelligent, sophisticated design and constant iteration.

Maintaining safeguards on the biomechanical and other sensitive physiological data of the patient is also a critical issue. Safe guards such as strong encryption and secure storage systems in accordance to health regulations such as HIPAA and GDPR are critical. Finally, the integration of a system in a hospital setting is a challenge that is both respective and cultural.

The integration of AI systems into common use depends on their compatibility with current clinical routines, endorsement from healthcare practitioners and inclusion in policy reimbursements (Thomas and Yadav 2024). The absence of these criteria will hinder adoption regardless of the technological sophistication of the systems.

Finally, the cooperation of users is the last of the factors that AI systems depend on. Wearable systems use comfort, ease of use and unobtrusiveness as their main strategy to capture audience cooperation. The comfort, ease and unobtrusiveness must be present in the wearables. Otherwise, the goal of capturing data to train AI through machine feedback loops will be lost.

10.8. Future directions

Advanced technology, particularly the integration of AI, has post-radiation therapy rehabilitation. Robotic-assisted recovery systems led the changes using AI algorithms that customize movement therapy in real-time (Elhaddad and Hamam 2024). These systems account for individual patient biomechanics and progress, providing responsive assistance that modifies based on predetermined performance and biological parameters. This individualized focus for robotic systems goes a long way in motor function rehabilitation and recovery period shortening.

AI systems using multiple technology systems in parallel (multi-modal systems) perform this function effectively. More AI systems also incorporate wearable sensors, diagnostic imaging and electronic health records to form a complete picture of a patient and then synthesize that data for more accurate predicting recovery and functionality (Nizamis et al. 2021). Appropriate data-integrated AI systems also created accurate, real-time data predictions that promptly assist clinicians in adjusting therapy based on attained functional milestones.

Scalable AI systems that use federated learning can analyze an AI built on multiple decentralized data sources (Jabarulla and Lee 2021). AI that runs on decentralized systems can maintain privacy while providing large datasets to AI models for training. The widespread use of these enhanced AI systems also relies on the issuance of standardized clinical protocols.

Development of evidence-based guidelines facilitates compliance, improves collaboration among institutions and facilitates review procedures related to regulation. When combined, these advances suggest that more intelligent and adaptive pathways of recovery of safety are used in patients undergoing radiation therapy (Gudur et al. 2024).

10.9. Conclusion

Combining AI and biomechanical feedback offers novel methods of rehabilitation in the post-radiation rehabilitation. The systems use real-time evaluation of rehabilitation exercises coupled with AI-intensive predictive analytics to adjust to the individual. They provide real-time insights on patients and enable them to be active in improving their recovery. The systems are capable of detecting inefficient patterns, proposing a recovery schedule and changing the exercise parameters to optimize the effectiveness of rehabilitation. New opportunities are provided by AI in rehabilitation, namely, AR and VR therapies and assisted robotics. There are a number of issues in automation. Mainstream use should not ignore the consequences of bad-quality data, clinical usefulness of loose integration, data secrecy and workflow accountability. Usability should enhance compliance, both system-wise and patient-wise. The optimization of the biomechanical feedback systems will involve the AI developments of real-time predictive analytics and data fusion. These technologies can be widely integrated into post-radiation theraputic care as they are highly validated, if there is interdisciplinary cooperation and ethical consideration.

10.10. References

Ali, H. (2022). Reinforcement learning in healthcare: Optimizing treatment strategies, dynamic resource allocation, and adaptive clinical decision-making. *International Journal of Computer Applications Technology and Research*, 11(3), 88–104.

Alsunaidi, B., Althobaiti, M., Tamal, M., Albaker, W., Al-Naib, I. (2021). A review of non-invasive optical systems for continuous blood glucose monitoring. *Sensors*, 21(20), 6820.

Amiri, Z., Heidari, A., Navimipour, N.J., Unal, M., Mousavi, A. (2024). Adventures in data analysis: A systematic review of deep learning techniques for pattern recognition in cyber-physical-social systems. *Multimedia Tools and Applications*, 83(8), 22909–22973.

Dianti, J. (n.d.). Monitoring and controlling respiratory effort in acute hypoxemic respiratory failure. Master's Thesis, University of Toronto, Canada.

Díaz, S., Stephenson, J.B., Labrador, M.A. (2019). Use of wearable sensor technology in gait, balance, and range of motion analysis. *Applied Sciences*, 10(1), 234.

Elhaddad, M. and Hamam, S. (2024). AI-driven clinical decision support systems: An ongoing pursuit of potential. *Cureus*, 16(4), e57728. doi: 10.7759/cureus.57728.

Emerzian, S.R., Wu, T., Vaidya, R., Tang, S.Y., Abergel, R.J., Keaveny, T.M. (2023). Relative effects of radiation-induced changes in bone mass, structure, and tissue material on vertebral strength in a rat model. *Journal of Bone and Mineral Research*, 38(7), 1032–1042.

Erdoğan, K., Ceylan, R., Albayrak Gezer, İ. (2023). The use of augmented reality in the teaching and training of basic exercises involved in the non-surgical treatment of neck pain. *Virtual Reality*, 27(1), 481–492.

Fang, J., Lee, V.C., Ji, H., Wang, H. (2024). Enhancing digital health services: A machine learning approach to personalized exercise goal setting. *Digital Health*, 10, 20552076241233247.

Fatunmbi, T.O. (n.d.). Predicting precision-based treatment plans using artificial intelligence and machine learning in complex medical scenarios. *World Journal of Advanced Engineering Technology and Sciences*, 13(01), 1069–1088.

Grosu-Bularda, A., Lita, F.F., Hodea, F.V., Bordeanu-Diaconescu, E.M., Cretu, A., Dumitru, C.S., Cacior, S., Marinescu, B.M., Lascar, I., Hariga, C.S. (2024). Navigating the complexities of radiation injuries: Therapeutic principles and reconstructive strategies. *Journal of Personalized Medicine*, 14(11), 1100.

Gudur, A.K., Kale, S.R., Gudur, R.A., Bhosale, S.J., More, A.L., Datkhile, K.D. (2024). Single nucleotide polymorphisms in ape1, hogg1, rad51 genes and their association with radiotherapy induced toxicity among head and neck cancer patients. *Asian Pacific Journal of Cancer Prevention: APJCP*, 25(8), 2645.

Jabarulla, M.Y. and Lee, H.N. (2021). A blockchain and artificial intelligence-based, patient-centric healthcare system for combating the COVID-19 pandemic: Opportunities and applications. *Healthcare*, 9(8), 1019. MDPI.

Jubair, H., Chowdhury, R.T., Akter, F., Sami, I.S., Nawal, N., Arshi, S.T., Shil, P. (2025). Machine learning for real-time exercise correction and injury prevention: A systematic review. *Machine Learning*, 13, 1.

Kaul, S.M. and Maurya, N.K. (2025). *Integrating Technology in Physiotherapy: From Virtual Rehab to Wearable Devices*. Chyren Publication, Haryana.

Kumar, A., Batta, P., Ahuja, S., Rathore, P.S. (eds) (2026a). *AI-driven Innovations in Physiotherapy and Oncology 1*. ISTE Ltd, London and Wiley, Hoboken.

Kumar, A., Batta, P., Ahuja, S., Rathore, P.S. (eds) (2026b). *AI-driven Innovations in Physiotherapy and Oncology 2*. ISTE Ltd, London and Wiley, Hoboken.

Li, X., Fan, D., Feng, J., Lei, Y., Cheng, C., Li, X. (2025). Systematic review of motion capture in virtual reality: Enhancing the precision of sports training. *Journal of Ambient Intelligence and Smart Environments*, 17(1), 5–27.

Liang, W., Tadesse, G.A., Ho, D., Fei-Fei, L., Zaharia, M., Zhang, C., Zou, J. (2022). Advances, challenges and opportunities in creating data for trustworthy AI. *Nature Machine Intelligence*, 4(8), 669–677.

Mallawatantri, S. and Fernando, P. (2024). Enhancing adherence and completion rates through gamification and micro AI interventions in cognitive behavioral therapy. In *2024 International Conference on Computer and Applications (ICCA)*, 17 December, 1–7. IEEE.

Nascimento, L.M., Bonfati, L.V., Freitas, M.L., Mendes Junior, J.J., Siqueira, H.V., Stevan Jr, S.L. (2020). Sensors and systems for physical rehabilitation and health monitoring–A review. *Sensors*, 20(15), 4063.

Nizamis, K., Athanasiou, A., Almpani, S., Dimitrousis, C., Astaras, A. (2021). Converging robotic technologies in targeted neural rehabilitation: A review of emerging solutions and challenges. *Sensors*, 21(6), 2084.

Patil, V., Narayan, J., Sandhu, K., Dwivedy, S.K. (2022). Integration of virtual reality and augmented reality in physical rehabilitation: A state-of-the-art review. In *Revolutions in Product Design for Healthcare: Advances in Product Design and Design Methods for Healthcare*, Subburaj, K., Sandhu, K., Ćuković, S. (eds). Springer, Singapore, 177–205.

Petkevičius, M. (2023). The impact of gamification elements on the intention to use health related applications. Thesis, Vilniaus universitetas, Vilnius.

Seçkin, A.Ç., Ateş, B., Seçkin, M. (2023). Review on wearable technology in sports: Concepts, challenges and opportunities. *Applied Sciences*, 13(18), 10399.

Shah, K.N., Gami, S.J., Trehan, A. (2024). An intelligent approach to data quality management AI-Powered quality monitoring in analytics. *International Journal of Advanced Research in Science Communication and Technology*, 4(3), 109–119.

Singh, K. (2023). Artificial intelligence & cloud in healthcare: Analyzing challenges and solutions within regulatory boundaries. *SSRG International Journal of Computer Science and Engineering*, 10(9), 1–9.

Singh, S., Masood, H., Singh, D., Ansari, A.H., Pratap, R. (2024). Neurological injury: Consequences and restorative and rehabilitation strategies. In *Evidence-Based Neurological Disorders*, Upadhyay, S.K., Dan, S., Ali, S.A. (eds). Jenny Stanford Publishing, New Delhi, 357–381.

Tanaka, S. and Yamamoto, H. (2020). Biomechanical modeling in physical therapy: Advancing the science of movement to improve patient outcomes. *International Journal of AI, BigData, Computational and Management Studies*, 1(2), 11–18.

Thomas, M.S. and Yadav, T. (2024). Prevalence of upper cross syndrome in multipara women. *Journal of Ecophysiology and Occupational Health*, 24(1), 67–71.

Voiculescu, Ş. (2021). Long-term complications and management. In *Breast Cancer Essentials: Perspectives for Surgeons*, Rezai, M., Kocdor, M.A., Canterk, N.Z. (eds). Springer, Cham, 599–614.

Wang, X., Fu, Y., Ye, B., Babineau, J., Ding, Y., Mihailidis, A. (2022). Technology-based compensation assessment and detection of upper extremity activities of stroke survivors: Systematic review. *Journal of Medical Internet Research*. 24(6), e34307.

Wei, S. and Wu, Z. (2023). The application of wearable sensors and machine learning algorithms in rehabilitation training: A systematic review. *Sensors*, 23(18), 7667.

11

Computer Vision for Real-time Postural Correction in Cancer Physiotherapy

Cancer rehabilitation is becoming more integrated into survivorship care due to, fatigue and musculoskeletal complications after treatment. Survivors underappreciate the functional, adaptive and complex postural problems that they may face. The consequences of mastectomies, post-radiation fibrosis, neuropathic chemotherapy or sarcopenia-activating treatments are systemically underappreciated and, in turn, directly exacerbate psychosocial postural functional problems and overall quality of life. Computer vision (CV) technologies are beginning to receive more attention in their capacity to non-invasively and instantaneously recognize and potentially remedy postural problems, thus improving the precision and safety of rehabilitation. This chapter focuses on the expanding use of CV systems in real-time postural correction in cancer physiotherapy, including marker-based and markerless systems. The chapter describes the progression of systems from lab prototypes to clinically useful systems in providing consistent biomechanical evaluation, automated feedback and increased patient adherence. Other integrations of CV systems include augmented reality (AR), wearable inertial sensors and digital twins. This describes a newly formed ecosystem of CV and postural rehabilitation. The chapter describes this ecosystem, and the critique of the clinical validation attempts, potential for engagement and usability of the system consider the unique challenges of privacy of data, biases in algorithms and problems of scale and access. Lastly, the future directions of this research focus on the importance of cross-discipline research, design for the user and extensive clinical studies to advance personalized, tech-based rehabilitative cancer care.

Chapter written by Anand GUDUR and Rahul PETHE.

11.1. Introduction

11.1.1. *Background*

As healthcare systems evolve, cancer survivorship is increasingly prioritized and the after-effects of cancer care are being examined more closely. For cancer survivors, physical and functional impairments resulting from surgical procedures, chemotherapy, hormonal therapy and radiation therapy can be widespread. Although these treatments save lives, they can be associated with a complex spectrum of complications in the musculoskeletal and neuromuscular systems. Side effects reported by cancer survivors include muscle and joint restrictions, postural balance impairments, neuropathy, fatigue and lymphedema. Consequently, these effects can hinder a person's performance in the activities of daily living, social interactions and work, thus reducing overall quality of life (Catt et al. 2017).

In fact, postural dysfunction can be severe. Examples include women with mastectomies or lumpectomies, who manifest postural changes of rounded shoulders, winged scapula and kyphosis. Beyond aesthetics, these postural changes can be the cause of chronic pain, pulmonary dysfunction and movement restriction. Survivors of head and neck cancers can have abnormal cervical muscle tone, and pelvic cancer patients, especially those who have had radiation therapy, have common postural deficits of the pelvis and abnormal walking patterns (Sürmeli 2025). Falls, fractures and other injuries may be severe problems caused by biomechanical dysfunction.

Physiotherapy can significantly help lessen these impacts. Part of the rehabilitation caring plan involves therapist exercises, which aim to strengthen, stretch, balance and retrain some motion patterns as well. Yet, some of the problems of older physiotherapy approaches do seem to persist. Programs still use a lot of in-person sessions, which means those in remote regions or those with no means of travel/mobility issues may not be able to take part (Paul et al. 2023). Even these therapist evaluations tend to be subjective with no real consistent balance. There can be problems with motivation and adherence to a plan when exercises have to be done alone at home, unsupervised, meaning results can be less than ideal.

The increasing demands of cancer survivorship, especially when combined with the digital physiotherapy world gap, has created a new focus in the area of digital health, in particular the use of artificial intelligence (AI) and sensorless remote physiotherapy. Within sensorless remote physiotherapy, computer vision (CV) stands out as a unique digital health tool with the capability of transforming rehabilitation.

11.1.2. *Rise of CV in physiotherapy*

CV refers to the field of AI that enables machines to interpret and understand visual information from the physical world. In physiotherapy, CV systems are used to analyze human movement captured on regular RGB or depth cameras. They evaluate posture, the angle of joints, range of motion and symmetry of movement. This is performed using complex algorithms, such as pose-estimation models, which move anatomical landmarks (e.g. the shoulder, hip or knee) in two or three dimensions (Kumar et al. 2026a).

The efficiency and accuracy of pose estimation have significantly increased due to the recent developments in deep learning (DL) and computing capabilities. With only a smartphone or web camera, open-source skeletal key point detectors such as OpenPose, MediaPipe and BlazePose can identify skeletal key points in real time without the use of an expensive motion capture system or wearable sensor (Kumar et al. 2026b). The remote and interactive nature of physiotherapy is scalable and cost-effective with these innovations.

In contrast to the conventional video recording or therapist-based observations, CV allows objective, continuous and quantitative measurement of patient movement. For example, a CV system can estimate joint angles, identify the pattern of compensations and identify whether an exercise is done correctly. More technology-oriented systems are able to offer feedback in real time with on-screen alerts, auditory signals or gamified interfaces to assist patients in better posture and maintaining proper form when the rehabilitation process is underway.

11.1.3. *Relevance in oncology rehabilitation*

CV technology is important for oncology rehabilitation, especially considering that cancer survivors often have postural and functional limitations. Some cancers and their treatments lead to specific biomechanical challenges and CV-based tools can monitor and help address these challenges.

For example, after breast cancer surgery or radiation, survivors often have functional issues. Research indicates that up to 60% of women post-mastectomy suffer from shoulder impingement and have a limited range of motion, often performing abnormal scapular movements (Lang 2020). CV systems can monitor and track the movement of the scapula and humerus in real time during rehabilitation to help guide survivors, prevent further injury and maintain the correct alignment.

Head and neck cancer survivors may have malpositioned postures due to radiation-induced fibrosis and neck dissection. CV technology can help assess and monitor the neck's range of motion, tell the patient when they need to realign their neck to engage specific torso muscles and help track their progress to functional recovery while lowering pain.

Contributing factors to limited function post-surgery and radiation for pelvic cancer can include morphologic malposition, gait dysfunction and lower limb weaknesses. CV enables the detection of changes in stride length, cadence and limb loading, providing the basis for longitudinal tracking of the progress of customized therapeutic interventions (Hulleck et al. 2022).

Beyond the rehabilitation of musculoskeletal conditions, the immediate delivery of visual and auditory feedback is another considerable advantage of CV systems in rehabilitation oncology. Patients become more active participants in the recovery process, which enhances motivation and motor learning.

CV system rehabilitative exercises are gamified, and monotonous tasks are replaced by interactive CV tasks, improving compliance and the overall effect of the rehabilitation exercises. Clinicians can conduct longitudinal assessments and revise rehabilitation programs using stored quantitative movement data. This is especially important in oncology rehabilitation, as unaddressed progressive cancer, aging and late sequelae can change the rehabilitation demands of the physically debilitating cancer.

11.1.4. *Aim of the chapter*

In conjunction with the potential of CV in the physiotherapy domain, especially oncology, the chapter systematically evaluates the extent to which this CV technology has been researched and implemented in clinic practice. It integrates the available literature related to CV technology systems designed for postural correction during the rehabilitation of patients with cancer, as well as the technology that powers the systems and their uses in practice, along with their advantages and disadvantages. This chapter evaluates the systems in order to understand the optimal use, available research and gaps that indicate possible future research. The chapter ultimately assesses the potential for CV technology to meet the needs of scalable customization and results-oriented cancer rehabilitation.

11.2. CV technologies for postural correction

11.2.1. *Marker-based motion capture systems*

The earliest CV applications in biomechanics and rehabilitation used marker-based motion capture systems, which remain a gold standard for accuracy in motion analysis. These systems, such as Vicon, OptiTrack and Qualisys, use reflective markers attached to anatomical landmarks, captured by multiple synchronized infrared cameras to triangulate and reconstruct 3D joint trajectories. Their sub-millimeter precision has been instrumental in validating musculoskeletal models and establishing normative movement databases. However, these systems are expensive, require dedicated laboratory space, controlled lighting and skilled technicians to operate and interpret the data (National Research Council 2011). As such, their use is largely restricted to research labs and elite clinical centers, making them inaccessible for routine cancer rehabilitation, especially in low-resource or home-based settings. Nevertheless, their contribution to the development and training of machine learning (ML) algorithms for posture estimation and movement analysis remains critical, serving as the ground truth for evaluating the accuracy of emerging markerless systems. Table 11.1 shows a comparative overview of marker-based and markerless CV systems for postural correction in cancer physiotherapy.

Parameter	Marker-based systems	Markerless systems	Relevance in cancer physiotherapy
Technology used	Infrared cameras with reflective body markers (e.g. Vicon, OptiTrack, Qualisys, etc.)	RGB, RGB-D or depth cameras using AI-based pose estimation (e.g. OpenPose, MediaPipe, BlazePose, etc.)	Markerless systems suitable for both clinics and home rehabilitation
Setup requirements	Requires controlled lighting, calibrated cameras and technical expertise	Can operate with standard webcams, tablets or smartphones	Reduces cost and increases accessibility for patients
Accuracy	Sub-millimeter precision for 3D motion tracking	Slightly lower accuracy (2–3 cm joint error) but clinically acceptable	Sufficient for posture correction and range-of-motion monitoring
Cost and infrastructure	High equipment and maintenance cost	Low-cost, scalable and mobile-friendly	Ideal for remote and tele-rehabilitation
Data processing	High-end computer and specialized software	Real-time AI inference using edge computing or mobile processing	Enables on-device privacy-preserving correction
Clinical applicability	Limited to research and elite labs	Widely adaptable for home and outpatient settings	Promotes continuity of care and self-guided physiotherapy
Limitations	Requires markers; time-consuming setup	May struggle with occlusions, poor lighting and depth ambiguity	Continued AI improvement needed for precision and reliability

Table 11.1. *Comparative overview of marker-based and markerless CV systems for postural correction in cancer physiotherapy*

11.2.2. *Markerless vision systems*

With advances in DL and CV, a shift towards markerless pose estimation has enabled broader applications in real-world clinical and home environments. These systems rely on standard 2D RGB cameras, depth sensors or stereovision setups, eliminating the need for physical markers (Avogaro et al. 2023). Deep neural networks (DNNs) that have been trained on large human movement datasets suggest the skeletal key points and angle of joints in real time. Major models and frameworks include, but are not limited to:

– *OpenPose*: developed by the Carnegie Mellon Perceptual Computing Lab, OpenPose extracts 2D skeletal key points using multi-stage convolutional neural networks (CNNs) and is widely used for activity recognition and rehabilitation research.

– *MediaPipe*: developed by Google, MediaPipe is a lightweight, cross-platform library optimized for mobile and web applications, making it highly suitable for tele-rehabilitation.

– *AlphaPose*: known for high-precision multi-person pose estimation, this is useful in group therapy or community exercise settings where multiple users are present.

– *PoseNet and BlazePose*: these models offer real-time pose detection optimized for augmented reality (AR) and virtual reality (VR) environments, with applications in interactive physiotherapy.

11.2.3. *Integration with AR and visual feedback*

Modern CV systems stand out because of their ability to interactively visualize feedback through AR overlays. In rehabilitation, these systems and screens show patients posture correction cues in real time as patients execute their prescribed movements (Maskeliūnas et al. 2023). For instance, during a shoulder abduction exercise, if a patient makes a compensatory lean at the trunk, an animated posture guide or visual alert such as "detected misalignment" realigns the patient's posture.

Biofeedback loops are known to be effective in the reinforcement of correct functional movement, thereby increasing motor learning. More neurocognitive involvement, motivation and compliance with therapy regimens can be achieved by adding interface gamification (metaphoric display, such as "keep your wings level" or following the "glowing path"). This is particularly relevant to cancer rehabilitation, whereby patients are usually affected by cognitive fatigue, motivational difficulties and psychosocial distress.

11.2.4. *Real-time processing and edge computing*

Posture correction technology based on CV has been enhanced by edge computing. Real-time neural-network inference can be done on-the-fly by devices as small as the NVIDIA Jetson Nano, Raspberry Pi 5 with Coral TPU and even by the latest smartphones. This does not require the use of cloud computing and feedback is provided, which is essential to successful physiotherapy.

In the case of healthcare applications, privacy and security are the most important issues. Sensitive data are processed locally, which minimizes the possible risks. Another advantage of edge computing is the offline operation, which reduces the need to have high-bandwidth networks and external servers and makes it ideal in underdeveloped and rural regions (Kong et al. 2022).

11.3. Applications in cancer physiotherapy

11.3.1. *Breast cancer rehabilitation*

Postural asymmetries, rounded shoulders and frozen shoulders are common problems that breast cancer survivors experience after a mastectomy, the removal of lymph nodes and radiation therapy, and this may limit the movement of shoulders, create compensatory torso movements and lead to chronic pain. CV systems are under investigation to help in rehabilitation through real-time analysis and feedback on correction. They are able to evaluate the scapular position during shoulder elevation, monitor shoulder range of motion exercises, such as wall slides or pulley movements, and identify the abnormal scapular winging or the lean of the trunk (Sanchez and Sanchez 2018). OpenPose and Microsoft Kinetop pilot studies have demonstrated more than 85% accuracy in detecting these problems and are capable of providing visual and audio feedback to therapists and patients. It is also a technology that enhances exercise adherence particularly where home-based programs are involved.

11.3.2. *Head and neck cancer*

The radiation treatment of cancers of the head and neck can frequently result in the development of fibrosis, cervical muscle rigidity and head forward posture. The changes complicate the movements of the neck and may affect swallowing and breathing. CV systems are able to measure cervical range of movement flexion, extension and rotation objectively. As an illustration, in chin-tuck or neck-retraction exercises, the system has the ability to provide real-time feedback so that there is proper alignment and to avoid thoracic flexion (Crane 2024). Vision-based equipment is also being considered to scan one-sided neck and jaw movements in swallowing therapy, and this is very important in managing dysphagia.

11.3.3. *Pelvic and colorectal cancer*

Gait issues, hip displacement and loss of core stability are common side effects of pelvic surgeries and radiation therapy for colorectal and gynecological tumors. These incontinences can impair balance, daily mobility and control of the pelvic floor. During exercise, CV systems are able to monitor the angles of the pelvis, lumbar lordosis and hip (squats, bridges and lunges). They are also able to note asymmetrical weight distribution or compensatory tendencies such as lateral trunk lean during the movement. Online feedback on step length and symmetry can be provided with the help of vision to facilitate neuromuscular reeducation and safer gait (Esposito et al. 2021).

11.3.4. *Cancer-related fatigue and sarcopenia*

Most cancer survivors experience a lack of endurance, loss of muscle (sarcopenia) and poor coordination due to chemotherapy and inactivity. These changes manifest as slouched posture, slower movements and biomechanical inefficiencies. It is possible for CV systems to monitor for changes to individual kinematics and detect signs of fatigue and recommend breaks or changes in posture. These breaks can conserve energy and reduce the risk of overexertion and promote movement efficiency aligned to individual capabilities (Baghdadi et al. 2018).

11.4. Validation and clinical studies

11.4.1. *Accuracy and reliability*

The clinical utility of CV systems in cancer physiotherapy largely depends on their ability to accurately and reliably assess human movement. Multiple validation studies have demonstrated strong agreement between CV-based pose estimation algorithms and gold-standard motion capture systems or expert physiotherapist assessments. For instance, open-source tools such as OpenPose and MediaPipe have shown mean joint position errors of less than 2.5 cm in rehabilitation-specific environments. These low error margins are generally acceptable for gross motor assessments such as range-of-motion (ROM) evaluations and postural analysis in physiotherapy.

Inter-session consistency (pose estimation reliability) in gross motor tasks such as shoulder flexion, abduction or cervical rotation has shown more than 90% inter-session reliability in a number of pilot studies. This suggests that CV systems have the ability to offer repeatable, objective data over various sessions and lighting conditions, as required in long-term rehabilitation monitoring (Nascimento et al. 2020). Also, CV lessens inter-ratio variability, which is inherent to manual physiotherapy measurements, and this is therefore useful in a resource-constrained environment or at home.

11.4.2. *User acceptance*

One of the aspects of CV-based rehabilitation deployment is user acceptance. Research on breast cancer survivors where Microsoft Kinect was applied to provide real-time visual and audio feedback to patients during exercise have indicated a high degree of patient satisfaction, particularly where exercise programs are customized and contain some elements of gamification (Su et al. 2024).

It was also indicated that there were significant changes in self-reported pain and functional limitations in the shoulders of the participants who used CV-guided programs. CV-enhanced protocols exhibited better adherence than the exercise protocols based on the use of standard video or guided by a book, which is likely explained by the engaging feedback and visualization of the progress. This increased compliance is crucial in cancer rehabilitation where motivation and perseverance is challenged by fatigue and emotional distress.

11.4.3. *Limitations in clinical trials*

Although the outcomes are promising, the evidence of CV systems in oncology rehabilitation is constrained by methodological flaws. Small studies are usually feasibility studies, which may not always have control or standard outcome measures. Additionally, the trials are on healthy subjects or general musculoskeletal patients, rather than cancer survivors that have distinct physiological and psychological difficulties.

There are limited studies that are directly based on oncology populations. One interesting case is an Iranian trial in 2022 that trained MediaPipe on the postural correction of breast cancer survivors. Although the study presented more favorable results in terms of improved scapular symmetry and pain alleviation, the results require bigger randomized controlled trials (RCT) to verify the findings and lay the appropriate clinical guidelines.

11.5. Challenges and barriers

11.5.1. *Technical limitations*

Although the use of CV systems has potential in postural assessment and rehabilitation, there are a number of technical difficulties to overcome. A significant drawback is that there is less accuracy when sections of the body are obscured (i.e. a limb obstructing the torso or only half of the body being visible to the camera). This is unsuitable for skeletal key point detection, particularly when looking off-side or when performing complicated shoulder or scapular turns. Consequently, real-time feedback may be lost, which is unfavorable in the case of clinical correction, which requires a strong degree of accuracy.

Moreover, pose-estimation algorithms can be undermined by a difference in lighting, the patterns of clothing and the presence of background clutter. Baggy or dark clothes blur the lines of the body, and it is difficult to make 2D models differentiate joint parts. Even though some resistance to these changes is offered by sophisticated frameworks such as OpenPose or MediaPipe, their work can still continue to deteriorate when applied to real-world, less-controlled settings such as in the patients' homes (Ram et al. 2024).

Another urgent problem is that in many 2D-only CV systems, information on depth is not given. This limitation is critical with regard to spine and scapular tracking in which even minor anterior–posterior deviations (e.g. kyphosis or scapular winging) are critical in diagnosis and treatment. Although this is partially met by RGB-D cameras (e.g. Kinect or Intel RealSense) that capture depth data, these come at the cost of trade-offs in price, complexity of set-up and compatibility with mobile devices.

11.5.2. *Clinical validation*

Evidence and standardization of the clinical implementation of CV systems to aid cancer rehabilitation is flawed. Nowadays, no universal guidelines on the classification of posture and the detection of compensatory movements are established. This inconsistency restricts the extrapolability of the results of the studies as well as their application to broader clinical populations.

Additionally, there are limited large-scale, multicenter RCTs that prove the usefulness of CV tools in the context of oncology. The available literature consists of pilot studies or small-cohort validation which have little overall validity. This means that there is a lack of high-quality information that can justify regulatory approval or massive reimbursement of such technologies.

Lastly, CV systems have not been integrated with electronic health records (EHRs) or physiotherapy dashboards. In the absence of seamless interoperability, it is hard to embed CV outputs in the workflows or to monitor the long-term patient progress. This limitation constrains the use of such systems in contexts other than academic or research settings as well as their application in regular care provision.

11.5.3. *Privacy and ethics*

One of the key issues of the CV in home rehabilitation is privacy and data security. The concept of observing patients at all times at home may be intrusive to the patient as well as to the caregivers. They include issues such as the accidental capture of faces of individuals who are not program participants, anxiety or discomfort arising from being observed, and the potential misuse of recorded footage (Salehzadeh Niksirat et al. 2024).

These systems are required to comply with data protection laws, including the General Data Protection Regulation (GDPR) in Europe and the Health Insurance Portability and Accountability Act (HIPAA) in the United States, in order to remain legal. The principles dictate that data must be kept safely and transferred through encryption, explicit informed consent must be given and participants should be allowed to withdraw.

Edge computing is one method of reducing the risk, as the data processing happens on the device of the user rather than on the cloud. This can maintain the data privacy more, but can restrict the capability of clinicians to audit remotely, intervene in real-time or extract aggregate information to use in research.

11.5.4. *Equity and accessibility*

Equal utilization of CV systems implies bridging the divide in access to technology and digital literacy. Several of the solutions require the use of high quality phones, tablets or webcams which may be lacking among older adults, individuals with low income or rural dwellers. Sporadic Internet access may also impair the transfer of data and real-time feedback.

Engagement may also decrease due to language barriers and poor digital literacy, particularly among older people or non-native speaking cancer survivors. The developers can also create products that cater to all people, such as making them in multiple languages, with easy navigation and cheap hardware (Srivastava et al. 2021) so that more individuals can use these products.

Cost is still a very major obstacle. Although CV has the potential to reduce physical visits, the initial cost of set-up and purchase of the devices may be prohibitive to underserved communities. The solution to these issues is necessary in order that even those who do not enjoy the benefits of CV with regard to posture correction will benefit.

11.6. Innovations and integrations

11.6.1. *Integration with digital twins*

Oncology rehabilitation has been implemented with the application of digital twins which are virtual versions of real objects. They include CV information in the dynamic models of the musculoskeletal system of a patient. Real-time posture data captured on CV cameras enables clinicians and physiotherapists to simulate and visualize the impact of various interventions on a patient in real time and over a period of time on their posture and function. These computer-based models allow practitioners to perform predictive simulations of remedial interventions such as stretching, strengthening or orthosis (Huang et al. 2023). As an example, a breast cancer survivor with post-mastectomy scapular asymmetry can determine how shoulder-kinesis training, ROM training and muscle activation sequences will perform by using a digital twin. Digital twins also give us a historical perspective of movement deficits, which allow the patient to see progress, identify plateaus and identify relapses. This will improve the user experience, decision-making, support adaptive rehabilitation plans and minimize the use of guesswork.

11.6.2. *Sensor fusion*

CV systems have the benefit of being scalable and contactless, but can have issues with accuracy when in specific postures or under certain lighting conditions. Sensor fusion is a combination of CV and wearable IMUs, gyroscopes and surface electromyography sensors. This is a multimodal method that enhances spatial and temporal tracking of posture. Adaptively, scapular motion, which is of paramount importance in the rehabilitation of the shoulders, cannot be accurately measured by vision due to overlapping tissues with respect to small changes in angles. With CV data plus IMU orientation tracking and sEMG signals, which can be used to indicate muscle activity, clinicians can have a better understanding of scapular kinematics. This feedback is multidimensional and particularly invaluable in cases of post-radiation fibrosis whereby soft tissues retain compensatory patterns that require multi-angle minor corrections. Sensor fusion also aids in the detection of micro-movements, asymmetries and dynamic compensations in complex daily movements, such as reaching and twisting. These movements are often difficult to assess using a single modality (Huang et al. 2023).

11.6.3. *Gamification and AR/VR*

Using gamification in healthcare has been effective in focusing patients in rehabilitation, especially with younger patients in pediatric or adolescent oncology. CV systems integrated with AR and VR technologies provide interactive real-time feedback systems which promote motivation and encourage compliance. During therapeutic exercises, patients may score points for holding a therapeutic symmetrical posture. An AR avatar will mimic the patient and provide posture correction cues while guiding them through exercises (Kaul and Maurya 2025).

For example, a child with leukemia-related deconditioning may use VR posture correction exercises while having therapy and interact with virtual characters and animated rehabilitation progress to gamify their therapeutic goals. Immersive VR therapy reduces the boredom associated with rehabilitation therapy's repetitive movements, improving consistency and encouraging longer sessions. In addition to the AR-enhanced VR environment therapy sessions, AR control overlays provide intuitive posture correction feedback.

11.6.4. *Personalized AI coaching*

Once a CV system starts accruing user-specific data, it can be paired with AI tools such as large language models and reinforcement learning agents to provide real-time, personalized mentoring. AI systems can evaluate and recognize patterns in a user's posture and adherence, and plan rehabilitation activities accordingly.

Suppose a patient exhibits anterior pelvic tilt during squats, the system would be able to provide real-time verbal commands such as "engage your core" or "shift weight back slightly", akin to a human therapist (Patel et al. 2024). With reinforcement learning, the system adjusts coaching and feedback in real time and over time, improves the relevance and timing of feedback. These AI coaches can provide motivational support by monitoring and analyzing emotional cues and adjusting the interaction to help sustain engagement, offering emotional support and guidance during rehabilitation. Such a system could be the foundation of responsive virtual rehabilitation assistants. These systems would be empowered with emotional intelligence. Designed to help cancer rehab patients, they would respond to the patient's emotional and motivational cues to support rehabilitation (Lin et al. 2020).

11.7. Future directions

11.7.1. *Clinical trials*

Researchers have begun multiple nationwide RCTs involving CV systems in cancer rehabilitation with a special focus on post-therapeutic mastectomy, head-and-neck resections and post-prostate cancer surgery rehabilitation (Unger et al. 2019). These trials focus on evaluating the functional expected outcomes (e.g. ROM and postural alignment) along with the subjective outcomes to the patient (e.g. pain, quality of life, etc.). The stratified trial designs have now considered the demographical and clinical majorities: age, sex, cancer type, stage of treatment and comorbid conditions, such as lymphedema or diabetes (Sheu and Pardeshi 2022). This guarantees individual rehabilitation procedures and renders the findings more representative in a variety of patients.

11.7.2. *Model improvement*

The model is enhanced in terms of robustness and clinical relevance. Depth-aware pose estimation is an amalgamation of both monocular RGB data and AI-based depth estimation to estimate the 3D joint locations without any special hardware. This is particularly applicable to spine and scapular tracking, which cannot be reliably achieved using 2D data alone. Explainable AI (XAI) is another trend that can be explained by the visual on-the-fly, highlighting the rationale behind corrective suggestions. This openness instills confidence in both patients and clinicians (Nanduri 2024).

11.7.3. *Accessibility initiatives*

Efforts to make technology-assisted rehabilitation democratic come through accessibility efforts. Open-source CV apps are usually low bandwidth with low processing requirements, and thus can be implemented on simple smartphones or tablets in low-resource environments. There is the development of regional language support and culturally modified training resources such as voice-assisted videos to increase the engagement and compliance of various mass language and education groups (Marshall et al. 2021). These initiatives attempt to close the digital divide and offer universal rehabilitation globally.

11.7.4. *Regulatory guidelines*

As cardiovascular rehabilitation devices move from purely research-based tools to rehabilitation research clinical tools, the integration of CV rehabilitation devices has provided the rehabilitation research community opportunities to refine the clinical CV rehabilitation pathway for patients with poor and compromised mobility. Correspondingly, establishing a clinical CV rehabilitation pathway promotes the integration of CV-based devices into routine rehabilitation practice (Gudur et al. 2024). Consequently, the integration of CV rehabilitation devices into clinical pathways has prompted the incorporation of other devices to mitigate any possible risks and errors in the clinical CV rehabilitation pathway, specifically concerning self-use and feedback mechanisms.

11.8. Conclusion

CV can be revolutionary in the monitoring and real-time correction of patient movement in cancer rehabilitation physiotherapy. It offers scalable and objective ways to assess and improve patient movement. For instance, in the rehabilitation of breast cancer patients who exhibit shoulder dysfunction, scapular winging and restricted movement in the shoulder, CV can be used to detect compensatory movements and offer corrective instructions in real time. CV is also capable of tracking the pattern of movements in prescribed physiotherapy of survivors of head, neck or prostate cancer who suffer from fatigue, postural collapse and muscle imbalance. CV is creating the future of customized rehabilitation, coupled with AI, digital twins and AR. AI-powered pose estimation algorithms track the joints and posture of a patient with increasing accuracy. Digital twins of the patient's musculoskeletal system help visualize the functional outcomes of certain rehabilitation interventions. AR is a distraction and encourages passive patient participation by overlaying corrective guidance during movement tasks. Still, the use of CV in physiotherapy rehabilitation of cancer patients is faced with the issues of occlusion and poor lighting. To achieve the full potential of CV in the treatment of cancer physiotherapy, it is necessary to implement strong clinical trials, provisions of equal access and interdisciplinary teams composed of engineers, clinicians and rehabilitation professionals. Having dealt with these challenges, the strength and the formalizability of the approach will be maximized.

11.9. References

Avogaro, A., Cunico, F., Rosenhahn, B., Setti, F. (2023). Markerless human pose estimation for biomedical applications: A survey. *Frontiers in Computer Science*, 5, 1153160.

Baghdadi, A., Megahed, F.M., Esfahani, E.T., Cavuoto, L.A. (2018). A machine learning approach to detect changes in gait parameters following a fatiguing occupational task. *Ergonomics*, 61(8), 1116–1129.

Catt, S., Starkings, R., Shilling, V., Fallowfield, L. (2017). Patient-reported outcome measures of the impact of cancer on patients' everyday lives: A systematic review. *Journal of Cancer Survivorship*, 11(2), 211–232.

Crane, P.A. (2024). Management of cervical and thoracic spine orthopedic conditions. In *Principles of Therapeutic Exercise for the Physical Therapist Assistant*, Kopack, J. and Cascardi, K. (eds). Routledge, New York.

Esposito, D., Centracchio, J., Andreozzi, E., Gargiulo, G.D., Naik, G.R., Bifulco, P. (2021). Biosignal-based human–machine interfaces for assistance and rehabilitation: A survey. *Sensors*, 21(20), 6863.

Gudur, R.A., Bhosale, S.J., Gudur, A.K., Kale, S.R., More, A.L., Datkhile, K.D. (2024). Influence of (C1236T and C3435T) polymorphisms of ABCB1 gene on chemotherapy treatment outcome and toxicity in breast cancer patients. *Asian Pacific Journal of Cancer Prevention: APJCP*, 25(5), 1567.

Huang, X., Xue, Y., Ren, S., Wang, F. (2023). Sensor-based wearable systems for monitoring human motion and posture: A review. *Sensors*, 23(22), 9047.

Hulleck, A.A., Menoth Mohan, D., Abdallah, N., El Rich, M., Khalaf, K. (2022). Present and future of gait assessment in clinical practice: Towards the application of novel trends and technologies. *Frontiers in Medical Technology*, 4, 901331.

Kaul, S.M. and Maurya, N.K. (2025). *Integrating Technology in Physiotherapy: From Virtual Rehab to Wearable Devices*. Chyren Publication, Haryana.

Killen, B.A., Falisse, A., De Groote, F., Jonkers, I. (2020). In silico-enhanced treatment and rehabilitation planning for patients with musculoskeletal disorders: Can musculoskeletal modelling and dynamic simulations really impact current clinical practice? *Applied Sciences*, 10(20), 7255.

Kong, L., Tan, J., Huang, J., Chen, G., Wang, S., Jin, X., Zeng, P., Khan, M., Das, S.K. (2022). Edge-computing-driven internet of things: A survey. *ACM Computing Surveys*, 55(8), 1–41.

Kumar, A., Batta, P., Ahuja, S., Rathore, P.S. (eds) (2026a). *AI-driven Innovations in Physiotherapy and Oncology 1*. ISTE Ltd, London and Wiley, Hoboken, NJ.

Kumar, A., Batta, P., Ahuja, S., Rathore, P.S. (eds) (2026b). *AI-driven Innovations in Physiotherapy and Oncology 2*. ISTE Ltd, London and Wiley, Hoboken, NJ.

Lang, A.E. (2020). Towards improving clinical evaluation of the shoulder: Defining upper limb biomechanics of breast cancer survivors during functional evaluation tasks. Doctoral Dissertation, University of Saskatchewan.

Lin, J., Ma, Z., Gomez, R., Nakamura, K., He, B., Li, G. (2020). A review on interactive reinforcement learning from human social feedback. *IEEE Access*, 8, 120757–120765.

Marshall, J., Morrill, K., Gobbe, M., Blanchard, L. (2021). The difference between approval processes for medicinal products and medical devices in Europe. *Ophthalmologica*, 244(5), 368–378.

Maskeliūnas, R., Damaševičius, R., Blažauskas, T., Canbulut, C., Adomavičienė, A., Griškevičius, J. (2023). BiomacVR: A virtual reality-based system for precise human posture and motion analysis in rehabilitation exercises using depth sensors. *Electronics*, 12(2), 339.

Nanduri, D.K. (2024). Exploring the role of generative artificial intelligence in culturally relevant storytelling for native language learning among children. Master's Thesis, University of Maryland, College Park.

Nascimento, L.M., Bonfati, L.V., Freitas, M.L., Mendes Junior, J.J., Siqueira, H.V., Stevan Jr, S.L. (2020). Sensors and systems for physical rehabilitation and health monitoring – A review. *Sensors*, 20(15), 4063.

National Research Council (2011). *Prudent Practices in the Laboratory: Handling and Management of Chemical Hazards, Updated Version*. The National Academies Press, Washington, D.C.

Patel, D.J., Sharma, P., Gudur, R., Banerjee, D., Singh, J. (2024). Prognostic study to detect ailments, including cancer, during solid organ transplantation to prevent postoperative complications using perioperative antibiotics therapy. *Onkologia i Radioterapia*, 18(9), 001–007.

Paul, S.S., Hubbard, A., Johnson, J., Dennis, S.M. (2023). Transition to a virtual model of physiotherapy and exercise physiology in response to COVID-19 for people in a rural Australia: Is it a viable solution to increase access to allied health for rural populations? *PLoS One*, 18(1), e0280876.

Ram, B.G., Murgod, T.R., Chandan, H., Kumar, D.S. (2024). Real-time exercise assessment: MediaPipe & OpenCV approach. In *Computer Science Engineering*, Gururaj, H.L., Flammini, F., Srividhya, S. et al. (eds). CRC Press, London.

Salehzadeh Niksirat, K., Velykoivanenko, L., Zufferey, N., Cherubini, M., Huguenin, K., Humbert, M. (2024). Wearable activity trackers: A survey on utility, privacy, and security. *ACM Computing Surveys*, 56(7), 1–40.

Sanchez, H.M. and Sanchez, E.G. (2018). Scapular dyskinesis: Biomechanics, evaluation and treatment. *International Physical Medicine & Rehabilitation Journal*, 3(6), 514–520.

Sheu, R.K. and Pardeshi, M.S. (2022). A survey on medical explainable AI (XAI): Recent progress, explainability approach, human interaction and scoring system. *Sensors*, 22(20), 8068.

Srivastava, A., Kapania, S., Tuli, A., Singh, P. (2021). Actionable UI design guidelines for smartphone applications inclusive of low-literate users. *Proceedings of the ACM on Human-Computer Interaction*, 5(CSCW1), 1–30.

Su, Z., Zhang, L., Lian, X., Guan, M. (2024). Virtual reality-based exercise rehabilitation in cancer-related dysfunctions: Scoping review. *Journal of Medical Internet Research*, 26, e49312.

Sudhakar, A. (2025). *Applied Deep Learning for Clinical NLP and Radiomics: Architectures, Pipelines, and Case Studies*. Yashita Prakashan Private Limited, Delhi.

Sürmeli, M. (2025). Postural and balance problems in breast cancer survivors and managing options. In *Managing Side Effects of Breast Cancer Treatment*, Baker, Y. and Tuğral, A. (eds). Springer, Cham.

Unger, J.M., Hershman, D.L., Fleury, M.E., Vaidya, R. (2019). Association of patient comorbid conditions with cancer clinical trial participation. *JAMA Oncology*, 5(3), 326–333.

12

AI-Driven Remote Physiotherapy Platforms for Immunocompromised Cancer Patients

Patients with compromised immune systems who have cancer are the most disadvantaged of all in regard to accessing traditional in-person physiotherapy. This is due to the risk of infection, the fatigue that comes from traveling to a healthcare facility, the impact of scheduling around intensive treatment, the risk of hospital-acquired illness and general treatment vulnerability. Many of these patients do not get the rehabilitation they need, and this will, in turn, impact recovery, create a poor quality of life and invite other long-term health issues that can arise.

Artificial intelligence (AI) rehabilitation remote physiotherapy portals of care help these patients by offering individualized rehabilitation safely in the comfort of their homes. Patients can take advantage of the latest advancements in real-time motion tracking technologies and virtual exercise coaching, along with other predictive analytics, wearables and integration with AI video consultations with physiotherapists. Continuous machine learning can alter exercise regimens with active symptom management, while predictive algorithms ensure ongoing monitoring of safety and adherence.

This chapter discusses the remote physiotherapy sector, which is based on AI and its application to cancer patients. It looks at infection control and patient engagement as the essential clinical advantages and also evaluates data security, digital illiteracy and reliance on automated systems as well as raising awareness of potential risks. This chapter also talks about AI application among vulnerable groups ethically and gives practical solutions on how interdisciplinary collaboration, regulatory compliance and patient-centered design can be used. It also identifies the priorities of future research, such as the significance of clinical trials, feedback loops to refine the model and equal access to technology.

Chapter written by Rashmi GUDUR and Abhay KASHETWAR.

12.1. Introduction

The effectiveness of cancer treatment can be lessened due to its numerous side effects. This includes side effects such as fatigue and general deconditioning, as well as muscle atrophy, reduced range of motion, neuropathy and lymphedema. Physiotherapy therefore becomes important to help improve function, ease pain and enhance mobility and overall physical functioning, both during and after treatment. Yet, traditional, in-person physiotherapy becomes inaccessible and even risky to immunocompromised cancer patients, especially those in chemotherapy and radiotherapy or those who have blood cancers, bone marrow transplants or hematological malignancies. Infections and complications for these patients can be hazardous. Frequently visiting clinics or hospitals is very risky and logistically demanding, including traveling, time and physical exhaustion (Zheng et al. 2021). These difficulties tend to put rehabilitation assistance behind schedule. Remote physiotherapy accompanied by artificial intelligence (AI) can also prove useful in this case.

AI remote physiotherapy is a safe, home-based and personalized rehabilitation that reduces the burden experienced by immunocompromised patients. These platforms provide high-quality interventions using state-of-the-art AI, such as machine-learning models and real-time cloud data. The AI tools keep track of the movement patterns, provide immediate corrective feedback and modify the treatment plans according to the changing needs of each patient (Rasa 2024). AI is used in combination with human therapists in most instances, and the human therapists are involved in safety and efficacy monitoring, but they have fewer direct contacts.

AI goes beyond fundamental telemedicine in remote physiotherapy. It integrates video conferencing with automatic and interactive applications and smart decision-making. Android or iOS cameras or even simplistic motion-capturing devices such as wristbands or fitness trackers objectively measure the correct performance of exercises by patients (Yang et al. 2024). AI determines the bad form, fatigue, compensations or wrong posture, and it is able to forecast complications such as falls or strains and warn a clinician. This feature is particularly useful when it comes to complicated and delicate oncology cases that require particular attention.

AI oncology systems follow continuous rehabilitation in pre-habilitation, treatment, recovery and survivorship. Physiotherapy is normally interrupted by infections, hospitalization or fatigue caused by treatment in immunocompromised patients. Remote solutions that use AI can overcome these gaps and maintain continuity, compliance and long-term outcomes of care. They also improve the levels of patient engagement through the establishment of individualized objectives and the provision of gamified exercises and real-time feedback – crucial to the long-term motivation (Cascella et al. 2023).

The AI systems have the ability to seal the gaps in cancer care, particularly in the underserved or rural areas where there are no trained physiotherapists. The proliferation of mobile health technology and Internet access allows AI to provide real-time remote physiotherapy in any location, extending access to high-end rehabilitation. Such increased access enhances individual performance, decreases unnecessary hospitalization and readmission and democratizes healthcare.

The prospect of AI in distant physiotherapy is evident, but a few more areas should be mentioned. Among the challenges are privacy issues, fair access to technology, bias in the algorithm and inadequate regulation. It has to be technology that places the patient's well-being as a priority and makes sure that the human aspect of care remains intact (Bekbolatova et al. 2024).

This chapter is a literature review on the history, applications, benefits, challenges and future of AI-enabled remote physiotherapy in cancer patients with weakened immune systems. It demonstrates the ways in which smart technology is transforming cancer rehabilitation, providing safe, reliable, high-quality, and personalized care to some of the most vulnerable people.

12.2. Background: cancer rehabilitation needs and barriers

12.2.1. *Physical sequelae of cancer treatment*

Different types of cancer therapies such as chemotherapy, radiation, surgery and newer immunotherapies may have different physical sequelae consequences that target the functional abilities and the quality of life of a patient. The systemic adverse sequelae of chemotherapy are usually observed during and after the treatment (sarcopenia, fatigue and chemotherapy-induced peripheral neuropathy (CIPN)). CIPN causes feet and hand numbness, tingling and pain, as well as limiting mobility and coordination. Radiation treatment can result in focal tissue fibrosis, stiffness of joints and loss of range of motion, particularly when it is applied to musculoskeletal or thoracic areas. Depending on the extent and location, surgical procedures often lead to post-optical pain, adhesion of the scar or limited mobility (Fatehi Hassanabad et al. 2021). An example is the dysfunction of the shoulder and risks of lymphedema in breast cancer patients who have lymph node dissection.

Long-term cancer survivors are still faced with chronic fatigue, deconditioning, cardiopulmonary fitness loss and balance problems. These outcomes decrease the autonomy of activities of daily living (ADLs) and augment the chances of falls, readmission and even early death. As such, rehabilitation as part of survivorship care becomes the role of physiotherapy. Individualized interventions will help to regain muscle strength, increase cardiovascular capacity, reduce pain, improve flexibility and balance, and support the overall well-being. Early and regular rehabilitation would significantly enhance the quality of life, ease the reintegration to work and social roles and minimize the impact on caregivers and healthcare systems.

12.2.2. *Immunocompromised and access barriers*

Immunosuppression is another significant problem of cancer patients, and particularly active treatment patients. Treatments including chemotherapy, stem cell transplants and some targeted therapies lower the number of white blood cells and therefore leave patients open to dangerous infections, which means patients cannot go to physiotherapy clinics or hospitals for treatment because they could be exposed to infections in the environment and the clinics.

Even without the risk of infections, the sheer distance patients must travel for physiotherapy is exhausting. This is particularly true for patients who have fatigue, pain and limits in mobility, not to mention the patients who live in rural or underserved regions where it is particularly difficult to access specialized oncology rehabilitation services that are crucially important to mitigate treatment side effects. These access challenges often lead to a decrease in rehabilitation participation when it is most needed (Stinear et al. 2020). Addressing the challenges above to ensure safe and timely physiotherapy care for these patients is of utmost importance.

12.3. Telerehabilitation in oncology

12.3.1. *Evidence of feasibility and efficacy*

Research in the field has increasingly recognized the potential and effectiveness of telehealth home-based rehabilitation for cancer survivors. Several remote rehabilitation initiatives have yielded significant results, including, but not limited to, physical capacity, fatigue, pain and psychological wellness. Such programs are usually virtual meetings with physiotherapists, recommended exercises and a mix of mobile apps and wearables. Indeed, when live video monitoring and coaching incomitance with the activity trackers were introduced in the program, adherence and participation increased even more. Patients, even those who were surgically recovering or still receiving treatment, reported significant improvements in quality of life and other health metrics such as cardiopulmonary endurance and muscle strength. In addition, the absence of health risks associated with these remote programs has been documented for people of varied frailty (Li and Zhang 2025). As such, remote rehabilitation programs are an excellent option for patients with weakened immune systems and for those who cannot risk attending in-person rehabilitation classes.

12.3.2. *Limitations of conventional telerehabilitation*

When it comes to virtual rehabilitation, there are obstacles to overcome in order to provide successful and sustainable results, even when the projected outcomes are positive. In traditional care service models, there is an over-reliance on scheduled video consultations, which are attended by recordings of patients and data from wearables. As a result, the real-time engagement from practitioners is absent concerning feedback on the patient's exercise performance in terms of technique, posture, angles and other critical features. Moreover, there is a substantial risk of injury if exercises are executed poorly.

Most virtual rehabilitation systems also fail to integrate real-time responsiveness and personalization to evolving rehabilitation plans through automated machine learning approaches. Instead, the systems focus on retrospective logging, which is inefficient and serves to diminish responsiveness. The overall outcome of such systems is the absence of individual virtual rehabilitation, not to mention a sustainable, responsive system, which is a much bigger problem for cancer patients (Nuara et al. 2022).

12.4. Enabling technologies in AI-driven remote physiotherapy

12.4.1. *Computer vision and motion tracking*

The contemporary AI-based physiotherapy involves computer vision to monitor patient movements with their daily gadgets such as smartphones, pads, laptops and webcams. Such systems use pose estimation to study skeletal architecture and measure joint angles and limb positions that enable a detailed evaluation of posture and gait in exercises. The system will immediately alert the patient if the system notices some form of misalignment such as bending of the knee during a squat or uneven motions of the arm (Nataraj et al. 2023). These feedback systems maximize rehabilitation and risk of injury reduction. Above all, motion analysis in real time presents the option of tracking the progress, which enables patients and clinicians to determine the degree of rehabilitation and pace objectively.

12.4.2. *Machine learning (ML) for personalization*

Machine learning (ML) technology is the technology that generates physiotherapy plans that are most effective with the individual cancer patients. ML models examine various sources of information, demographics, clinical history, exercise diaries, compliance and subjective symptoms. From this, the system determines the optimal parameters for exercises: what the intensity, frequency, duration and type of exercises will be. For instance, if a patient is exhibiting fatigue and signs of stress to the heart, the system will reduce the length of the session and lower the degree of difficulty of the exercises to be performed. Moreover, predictive models will identify patients who are likely to be non-compliant and recovering slowly, allowing timely clinical actions (Giddings et al. 2024). This provides the system the opportunity to learn more and become more accurate, thereby personalizing dynamic advances in rehabilitation to patient expectations and clinical safe flexibility.

12.4.3. *AI-driven virtual assistants and coaching*

To improve user involvement and stay in touch with patients, some systems build AI virtual assistants whose role is to converse with patients while they are on the active rehabilitation route. These virtual assistants provide educational snippets about cancer rehabilitation, answer rehabilitation-related questions and remind patients about upcoming sessions. They are supportive, rewarding and redress the personal issues. These virtual assistants provide companionship to patients who might be isolated or have less contact because of the immunocompromised rehabilitation phase. They are particularly helpful when combined with ML and can be used in customizing communication and therapeutic messages to achieve a particular therapeutic objective (Virvou 2023).

12.4.4. *Remote patient monitoring and multimodal data integration*

Intelligent AI platforms are able to monitor exercises in addition to incorporating wearables, mobile applications and self-reporting devices. In one of the sessions, patients will be able to document pain, weakness, nausea and mood changes, and wearables will supply us with real-time data on heart rate, respiratory rate, oxygen saturation, sleep and activity. As AI algorithms process this mixed data, it becomes possible to detect essential changes and monitor recovery and to convert personalized programs, which are dynamically adjusted in the context of analysis. To illustrate, an AI can reduce the intensity of the next workout or notify a healthcare professional in case the real-time data indicates tiredness and irregular heart rate (Prieto-Avalos et al. 2022).

12.4.5. *Federated learning and privacy-preserving AI*

Due to the sensitive nature of oncology as well as rehabilitation, there is even more reason for the use of federated learning to improve privacy and performance. In federated learning, there is no need to send raw data to a central server, and instead, it is trained on decentralized data sources such as hospitals, clinics and even home devices. Only updates to the model and training results, which do not contain identifiable patient information, are sent to the central server and aggregated (Nguyen et al. 2022). This enables the construction of accurate, generalizable AI models, which is a critical need in the oncology AI space. The privacy of patients is also respected as data "hiding" such as differential privacy or computation cloaking is used. This provides a final layer of encryption to the data and is fully compliant with laws, regulations and ethical requirements in healthcare.

12.5. Commercial platforms and use cases

Remote physiotherapy systems using AI technologies provide various digital rehabilitation options to patients. Every platform will have its interface and specific features, yet the majority of them will have major similarities. They can apply motion-tracking technology to evaluate movement in real time, have virtual coaches to help students in their exercises and have dynamic and customized programs to update patient progress. Many of them also provide clinicians with dashboards containing performance-related data and vital monitoring at a distance. These devices make it possible to provide continuous and personalized care out of the clinic. They have been popular musculoskeletal and neurological rehabilitation tools; however, oncology has high potential. Such systems can be used to provide immunocompromised cancer patients with no access to face-to-face treatment with fatigue-reduction protocols, personalized workouts and additional cancer-specific safety measures to improve accessibility and safety.

Platform/technology	Core AI features	Clinical function/application	Key benefits for immunocompromised patients
Kaia health	Computer vision for posture correction and ML-based exercise personalization	Musculoskeletal and post-cancer rehabilitation	Contactless motion analysis; automated feedback; improved adherence

PhysiApp/Physitrack	Predictive analytics for activity tracking and adherence prediction	Virtual physiotherapy via mobile interface	Enables safe home exercise and integrates therapist supervision
ePhysio platform	Wearable-integrated AI analytics with motion sensors	Remote management of physical rehabilitation	Real-time tracking of limb motion and fatigue detection
RehabGuru AI	ML algorithms for progress monitoring and program adaptation	Cancer recovery and chronic care management	Personalized exercise intensity adjustment; scalable for remote use
AI-powered virtual assistants (e.g. Alexa health skills)	NLP-based conversational support and emotional engagement	Home-based coaching and reminder system	Reduces isolation and enhances motivation, supports therapy continuity
Federated learning frameworks (e.g. NVIDIA Clara FL)	Privacy-preserving distributed model training	Secure data collaboration across healthcare centers	Protects sensitive oncology data while improving model accuracy
Computer vision-based mobile apps (e.g. Move AI, Kaia motion coach)	Pose estimation and skeletal mapping using deep learning	Real-time exercise assessment	Contactless form correction and infection risk reduction

Table 12.1. *Overview of AI-driven remote physiotherapy platforms and technologies for immunocompromised cancer patients*

12.6. Benefits for immunocompromised cancer patients

12.6.1. *Clinical safety and reduced infection risk*

In severely weakened cancer patients with immunocompromised conditions (under chemotherapy, stem cell transplants or immunotherapy), it is crucial to minimize contact with germs (Cutino-Moguel et al. 2017). Remote physiotherapy powered by AI allows patients to recover completely at home, without having to visit clinics, which are also possible risk factors of infection. Such platforms also measure vital signs, such as heart rate, oxygen saturation, temperature in real time and symptoms. They will be able to identify the first signs of complications such as neutropenic fever, lymphedema outbreak or cardiopulmonary distress. Clinicians receive automatic alerts to enable swift response, even without visiting the patient.

12.6.2. *Personalized rehabilitation and feedback*

The current rehabilitation technology is based on AI, computer vision and ML, which provides personalized therapy. The rehab device has a camera that records the movement, including joint angles, posture, balance and range of motion. Based on that data, together with the functional level of the patient, their progress, and symptoms they report, ML modifies exercises in real-time (Fang et al. 2024). As an illustration, in the case where the patient reports that they experience joint pains, the system reduces the intensity level, alters the duration or replaces the exercises. Such an actual-time modification mitigates the risk of injuries, ensures safety and maintains patients on clinical standards.

12.6.3. *Improved access and adherence*

Remote physiotherapy based on AI allows patients to exercise at their own pace, not necessarily traveling to do so, and not necessarily feeling tired after it, which can be highly valuable in the case of cancer patients who have to fight cancer fatigue or low mobility. Motivational AI, virtual coaching, gamified features and progress images are also used on these platforms to capture the attention of users. They can increase the level of adherence and assist in entrenching self-care into daily life by making rehab an engaging and fun routine (Ambros-Antemate et al. 2023).

12.6.4. *Data-driven outcome tracking*

The track record measures the distance of movement, counts, exercise precision, heart rate, and breathing rate in real time. Such data can be used to gauge progress and to develop clinical plans. Dashboards also enable therapists and oncologists to evaluate the rehabilitation effect, plateau or decrease and modify care (Mlakar et al. 2021). In the long run, longitudinal data can demonstrate the reaction of patients to various treatments and justify individual therapy and evidence-based choices.

12.6.5. *Scalability and therapist leveraging*

Physiotherapy platforms enhanced by AI can be used with large populations and can address both the issue of workforce shortages and geographical constraints.

Assessment, feedback and data collection can be automated to allow a single therapist to monitor and serve large numbers of patients simultaneously. This efficiency comes particularly in resource-deficient or rural regions where the oncology rehab is not particularly scarce. The ability to extend expert care through virtual platforms not only improves equity in healthcare access but also reduces the overall cost burden on both healthcare systems and patients (Lakshminarayanan et al. 2023).

12.7. Challenges and risks

AI-driven remote physiotherapy platforms are highly promising for the enhancement of cancer rehabilitation. That said, however, numerous highly important issues and ethical questions are still likely to arise regarding the safe, fair and efficient use of such platforms.

12.7.1. *Technology access and digital literacy*

In the case of AI-enabled remote physiotherapy technology, access to the Internet or the technology needed to connect poses a challenge to adoption. Some patients, especially those who live in remote, less populated or poor regions of a country, do not have reliable, affordable or any Internet connection. They may also lack the necessary smartphones, tablets or computers. In addition, a considerable portion of the oncology demographic is older and may have less technological nimbleness with poor vision or manual dexterity that impedes navigation of the technology (Leung et al. 2025). To lessen these challenges, systems should have a minimum of a simplified user interface that includes voice commands and navigations along with prominent instructions or guides. In addition, the provision of basic digital training to patients along with caregivers and readily available technical support would promote engagement.

12.7.2. *Data privacy and security*

With AI physiotherapy systems, there are concerns related to privacy, confidentiality and cybersecurity when processing sensitive health data. These platforms are not only storing health data but also video recordings, biometrics and exercise logs. Health data require encryption, role-based access controls and Health Insurance Portability and Accountability Act (HIPAA), General Data Protection Regulation (GDPR) or equivalent local laws compliance to ensure health data are stored and transmitted securely. Individuals whose health data are mishandled are probably harmed, and the public's trust in digital health technologies will be negatively impacted. In addition to the confidentiality and integrity of the health data, platforms should also have transparency data policies, secure authentication and access, and regular audits (Kumar et al. 2026a).

12.7.3. *Algorithm transparency and trust*

The AI systems should be transparent to enable patients and clinicians to trust them. Users must know why the system is suggesting something when these platforms modify the exercise protocols or issue clinical alerts. Explainable artificial intelligence (XAI) systems can provide understandable explanations for every decision. As an illustration, when the AI reduces the intensity of exercise due to the unusual heart-rate patterns, it must give a reason in simple language (Kumar et al. 2026b). This type of transparency is crucial in joint decision-making, as it is needed when clinicians use AI outputs to guide the intricate decision-making of cancer survivors with comorbidities.

12.7.4. *Clinical appropriateness and oversight*

Even with the development of AI, the systems will not be able to substitute the clinical discretion of physiotherapists or oncologists. Remote physiotherapy platforms are not supposed to be independent solutions but rather supplementary ones. Human monitoring is necessary, especially with the unexpected symptoms, medical crises or complicated rehabilitation requirements (Vallati et al. 2018). AI can fail to recognize minor indications of deteriorating conditions, such as lymphedema or heart complications. Thus, teleconsultation, electronic health record (EHR) integration and physician alerts should be regular to improve clinical appropriateness and patient safety.

12.7.5. *Equity, bias and cultural adaptation*

The quality of the data that AI models are trained on is all that matters. Training data should not be homogenous in terms of age, gender, ethnicity or socioeconomic status because a given algorithm may recreate healthcare disparities. Underrepresented groups can be harmed or even rendered ineffective. The developers have to deploy inclusive datasets and bias testing among the demographics. It should also be culturally sensitive and be provided in various languages to serve various groups of patients. Individualizing instructions and interfaces to cultural practices, languages and health beliefs increases interest and minimizes inequalities.

12.8. Ethical, regulatory and implementation considerations

12.8.1. *Informed consent and autonomy*

Informed consent is essential in the implementation of AI-based physiotherapy of immunocompromised cancer patients. The issue of AI in the rehabilitation of patients should be properly explained to them in terms of how it monitors the movements, provides suggestions or modifies exercises. Openness on the type of data gathered, its storage and its intended use fosters confidence and advocates self-reliance (Krishna et al. 2024). The patients must have an option to enable or disable certain functions, such as sharing their data or AI decision support. There is also the need to ethically practice the implementation of AI by explaining the boundaries of the AI to the patients by making them aware that AI is a helping device, and clinical supervision is the primary focus.

12.8.2. *Regulatory compliance*

The AI-enhanced physiotherapy platforms should also be in line with the medical device laws and data protection laws in every region where they will be used. The AI software in the healthcare field can be regulated as a medical device in most jurisdictions, and it must comply with the safety and performance requirements provided by such authorities as the Food and Drug Administration (FDA) (U.S.), European Medicines Agency (EMA) (Europe) or Central Drugs Standard Control Organisation (CDSCO) (India). Social media platforms should exhibit both clinical effectiveness and real-world safety in addition to being open in their algorithm design. The adherence to data protection standards, including HIPAA or GDPR, is the key to protecting patient privacy and security (Shafik 2025).

12.8.3. *Interoperability and integration*

In order to enable the concept of holistic, patient-centered care, AI physiotherapy systems should be seamlessly integrated with oncology platforms, such as EHRs, teleconsultation systems and multidisciplinary care plans. Interoperability facilitates sharing of information between oncologists, physiotherapists and primary care providers such that they can provide coordinated care (Sargiotis 2024). Integration can be achieved by the use of standardized communication protocols and health data formats (e.g. Health Level Seven International (HL7), Fast Healthcare Interoperability Resource (FHIR)) and the risk of information silos that may undermine the quality and continuity of care can be reduced.

12.8.4. *Pricing and reimbursement model*

Reimbursement frameworks are extremely important in the long-term sustainability and fair access to AI-based remote physiotherapy. The insurance companies and national health systems should be able to see the worth of the virtual AI-enhanced rehabilitation and add relevant coverage codes. The preventive and cost-saving opportunities of these platforms should be reflected in payment models by setting them so that they minimize the number of hospital visits, complications and long-term disability (Osterman et al. 2020). Clear policies and codes of reimbursement are essential in order to create an incentive for the providers and promote universal adoption.

12.9. Future directions and research agenda

12.9.1. *Clinical trials comparison of AI-augmented and traditional physiotherapy*

It is necessary to conduct randomized controlled trials that can demonstrate that AI-driven physiotherapy is effective. The research works should compare AI-enhanced remote programs with traditional face-to-face physiotherapy among immunocompromised cancer patients (Mbata et al. 2024). Such outcomes as functional recovery, fatigue mitigation, quality of life, adherence rates and safety are key results to measure. The outcomes will justify the approval of regulation and promote clinical uptake.

12.9.2. *Explainable AI to transparent decision-making*

Mobile AI systems must develop confidence with patients and caregivers by being explainable. The system is supposed to explain the reasons for why it recommends one exercise, why it uses a given intensity and the reason it suggests the next steps. This openness gives clinicians more power and less mistrust of automation, as well as helps make ethical and informed decisions.

12.9.3. *Federated learning for advanced secured institutional cooperation*

While the performance of AI improves with diversity in datasets, the lack of data sharing due to privacy concerns complicates this. Federated learning solves this by performing decentralized model learning with multiple arms and no transfer of sensitive patient data (Ravi et al. 2025). This protects privacy while still garnering the benefits of learning from varied patient datasets on a considerable scale.

12.9.4. *Multimodal data integration for personalized AI-enhanced rehabilitation*

Future AI should consider and use rehabilitation genomic data, biometric data such as heart rate, oxygen saturation and patient-reported outcomes. This will help the AI assess the data streams and refine personalized rehabilitation regularly to a patient's neural rehabilitation, response to the current active treatment and psychosocial factors. Greater explanations enhance safety and efficacy of rehabilitation (Medenica et al. 2024).

12.9.5. *Cultural and linguistic adaptation to acceptable use*

To be used widely, AI platforms should be able to fit various languages, literacy and cultures. The interface must be easy to use, customizable and localized with local imagery and voice inputs. This minimizes the numerous feedback processes and makes the system user-friendly.

12.9.6. *Long-term evaluation of benefits and cost effects*

Short-term profitability is just one side of the coin. It requires long-term research to evaluate functional improvements, persistent therapy adherence and the cost-effectiveness in general. They must evaluate healthcare utilization, readmission rates and savings as opposed to regular care (Gudur et al. 2023).

12.9.7. *Ethics and human control*

AI in cancer physiotherapy has to be authorized by easily comprehensible moral standards. They are equity, mitigation of bias, informed consent, accountability of algorithms and human control. These kinds of protectors ensure fairness and safety.

12.10. Conclusion

Remote physiotherapy based on AI provides additional opportunities for the recovery of immunocompromised cancer patients. These patients have high chances of getting infected, being exhausted and having low mobility. Subsequently, the regular hospital visits become cumbersome and risky. Remote platforms can provide customized and flexible care to the patient at home, which is made possible by AI technology. Advanced computer vision with AI capabilities can monitor the movements and assess the exercises of a patient and provide real-time feedback. Exclusively owned AI algorithms define the exercise amount, alter the plan in the course of rehabilitation and customize exercises to the individual. They put into consideration side-effect mitigation, symptom monitoring and other modifications in the course of treatment. Telehealth functionality is also available on remote platforms: virtual consultations, asynchronous communication with clinicians and live chatbots. These characteristics make patients stimulated and motivated. The telehealth AI improves rehabilitation. Predictive algorithms identify complications in time and permit planned interventions in the event of non-adherence. New platforms embrace the best privacy, encrypted privacy and legalities technology, preserving sensitive health data. AI systems are also monitored by clinicians so as to be safe, yet flexible, easy to use, multilingual and low-bandwidth friendly.

12.11. References

Ambros-Antemate, J.F., Beristain-Colorado, M.D., Vargas-Treviño, M., Gutiérrez-Gutiérrez, J., Hernández-Cruz, P.A., Gallegos-Velasco, I.B., Moreno-Rodríguez, A. (2023). Improving adherence to physical therapy in the development of serious games: Conceptual framework design study. *JMIR Formative Research*, 7, e39838.

Bekbolatova, M., Mayer, J., Ong, C.W., Toma, M. (2024). Transformative potential of AI in healthcare: Definitions, applications, and navigating the ethical landscape and public perspectives. *Healthcare*, 12(2), 125.

Cascella, M., Cascella, A., Monaco, F., Shariff, M.N. (2023). Envisioning gamification in anesthesia, pain management, and critical care: Basic principles, integration of artificial intelligence, and simulation strategies. *Journal of Anesthesia, Analgesia and Critical Care*, 3(1), 33.

Cutino-Moguel, M.T., Eades, C., Rezvani, K., Armstrong-James, D. (2017). Immunotherapy for infectious diseases in haematological immunocompromise. *British Journal of Haematology*, 177(3), 348–356.

Fang, J., Lee, V.C., Ji, H., Wang, H. (2024). Enhancing digital health services: A machine learning approach to personalized exercise goal setting. *Digital Health*, 10, 20552076241233247.

Fatehi Hassanabad, A., Zarzycki, A.N., Jeon, K., Dundas, J.A., Vasanthan, V., Deniset, J.F., Fedak, P.W. (2021). Prevention of post-operative adhesions: A comprehensive review of present and emerging strategies. *Biomolecules*, 11(7), 1027.

Giddings, R., Joseph, A., Callender, T., Janes, S.M., Van der Schaar, M., Sheringham, J., Navani, N. (2024). Factors influencing clinician and patient interaction with machine learning-based risk prediction models: A systematic review. *The Lancet Digital Health*, 6(2), e131–e144.

Gudur, A., Pati, P., Garg, P., Sharma, N. (2023). Radiomics feature selection for lung cancer subtyping and prognosis prediction: A comparative study of ant colony optimization and simulated annealing. *International Journal of Intelligent Systems and Applications in Engineering*, 12(3s), 553–565.

Krishna, S.R., Gudur, A., Jain, S., Deivasigamani, S., Tiwari, M., Venkatesan, K.G. (2024). Deep learning for automatic diagnosis of skin cancer using dermoscopic images. *Journal of Artificial Intelligence and Technology*, 4(2), 114–123.

Kumar, A., Batta, P., Ahuja, S., Rathore, P.S. (eds) (2026a). *AI-driven Innovations in Physiotherapy and Oncology 1*. ISTE Ltd, London and Wiley, Hoboken, NJ.

Kumar, A., Batta, P., Ahuja, S., Rathore, P.S. (eds) (2026b). *AI-driven Innovations in Physiotherapy and Oncology 2*. ISTE Ltd, London and Wiley, Hoboken, NJ.

Lakshminarayanan, V., Ravikumar, A., Sriraman, H., Alla, S., Chattu, V.K. (2023). Health care equity through intelligent edge computing and augmented reality/virtual reality: A systematic review. *Journal of Multidisciplinary Healthcare*, 16, 2839–2859.

Leung, W.K., Yau, C.Y., Lam, S.C. (2025). Facilitators, barriers, and recommendations for mobile health applications among Chinese older populations: A scoping review. *BMC Geriatrics*, 25(1), 396.

Li, P. and Zhang, H.P. (2025). From surgery to recovery: Measuring success through quality of life and functional improvements after cardiac surgery. *World Journal of Cardiology*, 17(2), 100213.

Mbata, A.O., Soyege, O.S., Nwokedi, C.N., Tomoh, B.O., Mustapha, A.Y., Balogun, O.D., Forkuo, A.Y., Iguma, D.R. (2024). Preventative medicine and chronic disease management: Reducing healthcare costs and improving long-term public health. *International Journal of Multidisciplinary Research and Growth Evaluation*, 5(6), 1584–1600.

Medenica, V., Ivanovic, L., Milosevic, N. (2024). Applicability of artificial intelligence in neuropsychological rehabilitation of patients with brain injury. *Applied Neuropsychology: Adult*, 1–28. doi: 10.1080/23279095.2024.2364229.

Mlakar, I., Lin, S., Aleksandraviča, I., Arcimoviča, K., Eglītis, J., Leja, M., Salgado Barreira, A., Gómez, J.G., Salgado, M., Mata, J.G. et al. (2021). Patients-centered Survivorship care plan after cancer treatments based on big data and artificial intelligence technologies (PERSIST): A multicenter study protocol to evaluate efficacy of digital tools supporting cancer survivors. *BMC Medical Informatics and Decision Making*, 21(1), 243.

Nataraj, R., Sanford, S.P., Liu, M. (2023). Joint moment responses to different modes of augmented visual feedback of joint kinematics during two-legged squat training. *Biomechanics*, 3(3), 425–442.

Nguyen, D.C., Pham, Q.V., Pathirana, P.N., Ding, M., Seneviratne, A., Lin, Z., Dobre, O., Hwang, W.J. (2022). Federated learning for smart healthcare: A survey. *ACM Computing Surveys (Csur)*, 55(3), 1–37.

Nuara, A., Fabbri-Destro, M., Scalona, E., Lenzi, S.E., Rizzolatti, G., Avanzini, P. (2022). Telerehabilitation in response to constrained physical distance: An opportunity to rethink neurorehabilitative routines. *Journal of Neurology*, 269(2), 627–638.

Osterman, T.J., Terry, M., Miller, R.S. (2020). Improving cancer data interoperability: The promise of the minimal common oncology data elements (mCODE) initiative. *JCO Clinical Cancer Informatics*, 4, 993–1001.

Prieto-Avalos, G., Cruz-Ramos, N.A., Alor-Hernandez, G., Sánchez-Cervantes, J.L., Rodriguez-Mazahua, L., Guarneros-Nolasco, L.R. (2022). Wearable devices for physical monitoring of heart: A review. *Biosensors*, 12(5), 292.

Rasa, A.R. (2024). Artificial intelligence and its revolutionary role in physical and mental rehabilitation: A review of recent advancements. *BioMed Research International*, 2024(1), 9554590.

Ravi, S., Kumar, A., Rathore, P.S. (2025). Multiscale feature fusion for automated classification of digital breast tomosynthesis volumes using 3D deep learning. *Discover Computing*, 28(1), 300.

Reddy, K.J. (ed.) (2025). Patient perspectives and outcomes. In *Innovations in Neurocognitive Rehabilitation: Harnessing Technology for Effective Therapy*. Springer, Cham, 111–135.

Sargiotis, D. (ed.) (2024). Data security and privacy: Protecting sensitive information. In *Data Governance: A Guide*. Springer, Cham, 217–245.

Shafik, W. (2025). Patient-centered healthcare: Trust and transparency through technology. In *AI and Blockchain Applications for Privacy and Security in Smart Medical Systems*, Ahad, A. and Ullah, F. (eds). IGI Global, Hershey, PA, 309–338.

Stinear, C.M., Lang, C.E., Zeiler, S., Byblow, W.D. (2020). Advances and challenges in stroke rehabilitation. *The Lancet Neurology*, 19(4), 348–360.

Vallati, C., Virdis, A., Gesi, M., Carbonaro, N., Tognetti, A. (2018). ePhysio: A wearables-enabled platform for the remote management of musculoskeletal diseases. *Sensors*, 19(1), 2.

Virvou, M. (2023). Artificial intelligence and user experience in reciprocity: Contributions and state of the art. *Intelligent Decision Technologies*, 17(1), 73–125.

Yang, L., Amin, O., Shihada, B. (2024). Intelligent wearable systems: Opportunities and challenges in health and sports. *ACM Computing Surveys*, 56(7), 1–42.

Zheng, J.Y., Mixon, A.C., McLarney, M.D. (2021). Safety, precautions, and modalities in cancer rehabilitation: An updated review. *Current Physical Medicine and Rehabilitation Reports*, 9(3), 142–153.

13

Machine Learning for Early Detection of Mobility Decline in Oncology Patients

The decline in mobility of oncology patients (usually observed as slowed gait speed, poor balance, poorer strength, decreased physical activity) is a serious aspect that determines the tolerability of the treatment, functional independence and quality of life. The loss of mobility is a critical issue that requires early identification of mobility impairment so that timely interventions may be implemented to prevent falls and decrease the number of hospital admissions and assist in recovery. Older clinical instruments, including the timed up-and-go (TUG) test or the six-minute walk test, are infrequent and resource-consuming. Machine-learning (ML) strategies, in turn, can provide early and personalized mobility decline detection using the multimodal continuous data.

This chapter is a review of the new applications of both supervised and unsupervised ML models, such as support-vector machines, random forests and deep neural networks, in mobility deterioration prediction and monitoring in oncology patients. The range of input data consists of wearable accelerometers and inertial sensor input data, electronic-health-record data, imaging, laboratory and patient-reported outcome measure data. The most important applications investigated are the prediction of chemotherapy-induced decline (e.g. in the taxane or platinum-based regimen), the post-operative functional recovery in the field of surgical oncology, and real-time monitoring during radiation therapy. The technological systems such as the mobile-phone sensors and the telehealth integration make it more feasible and expand the scope of coverage. The key issues include small samples, unbalanced datasets, no standardization in defining outcomes, interpretability of models and data privacy. The future directions will involve the multimodal integration, the personalized models, the edge computing to make real-time inference and the integration of the ML tools into the clinical trial in order to implement the innovations into the usual practice. ML-based early diagnosis has immense potential to revolutionize the mobility monitoring and intervention among oncology patients.

Chapter written by Trupti YADAV and Sanjay BADJATE.

13.1. Introduction

Mobility, which includes gait, balance, strength, physical activity is the basis of patient independence and total health. Mobility is particularly of great significance in oncology, because cancer and its treatments can present significant physical and psychological stress. The mobility loss is a typical and clinically important complication, which may be caused by cancer-related fatigue, chemotherapy-related toxicity, peripheral neuropathy, cachexia, inactivity, pain and depression (Rodin 2022). The progressive deconditioning and musculoskeletal dysfunction are caused by these multifactorial influences.

The implications of mobility decline in oncology patients are considerable. Passing of time, coupled with the consequences of mobility decline, negatively impacts the ability to perform basic activities of daily living (ADLs), considerable loss of mobility stiffness and the decline of basic functions of daily living (Admi et al. 2015). It also increases the risk of falls, fractures and hospitalizations. Patients with reduced mobility frequently require longer rehabilitation periods and are more likely to suffer loss of independence and reduced quality of life. Moreover, loss of mobility is likely to signal broader systemic issues, such as decline in nutritional status, worsening burden of disease or impact of treatment and clinical disease intolerability. Finally, loss of mobility is concerning as it is a sensitive, late indicator of general and systemic decline of health.

Considering the above, the decline of mobility needs to be identified as a priority to allow timely interventions aimed at maintaining functional mobility and purposeful living.

Several of the clinically established direct assessments, such as timed up-and-go (TUG), six-minute walk test (6MWT), and handgrip strength, are already well-known, established and provide adequate clinical insights into the realm of physical functioning and physical performance. However, such assessments will likely be performed infrequently in the clinic; there is a need for trained personnel to perform assessments, and subtle or gradual changes may be missed in the time between encounters (Yorkston et al. 2010). Finally, such assessments are a logistical burden to the patients and healthcare system, potentially explaining the popularity of self-report questionnaires. One such well-known and widely used scale would be the EORTC QLQ-C30 physical function subscale.

Although they provide perspectives on a patient's subjective experience and are straightforward to implement, they are inherently constrained by biases, absence of objective validation and frequently insufficient sensitivity to identify early-stage mobility changes. The availability of data and growing computational power are enabling machine learning (ML) to offer a viable alternative (Rochester et al. 2020).

In contrast to conventional approaches, ML algorithms can handle multidimensional and large volumes of data and reveal nonlinear and sophisticated relationships, which allow sophisticated predictive patterns. In relation to mobility decline, wearable, electronic health record (EHR), imaging, laboratory and patient-reported outcome (PRO) data can be combined to develop early predictive analysis and individualized risk analysis (Wilson and Anwar 2024). Wearables generate close to real-time live feeds which dynamically measure gait speed, number of steps taken, balance, posture and activity. The combination of clinical and demographic data with ML models can predict the future mobility status and alert about the upcoming functional deterioration, which can be irreversible (Dobkin and Martinez 2018).

This chapter describes how modern ML methods can be used to predict mobility decline in cancer patients, including data collection, feature engineering, model construction and testing and the environment. It also identifies the data quality, interpretability, integration into clinical workflow, patient privacy and further research opportunities. It is aimed at informing the design of proactive and intelligent monitoring systems, which allow customized, timely and efficient intervention during oncology care.

13.2. Clinical context: mobility decline in oncology

13.2.1. *Etiology and prevalence*

Compromise in mobility is a frequent issue and of clinical importance to oncology patients. It is an outcome of complicated interaction between disease-related, treatment-induced and comorbidity-driven factors. Many mechanisms are involved in its etiology. Peripheral neuropathy as a side effect of chemotherapy (chemotherapy-induced peripheral neuropathy (CIPN)), particularly of taxanes, platinum compounds and vinca alkaloids is one of the largest causes. These medications interfere with nerve conduction (sensory and motor), leading to balance difficulties, weakness and deranged gait. Myopathy caused by steroids is also another common cause. The condition is normally due to long-term usage of corticosteroids in the treatment of cancer and causes proximal muscle wasting and functional impairments (Tsetlina et al. 2024).

The nerves, muscles and connective tissues can be damaged by radiation therapy, particularly where the radiation is directed to the pelvis, spine or brain. The effects of this damage are fibrosis, contractures and neuromuscular dysfunction. Surgical sequelae, such as amputation of limbs, lymphedema and pain post-surgery are also major contributors to the deterioration of mobility. The size of tumor itself, especially the tumor of the bones or the spine, can result in pain, pathological fracture or spinal cord compression, which only worsens mobility problems. Even in the absence of overt motor deficit, treatment-associated fatigue – a multifactorial, widespread symptom – has the potential to reduce the potential to engage in physical activity and functional endurance.

Mobility impairment is dependent upon the type of cancer, stage and the modality of treatment. It has been estimated that 40 to 60% of the patients having neurotoxic chemotherapy treatment suffer a degree of mobility disability. A high incidence of decreased physical functioning is also indicated in post-surgical patients, particularly those who have had orthopedic surgery, neurosurgical surgery and in patients who have had pelvic or cranial radiation (Thackeray and Miller 2019). It is also important to note that the decline in mobility can be short-lived in certain patients and permanent in others – or it can be progressive.

13.2.2. *Impact on outcomes*

The clinical implications of impaired mobility of cancer patients are far-reaching. Lack of mobility is directly associated with a higher risk of falls, as these conditions are a major determinant of injury-causing morbidity and mortality in this group of patients. Fractures (including hip fractures), especially in elderly patients, are major causes of functional loss and institutionalization. The inability to perform ADLs when it comes to dressing, bathing or walking dramatically affects the quality of life and overloads the caregiver.

A decline in mobility is also associated with increased hospital readmissions, increased inpatient stay and increased healthcare utilization. In gait speed, a simple and strong biomarker has been proven to predict adverse outcomes in surgical oncology. Precisely, a slower gait speed of less than 0.8 m/s is linked to an increased incidence of postoperative complications, slow recovery and overall survival in different kinds of malignancies such as colorectal cancer, lung cancer and gynecologic cancer (Latajka et al. 2023). Mobility decline results in physical inactivity and sedentary behavior, which are directly linked to cancer relapse, metabolic complications and all causes of mortality.

13.2.3. *Current assessment modalities*

The conventional method of measuring mobility in oncology practices is based on in-clinic functional performance tests:

– TUG test;

– gait speed over a standardized distance test;

– 6MWT;

– sit-to-stand test;

– balance tests, both the static and dynamic.

These balance tests are episodic, resource intensive and fail to detect early subtle decline, despite their validity and clinical usefulness. They are commonly done in the clinic, and therefore, they may not be representative of functional capacity or daily mobility changes.

Alongside, patient-reported outcome measures (PROMs) such as the PROMIS Physical Function scale can provide subjective information on a perceived physical ability of a patient. Nonetheless, recall bias, variable responses and ceiling/floor effects constrain the use of PROMs in the highly functioning or individuals with severe impairments, in particular (Stolwyk et al. 2006). They are also subject to mood, cognition and social context, which could diminish their effectiveness as independent mobility measures.

With such restrictions, remote, continuous, objective measures of mobility to be captured by means of wearable sensors, smart devices and digital health platforms are increasingly becoming of interest. A combination of these technologies and ML methods is able to identify subtle changes, predict earlier functional decline and make interventions more personalized than traditional methods. This paradigm shift pertains to a bright future in supportive oncology and geriatric oncology.

13.3. ML approaches

13.3.1. *ML paradigms*

ML would assist in identifying a deterioration in early mobility in cancer patients. There are various computational methods that are applicable to different types and aims of data.

– The most common is the *supervised learning*. It is predictive using a model that is trained on labeled data. With a decline in mobility, there are results that pass clinical thresholds, including a decrease in gait-speed by 10% or longer inactivity or transition to assisted walking. Support-vector machines (SVMs) and decision trees, logistic regression, random forests, gradient-boosting machines (e.g. XGBoost, LightGBM, etc.) and deep neural networks are all found to be effective in a variety of healthcare prediction tasks (Rufo et al. 2021). Quality and fineness of labeled data are essential to supervised learning and can be provided by clinical tests, wearable gadgets or validated scales.

– *Unsupervised learning* learns not only to find concealed patterns when these labels are not pre-defined. The methods of clustering (k-means, density-based spatial clustering of applications with noise (DBSCAN), hierarchical clustering, etc.) cluster patients together by similar sensor-based pattern of movement or clinical phenotype, which could be aggregating sub-populations at risk. The anomaly-detection algorithm such as Isolation Forests or autoencoders identifies evidence of deviation from the normality of activity based on a patient and indicate early signs that have yet to reach diagnostic levels.

– *Deep learning* and in particular convolutional neural networks (CNNs) and recurrent neural networks (RNNs) with long short-term memory (LSTM) cells are most effective with high-dimensional sequential data, such as wearable data. Such architectures can automatically extract complex temporal relationships and hierarchical characteristics of raw signals (e.g. accelerator data) without much feature engineering. CNNs are used on short-time Fourier transforms or spectrogram-based representations of motion data, combined with RNNs and transformer models for continuous, long-term monitoring and analysis (Alqudah and Moussavi 2025).

13.3.2. *Data modalities*

The effectiveness of ML in this area depends on the ability to combine a variety of multimodal sources of data representing various components of patient functioning and health condition:

– Wearables (e.g. accelerometers, gyroscopes and inertial measurement units) can supply objective and high-resolution evidence about the dynamics of gait, postural responses and general activity. These are devices that record such measures as number of steps, gait speed, stride frequency and balance measures in a real-life context.

– Passive monitoring can be provided using smartphones that have integrated sensors. The number of steps, time spent on the screen and movement patterns of a patient based on GPS can be used as proxies of patient activity and can be collected remotely and without being obtrusive to the patient (Kulkarni et al. 2022).

– EHRs and administrative data will provide valuable contextual information, such as demographics, treatment history (e.g. chemotherapy type, dose), comorbidities, laboratory findings and clinical occurrences. The combination of these variables can be used to tailor risk models depending on the medical history of a patient.

– PROs, including validated questionnaires on fatigue (e.g. FACIT-F), physical function (e.g. PROMIS PF) and pain severity, offer subjective but clinically relevant insights into how patients perceive and experience their functional limitations (Nordin et al. 2016).

– Natural language processing (NLP) may be used to mine imaging and clinical notes, the former of which has not been well used, and descriptors of functional decline (e.g. uses walker, unsteady gait, etc.) can be extracted from unstructured text or radiological findings, suggesting musculoskeletal impairment.

13.3.3. *Feature engineering*

Predictive modeling involves the transformation of raw sensor and clinical data into meaningful features. The high interpretability and performance are achieved due to the good engineering of features:

– Gait parameters, including stride length, cadence, stride variability and stride symmetry, are early indicators of neuromuscular coordination and primary indicators of functional impairment, especially in peripheral neuropathy as a result of chemotherapy.

– The measures of activity, such as total number of steps taken per day, time and frequency of sedentary periods, and movement fragmentation (spread of movement throughout the day), are proxy measures of total physical activity and stamina.

– Balance indicators, such as postural sway or metrics from sit-to-stand transitions, offer insights into fall risk and proprioceptive deficits.

– Physiological indicators based on laboratory results (e.g. hemoglobin, C-reactive protein, serum albumin) may indicate systemic degradation, inflammation or malnutrition, which frequently predetermines the degradation of functional capacity.

– Temporal dynamics: rolling averages, daily or weekly trends and unexpected variability of baseline identify subtle longitudinal changes that forewarn of clinically evident decline.

The fusion of these diverse features across time, sensor input and clinical data significantly strengthens the sensitivity and specificity of models to detect early mobility decline, allowing for proactive intervention and personalized rehabilitation strategies (Lobo et al. 2024).

13.4. Predictive models for mobility decline

13.4.1. *Model types and performance metrics*

Early symptoms of mobility loss in oncology patients are detected with the help of different ML models. The models have distinct benefits, which are related to the data structure, sample size, and clinical application. SVMs have shown good performance in high-dimensional space and can be applied where there is a small sample size, such as in clinical studies. They create the best boundaries of decision between classes (e.g. decline vs. no decline) and could change kernel functions to represent the nonlinear relationships (Du et al. 2024).

GBMs and random forests are ensemble techniques that are robust and interpretable. They not only capture the complex interactions between variables, but they also are very robust to noise and missing data, as is typical of clinical data. Random forests limit overfitting through the combination of many trees, and GBMs follow a balanced pattern of tree-growing, correcting the errors made by the previous trees and usually providing a high level of predictive power. Neural networks (with CNNs or LSTM networks) are suitable for time-series data, such as measurements from accelerators, pedometers and smartphone apps.

CNNs are highly effective with respect to the identification of spatial or temporal patterns. The LSTM networks process time-dependent data of a long-term nature. These tools together predict the mobility decay due to the breaking down of the changing mobility trends, and customized measures guarantee clinical significance.

Generally, classification metrics that are commonly used to summarize model performance are overall accuracy, recall (sensitivity) and specificity. In the case of an imbalanced dataset, AUC-ROC and PR curves prove to be invaluable in case we are interested in rare events, such as falls and severe declines (Naidu et al. 2023). In the case of time-to-event or longitudinal data, we estimate time-varying risk, such as through survival analysis tools, such as Cox regression extensions or deep survival models.

13.4.2. *Modeling pipelines*

The standard procedure begins with the baseline data, including demographics, clinical data and digital health measurements, at the beginning of the chemotherapy or some other therapeutic period. The wearables or EHRs are later used to derive mobility features, including the gait speed, activity patterns, balance metrics and the level of physical activity.

Outcomes of interest (such as a clinically significant decrease of mobility or a fall in a specified time period (four weeks)) are associated with features. Data is divided into training, validation and test sets, and commonly cross-validation is used to increase the robustness of data (Yates et al. 2023). When the model is being trained, the hyperparameters and the models are adjusted to reduce overfitting. The most effective model is then tested against unknown test data, and risk scores can be scaled to have more clinical interpretability.

13.4.3. *Representative studies (hypothetical examples)*

Good examples include case studies of the wide variety of approaches:

– The use of random forest model on accelerator data and lab biomarkers predicted an 82.04 drop in gait speed of over 15%.

– In a study by an LSTM model fitted on smartphone sequences of steps, there was an increased fall risk prediction two weeks before normal clinical examination.

– An SVM model based on PRO and dose-intensity of chemotherapy was found to predict, with an accuracy of 80%, post-surgical mobility complications.

Model type	Data sources	Key features/inputs	Outcome predicted	Performance metrics (reported in literature)
Support vector machine (SVM)	Wearable accelerometers; clinical assessments	Gait speed; step variability; postural sway	Binary classification: "decline" vs. "no decline"	Accuracy: 78–82%; AUC: 0.80
Random forest (RF)	EHR + lab biomarkers + activity sensors	Step count trends; CRP; hemoglobin; albumin	15% gait speed decline risk	AUC: 0.82
Gradient boosting machines (GBMs)	Wearables + PROs	Activity fragmentation; daily fatigue score	Fall risk or reduced ADL performance	Precision: 0.81; recall: 0.79
LSTM networks	Smartphone time-series step data	Sequential step counts; rest-to-active transitions	Early fall warning/movement deterioration	Sensitivity: 85%; specificity: 83%
CNNs	IMU signal spectrograms	Frequency patterns; stride symmetry	Neuropathy-induced gait alteration	AUC: 0.84
Clustering (unsupervised)	Wearables + EHR	Baseline mobility clusters	Identification of "at-risk" groups	N/A (unsupervised)

Table 13.1. *Machine learning models applied for early detection of mobility decline in oncology patients*

These are oncology mobility monitoring examples as diverse in model architecture and clinical use as the current broad spectrum of model architectures and clinical applications being investigated. The ML models have been provided in Table 13.1.

13.5. Applications in oncology

13.5.1. *Chemotherapy-induced decline*

Taxanes, platinums and vinca alkaloids are neurotoxic chemotherapy agents that are famous in the occurrence of CIPN. This condition may insidiously and gradually affect gait, balance and fine motor control, which eventually may manifest itself in a clinical picture. These early signs may not be detected by traditional assessments, particularly in outpatient care. The potential solution is ML models on continuous data collected by wearable sensors. Using the alterations in gait parameters, such as stride variability, stance time or gait asymmetry, ML algorithms can detect early, subclinical signs of CIPN (Marshall et al. 2016). Early diagnosis can result in interventions such as dose modification, prompt development of physical or occupational therapy or the adoption of neuroprotective measures, which may prevent long-time disability.

13.5.2. *The patient reports no pain or discomfort in their post-surgical period*

There are no limitations reported in the patient's functional activities (including sexual and social activities). Immobilization, pain, and systemic stress reactions result in a high risk of patients with major oncologic surgeries (thoracic surgery, debulking abdominal tumor, limb-sparing orthopedic surgeries, etc.) developing rapid functional decline. ML-based remote monitoring may process the wearable sensor data, such as the number of steps, changes in cadence or fluctuations in daily pattern of activities, to detect patients who are not improving according to planned curves. The deviation in mobility patterns can be identified within several days after the operation and early referral to the rehabilitation facilities or individual physical treatment plans can be suggested, which may positively impact the long-term recovery process and decrease the likelihood of readmission (Kumar et al. 2026a).

13.5.3. *Radiation and combined modality therapy*

The disease can be treated by use of radiation therapy and combined modality therapy (Sun 2009). A late musculoskeletal weakness, fibrosis and neurologic impairment may occur in radiation therapy due to radiation of the spine, pelvis or brain (Kumar et al. 2026b). All of these effects might be enhanced when used together with chemotherapy. ML models that are trained on multimodal inputs (e.g. wearable mobility measurements, symptom measurements (e.g. fatigue, pain, etc.) and biological measurements such as inflammatory measure) can be used to identify the patients that are likely to experience functional decline. The system of early warning by these predictions is able to implement proactive management, including supportive therapies and interventions that maintain mobility.

13.5.4. *Hematologic malignancies and bone marrow transplantation*

Hematologic cancer patients who have received bone marrow transplantation or who are receiving intensive chemotherapy have special mobility problems because of prolonged inpatient therapy, isolation, neuropathy and catabolic stress. The ML models capable of combining various datasets, including in-room movement sensors, wearable accelerators and PROs on a daily basis, have the power to respond to deconditioning in real time. These models are able to warn the care teams about the impending subtle deterioration and therefore intervene in the early stages with the help of physical therapy, nutrition or psychosocial resources; and eventually have positive effects on the quality of life and limiting the complications (Tappenden et al. 2013).

13.6. Technological platforms

13.6.1. *Wearable sensors*

The wearable sensors, including commercial-grade accelerators (e.g. ActiGraph), inertial measurement units (IMUs) and special research-grade motion trackers, are necessary in the process of recording continuous, objective mobility information. These devices are becoming smaller, efficient in their use of energy and comfortable, enabling them to be used on a long-term outpatient basis without causing any inconvenience in the daily routine. Their ability to detect and measure gait speed, number of steps, transition of posture and other functional parameters that are essential in the early indication of mobility loss among oncology patients, is also possible.

13.6.2. *Smartphones and consumer devices*

Access is increased by smartphones and consumer-grade wearables (e.g. smartwatches, fitness bands). Being cheaply available, passive data collection methods through embedded accelerometers and gyroscopes mean most patients already carry such devices, making them a low-cost method of collecting data (Ciuti et al. 2015). These tools have the ability to monitor walking patterns, activity levels and behavioral rhythms, which can vary inconspicuously with health deterioration, with user consent.

13.6.3. *Telehealth integration*

The interface with telehealth enables remote monitoring where sensor data is sent to cloud-based systems to be analyzed via the use of ML. These insights fill clinician dashboards, generate an alert to facilitate early intervention and make timely referrals to physical therapy or rehabilitation by using telemedicine.

13.6.4. *Edge versus cloud analytics*

The processing can be done on the edge (analytical processing on the device) or on the cloud. Edge models are more focused on privacy, low-latency and offline capability, whereas cloud-based models can be used to perform more sophisticated analytics on larger datasets using deep learning architectures and centralized data integration (Joshi et al. 2023).

13.7. Challenges and limitations

13.7.1. *Data quality and heterogeneity*

The sensor data may vary due to inconsistency in the location of devices, patients and signal noise (Gopalakrisnan et al. 2024). Heterogeneity is further compounded by the variations in sensor hardware, firmware and data resolution among platforms, causing further problems in generalization by models.

13.7.2. *Sample size and labeling*

Generally, oncology-related data is small, which increases the risk of overfitting and restricts the robustness of the model. Also, no standard definition of mobility decline has been used across studies; the boundaries and outcomes definitions are different, and such a problem makes it difficult to label them consistently.

13.7.3. *Model interpretability*

Deep learning models are good, but they tend to be black boxes, hence restricting interpretability. Trust and understanding are essential in a clinical setting. A technique of explainable AI can enhance transparency and assist clinicians in interacting with the model, such as SHAP values, attention maps or scores of feature importance (Tucker et al. 2015).

13.7.4. *Clinical integration*

ML results need to seamlessly integrate into clinical practice to prevent disturbance. The insights cannot be timely and user-friendly and that is why they can be neglected. Clinicians are not convinced without solid evidence of how the insights of ML can enhance decision-making and patient outcomes.

13.7.5. *Ethical, privacy and regulatory concerns*

Constant patient monitoring has significant issues of informed consent, data privacy and autonomy. ML tools can be controlled as medical equipment (e.g. the FDA or CE), and they should be strictly validated. There is also the issue of equity with respect to training models with non-representative population, where the model may have a bias and exacerbate the disparities in health (Tuan 2024).

13.8. Future directions

13.8.1. *Multimodal data fusion*

In the multimodal data integration, the data is integrated into a hybrid format that is capable of supporting extra capabilities. Integrating a number of data streams, including wearable sensor streams, PROs, medical imaging, genomic profiles, and laboratory biomarkers, helps to improve early detection models and make them more predictive and accurate.

These multimodal, synthetic techniques provide a comprehensive picture of the condition of a patient both physiologically and functionally. As an illustration, sensor-based gait analysis combined with inflammatory biomarkers or genomic data indicating risk of neurotoxicity can be used to detect high-risk patients before they exhibit clinically apparent deterioration in mobility rates (Banerjee and Bhat 2025).

13.8.2. *Individualized predictive models*

Conventional models use the population averages, and it is possible that there are subtle changes in individuals. Individual models, however, are adjusted to the functional base of each patient, medical history, treatments taken and the reactions of the patients (Schniepp et al. 2021). Adaptive algorithms identify patterns of individuality and are able to detect early deteriorative behavior at an abnormal level, making them more sensitive and clinically relevant.

13.8.3. *Adaptive and real-time analytics*

The edge-enabled wearable devices can be used to monitor mobility metrics, including the number of steps, gait speed and posture changeover in real-time. Real-time analytics will offer minute-by-minute analytics and may raise early alarms in the event of an unforeseen shift in the expected pattern (Scheuer and Wilson 2004). This immediate feedback facilitates the timely behavioral changes that initiate in avoiding hospitalization, falls and other related deteriorations.

13.8.4. *Federated learning*

Federated learning provides an opportunity to allow several institutions to train ML models without accessing raw-patient data. This increases the variety of datasets, which is a precondition of model generalizability to other populations and kinds of cancer.

13.8.5. *Implementation research and clinical trials*

To ensure the real-world effect of ML tools, these tools should be introduced in prospective oncology trials (Williamson and Prybutok 2024). Effectiveness can be validated through outcomes such as a decrease in the fall rate, medical resource conservation and sustainability of function, which will inform the strategies of implementing them.

13.8.6. *Regulatory pathways and ethical design*

Responsible implementation must have clear, ethically sound frameworks that entail validation measures, explainability, bias surveillance and patient-centered design. It is these elements that need to be accepted in medicine (Gudur et al. 2024).

13.9. Conclusion

The technology of ML is introducing new opportunities to detect and continuously track the mobility deterioration in cancer patients. The conventional clinical examinations are too sporadic and usually overlook the more imperceptible and gradual deteriorations in physical functioning that antecedent actuality. ML algorithms are capable of processing high-frequency, multimodal wearable sensor data (gait speed, balance, step variability), PROs, EHRs, imaging and laboratory data in order to alert about the early signs of functional deterioration. These algorithms allow timely and accurate mobility-saving measures, such as preventing patients falling, modifying care and maintaining autonomy. The heterogeneity of data, noise by sensors and disparagingly labeling between studies contribute to impeding the training of models. ML also has a black-box design that compromises clinical utility and trust. The solution to inequity, opaque and ineffective workflow integration involves interdisciplinary collaboration. The next directions are individual, adaptive, real-time mobility analytics, cross-institution model training on federated learning and future clinical trial validation. These innovations can balance the technological advancement with patient-centered care with smart design, friendly interfaces and consultation with the healthcare professional and ethical considerations. Ultimately, ML may significantly contribute to the independence and quality of life of cancer patients in the course of treatment.

13.10. References

Admi, H., Shadmi, E., Baruch, H., Zisberg, A. (2015). From research to reality: Minimizing the effects of hospitalization on older adults. *Rambam Maimonides Medical Journal*, 6(2), e0017.

Alqudah, A.M. and Moussavi, Z. (2025). A review of deep learning for biomedical signals: Current applications, advancements, future prospects, interpretation, and challenges. *Computers, Materials & Continua*, 83(3), 3753–3841.

Banerjee, P. and Bhat, A. (2025). The role of artificial intelligence in solving bias: Transforming equity and inclusion through data-driven insights. *Archives of Engineering and Technology*, 2, 00340.

Ciuti, G., Ricotti, L., Menciassi, A., Dario, P. (2015). MEMS sensor technologies for human centred applications in healthcare, physical activities, safety and environmental sensing: A review on research activities in Italy. *Sensors*, 15(3), 6441–6468.

Dobkin, B.H. and Martinez, C. (2018). Wearable sensors to monitor, enable feedback, and measure outcomes of activity and practice. *Current Neurology and Neuroscience Reports*, 18(12), 87.

Du, K.L., Jiang, B., Lu, J., Hua, J., Swamy, M.N. (2024). Exploring kernel machines and support vector machines: Principles, techniques, and future directions. *Mathematics*, 12(24), 3935.

Gopalakrisnan, P., Parmar, Y., Samala, R.K., Sidhu, J., Upadhye, V., Gudur, A. (2024). Assessment of primary lung cancer survival rates in relation to the number of thoracoscopic lobectomies performed in hospitals. *Onkologia i Radioterapia*, 18(8), 001–007.

Gudur, R.A., Bhosale, S.J., Gudur, A.K., Datkhile, K.D. (2024). Genetic polymorphisms in glutathione S-transferase (GST) gene and their correlation with toxicity of chemotherapy in breast cancer patients. *Asian Pacific Journal of Cancer Prevention*, 25(7), 2271.

Joshi, P., Hasanuzzaman, M., Thapa, C., Afli, H., Scully, T. (2023). Enabling all in-edge deep learning: A literature review. *IEEE Access*, 11, 3431–3460.

Kulkarni, P., Kirkham, R., McNaney, R. (2022). Opportunities for smartphone sensing in e-health research: A narrative review. *Sensors*, 22(10), 3893.

Kumar, A., Batta, P., Ahuja, S., Rathore, P.S. (eds) (2026a). *AI-driven Innovations in Physiotherapy and Oncology 1*. ISTE Ltd, London and Wiley, Hoboken, NJ.

Kumar, A., Batta, P., Ahuja, S., Rathore, P.S. (eds) (2026b). *AI-driven Innovations in Physiotherapy and Oncology 2*. ISTE Ltd, London and Wiley, Hoboken, NJ.

Latajka, A., Stefańska, M., Woźniewski, M., Malicka, I. (2023). Walking speed and risk of falling patients operated for selected malignant tumors. *Healthcare*, 11(23), 3069.

Lobo, P., Morais, P., Murray, P., Vilaça, J.L. (2024). Trends and innovations in wearable technology for motor rehabilitation, prediction, and monitoring: A comprehensive review. *Sensors*, 24(24), 7973.

Marshall, T.F., Zipp, G.P., Battaglia, F., Moss, R., Bryan, S. (2016). Effects of chemotherapy-induced-peripheral-neuropathy on spatiotemporal gait parameters and fall risk in cancer patients after the completion of chemotherapy drug treatment. *Archives of Physical Medicine and Rehabilitation*, 97(12), 2027–2238.

Naidu, G., Zuva, T., Sibanda, E.M. (2023). A review of evaluation metrics in machine learning algorithms. In *Computer Science On-line Conference*. Springer, Cham, 15–25.

Nordin, Å., Taft, C., Lundgren-Nilsson, Å., Dencker, A. (2016). Minimal important differences for fatigue patient reported outcome measures: A systematic review. *BMC Medical Research Methodology*, 16(1), 62.

Rochester, L., Mazzà, C., Mueller, A., Caulfield, B., McCarthy, M., Becker, C., Miller, R., Piraino, P., Viceconti, M., Dartee, W.P. et al. (2020). A roadmap to inform development, validation and approval of digital mobility outcomes: The Mobilise-D approach. *Digital Biomarkers*, 4(1), 13–27.

Rodin, M.B. (2022). Maintaining functional status among older adult cancer patients and survivors. In *Pathy's Principles and Practice of Geriatric Medicine*, Sinclair, A.J., Morley, J.E., Vellas, B. et al. (eds). Wiley-Blackwell, Hoboken, NJ, 1208–1227.

Rufo, D.D., Debelee, T.G., Ibenthal, A., Negera, W.G. (2021). Diagnosis of diabetes mellitus using gradient boosting machine (LightGBM). *Diagnostics*, 11(9), 1714.

Scheuer, M.L. and Wilson, S.B. (2004). Data analysis for continuous EEG monitoring in the ICU: Seeing the forest and the trees. *Journal of Clinical Neurophysiology*, 21(5), 353–378.

Schniepp, R., Huppert, A., Decker, J., Schenkel, F., Schlick, C., Rasoul, A., Dieterich, M., Brandt, T., Jahn, K., Wuehr, M. (2021). Fall prediction in neurological gait disorders: Differential contributions from clinical assessment, gait analysis, and daily-life mobility monitoring. *Journal of Neurology*, 268(9), 3421–3434.

Stolwyk, R.J., Charlton, J.L., Triggs, T.J., Iansek, R., Bradshaw, J.L. (2006). Neuropsychological function and driving ability in people with Parkinson's disease. *Journal of Clinical and Experimental Neuropsychology*, 28(6), 898–913.

Sun, C.C. (2009). Materials science tetrahedron: A useful tool for pharmaceutical research and development. *Journal of Pharmaceutical Sciences*, 98(5), 1671–1687.

Tappenden, K.A., Quatrara, B., Parkhurst, M.L., Malone, A.M., Fanjiang, G., Ziegler, T.R. (2013). Critical role of nutrition in improving quality of care: An interdisciplinary call to action to address adult hospital malnutrition. *Journal of the Academy of Nutrition and Dietetics*, 113(9), 1219–1237.

Thackeray, A. and Miller, C. (2019). The management of post-surgical orthopedic conditions in the older adult. In *Guccione's Geriatric Physical Therapy*, Avers, D. and Wong, R. (eds). Mosby, Maryland Heights, MO.

Tsetlina, V., Stanford, R.A., Syrkin, G., Ibanez, K. (2024). Steroid myopathy and rehabilitation in patients with cancer. *PM&R*, 16(8), 908–918.

Tuan, D.A. (2024). Bridging the gap between black box AI and clinical practice: Advancing explainable AI for trust, ethics, and personalized healthcare diagnostics. *Preprints*. doi: 10.20944/preprints202409.1974.v2.

Tucker, C.S., Behoora, I., Nembhard, H.B., Lewis, M., Sterling, N.W., Huang, X. (2015). Machine learning classification of medication adherence in patients with movement disorders using non-wearable sensors. *Computers in Biology and Medicine*, 66, 120–134.

Williamson, S.M. and Prybutok, V. (2024). Balancing privacy and progress: A review of privacy challenges, systemic oversight, and patient perceptions in AI-driven healthcare. *Applied Sciences*, 14(2), 675.

Wilson, A. and Anwar, M.R. (2024). The future of adaptive machine learning algorithms in high-dimensional data processing. *International Transactions on Artificial Intelligence*, 3(1), 97–107.

Yates, L.A., Aandahl, Z., Richards, S.A., Brook, B.W. (2023). Cross validation for model selection: A review with examples from ecology. *Ecological Monographs*, 93(1), e1557.

Yorkston, K.M., Bourgeois, M.S., Baylor, C.R. (2010). Communication and aging. *Physical Medicine and Rehabilitation Clinics of North America*, 21(2), 309.

14

Predictive Analytics for Return-to-Function Timelines in Cancer Survivors

Oncological care now incorporates recovery and quality of life along with disease eradication as world cancer survival rates rise. Predictive analytics, using machine learning, artificial intelligence (AI), and big data, may help clinicians forecast cancer survivors' return-to-function timelines. Predicting recovery in the physical, cognitive and psychosocial domains will allow clinicians to customize rehabilitation and track recoveries, optimize resources and increase patient outcomes. This chapter discusses predictive analytics to model RTF cancer survivors' potential and current global advances. Predictive modeling as a theory, data sources, electronic health records, wearables, patient-reported outcomes and different methods such as regression, decision trees and deep learning are discussed. Predictive modeling across the musculoskeletal, cardiopulmonary and neurological areas is also covered, along with implementation and the ethical issues of data ownership, model vagueness and clinical realization. This chapter additionally discusses population heterogeneity, the absence of external validation and other gaps as critical weaknesses. Multimodal data fusion, real-time adaptive systems and federated learning, along with the rising need for equity design and explainable predictive analytics, are able to predict and determine recovery timelines through timely interventions. When developed ethically and applied responsibly, these technologies have the ability to change the rehabilitation pathways and provide individualized, improved and inclusive cancer survivorship care.

14.1. Introduction

The rise in cancer survivorship is a global phenomenon made possible by advances in early detection treatment modalities and supportive care. Improvements in survival rates are gaining more attention for survivors being able to live a functional and fulfilling life. The return-to-function (RTF) timelines consist of a multitude of dimensions and include recovery in the physical, cognitive,

---

Chapter written by Dhairyasheel PATIL and Pankaj THOTE.

psychosocial, occupational and social spheres. For clinicians, patients and caregivers, the unpredictability of these recovery pathways can pose challenges to psychological preparedness, planning rehabilitation and the allocation of healthcare resources (Shaik et al. 2024).

Predictive analytics is the relatively new interdisciplinary field of study, which combines statistics, data science and artificial intelligence (AI). Predictive analytics sits at the intersection of AI and healthcare predictive forecasting models. In cancer survivors, predictive analytics can be used to forecast recovery timelines, inform personalized rehabilitation plans, identify patients likely to experience prolonged recovery, and adjust healthcare resources to meet anticipated demand. Predictive analytics in oncology has been shown to be useful in estimating the likelihood of recurrence, modeling treatment responses and survivorship risk assessment (Obeagu and Obeagu 2024). For forecasting RTF, it is a necessary and logical step to build predictive models. The wide variety of types of cancer, treatment plans, patient ages and comorbidities make predictive analytics indispensable.

As an illustration, a young adult who has fought Hodgkin lymphoma and an elderly person with breast cancer undergoing chemotherapy will likely have very different recovery paths. Generic recovery timelines used in traditional rehabilitation methods tend to be based on arbitrary clinician judgments. Predictive models could integrate electronic health records (EHRs), genomics, data from wearables, patient-reported outcomes (PROs) and other variables to approximate personalized recovery timelines with far greater accuracy (Adeniran et al. 2024).

Quite uniquely, post-intervention periods can allow predictive analytics to attempt to forecast rather than respond to recovery challenges, such as fatigue, diminished mobility, cognitive issues and mental and social discomfort (Balaji et al. 2023). If predictive models identify and measure the risk of sustained fatigue depression in a specific group of colorectal cancer survivors, for instance, more aggressive exercise and diet strategies could be put in place to help manage fatigue when cancer survivors are likely to be most inactive. Predicting, for example, delays in social reintegration enables timely social support and recovery services.

There are, however, still very few predictive models in cancer rehabilitation. Most work tends to be siloed, covering narrow scopes, lacking a diversity of samples and broad, real-world context. Challenges in methodology include combining different types of data, ensuring that models can be understood and tackling ethical implications, such as privacy, equity and consent (Dinov 2016). That is why numerous models, particularly in primary care, continue to be experimental and lack clinical readiness for broad application.

This chapter studies new research on the application of predictive analytics to the RTF timelines of cancer survivors. It examines the functional recovery courses, predictive modeling types used, how the models are actually used and the ethical and methodological concerns of the models. The goal of this chapter is to aid in making predictive analytics a part of individualized rehabilitation of cancer survivors by listing the existing practice and the research gaps and making integration possible.

14.2. Scope and definitions

To interpret and predict the timelines of return-to-function (RTF) in cancer survivors, a definition is necessary and the numerous aspects of the concept of function need to be appreciated. Post-cancer functionality encompasses physical, cognitive, emotional and social functions. Recovery may differ in these areas and could depend on the type of cancer, the mode of treatment, age, comorbidities and psychosocial environment (Post 2019). Predictive analytics should thus be based on a broad, clear-cut model of operation to provide clinically useful predictions.

14.2.1. *Dimensions of function in cancer survivorship*

Physical ability is the capacity to carry out activities of daily living (ADLs), to move (walking, taking stairs) and to perform more complicated tasks (shopping, cooking). The frequent issues are fatigue, muscular weakness, nerve damage and heart-pulmonary deconditioning, particularly following surgery, chemotherapy or radiation (Stubblefield et al. 2006).

The cognitive function includes memory, attention, executive speed and speed of processing. Cognitive impairment, or so-called chemo brain, is a condition experienced by a significant number of survivors and may persist for months and even years. Its prevalence is dependent on treatment, age and genes.

Psychosocial functioning consists of emotional well-being, social integration and returning to work. Most survivors are anxious, depressed, have body-image problems and suffer from PTSD. Going back to work is not only important in terms of finances, but it is also a good sign of recovery, which is both economically and psychologically positive.

Reintegration into the job and into the larger society – returning to work, caregiving responsibilities and social life – demonstrates complete restoration of independence in a survivor. The outcomes can only be predicted by having clinical data as well as socioeconomic and occupation data (Wood et al. 2010).

14.2.2. *Timeline definitions: acute versus long-term recovery*

The timelines on RTF may be split to include acute, subacute and long-term:

– *Acute phase (3 months after treatment)*: elimination of the acute side effects (e.g. nausea, acute fatigue) and early physical rehabilitation.

– *Sub-acute phase (3–12 months)*: phase marked by gradual physical and mental recovery, employment and normalization of ADLs.

– *Long-term phase (12 months or above)*: check on any long-term impairments that have persisted, risk of late consequences and whether or not functional improvements persist.

Each stage may be a specific goal of predictive modeling, and can be based on the clinical or research question.

14.2.3. *Data sources for function assessment*

In order to predict RTF, the information should be precise about functioning on the specified levels. The following sources are commonly used:

– *Patient-reported outcome measures (PROMs)*: tools such as the EORTC QLQ-C30, FACT-G and PROMIS capture self-assessed physical and emotional function.

– *Performance-based measures*: timed up and go (TUG), six-minute walk test (6MWT), grip strength and balance measures are used to bring up objective physical functioning.

– *Wearables and digital health tools*: step counts, heart rate variability (HRV), and sleep patterns collected via smartphones or fitness trackers offer continuous, passive function monitoring.

– *Clinical records and imaging*: tumor staging, treatment details, comorbidities, and complications are essential inputs for contextualizing function (Sarfati et al. 2016).

– *Administrative and employment data*: return-to-work timing, disability claims and workplace accommodations serve as proxies for societal reintegration.

14.2.4. *Population-specific considerations*

The expectation of post-treatment and baseline functions of different populations of patients who are survivors of cancer differ. For instance:

– Older adults usually possess fewer physical reserves and are more susceptible to deconditioning.

– Pediatric survivors are also likely to have impaired developmental milestones.

– Young adult survivors are significantly exposed to special vocational and psychosocial problems when encountering the transition into life.

In prediction, stratification of the predictive models based on the population sub-group is relevant and accurate.

14.3. Current evidence on functional recovery in cancer survivors

The whole spectrum of post-invasive cancer survivorship implies a resurgence of the level of functioning that is essential but fluctuating. In the last 20 years, there has been increasing longitudinal research and cohort studies that have reported functional trends among various survivor groups (Feliciano et al. 2023). This research shows that the recovery patterns are diverse and unpredictable and that predictive tools should be used in order to tailor the rehabilitation process and teach patients about their expectations.

14.3.1. *Physical function recovery*

The most common sequelae of cancer treatment reported include physical impairments. A historic investigation of elderly breast-cancer survivors established that almost 47% who underwent a physical functioning decrease throughout treatment never went back to their pre-treatment level, even 12 months post-therapy. The predictors of poor recovery were old age, high burden of comorbidity and adjuvant chemotherapy. This is observed in the survivors of lung, colorectal and prostate cancer.

Other authors have found the long-term effects of cancer treatment (persistent fatigue, sarcopenia, problems with balance and neuropathy) to be more common among individuals subjected to neurotoxic chemotherapy (e.g. platinum-based regimens) or those subjected to invasive surgeries. As an example, the survivors of colorectal cancer often complain of mobility and bowel malfunction several years after the intervention, whereas survivors of head and neck cancers may experience trismus, dysphagia and low exercise tolerance.

Functional improvement has partially been a successful result of exercise interventions and structured rehabilitation programs, although there is a huge variation in outcomes. Such a response heterogeneity further supports the need to predict high-risk individuals as the subject of predictive analytics (Zaree et al. 2023).

14.3.2. *Cognitive and neuropsychological recovery*

Cognitive dysfunction (also known as chemobrain) occurs in 17–75% of cancer survivors, depending on the population under investigation and the measurement instrument. The survivors of breast cancer are particularly well researched, due to their great numbers and high incidences of systemic therapy. Deficits such as memory lapses, decrease in attention span, slow processing speed and impaired executive functions are common.

In 2024, a study in the *Journal of Cancer Survivorship* established the significant negative change in the cognitive ability of older breast cancer survivors during the 24 months following chemotherapy. Notably, the individuals who did not have any deficits at the baseline acquired new impairments when under follow-up, which shows that it was delayed in some individuals. These findings can be used to question the linearity and predictability of recovery and emphasize the importance of time-sensitive prediction models (Yang et al. 2024).

In addition to chemotherapy, other causative cognitive side effects include radiation therapy to the brain, targeted therapies and hormonal interventions. Neurocognitive issues can result in long-term effects on academic performance and quality of life in adulthood in pediatric cancer survivors, including survivors of leukemia or brain tumors.

14.3.3. *Return to work and societal reintegration*

Return to work (RTW) is also regarded as one of the most important milestones of survivorship, for both physical and psychosocial recovery. However, numerous survivors face enormous obstacles to work: exhaustion, cognitive disabilities, physical difficulties, discrimination in the workplace and loss of self-confidence are often reported.

Recent results of survival of early-onset colorectal cancer patients in China used logistic regression to produce the predictive model of RTW at one and two years of follow-up. The models were well-discriminated (area under the curves (AUCs) 0.70 to 0.89) and were acceptable in their calibration, with education level, modality of treatment, tumor stage and psychological readiness playing a significant role in therapy (Li et al. 2025). These results support the practicability of predictive methods and indicate that they should be validated in other contexts.

The same trends can be observed among breast and prostate cancer survivors and the RTW rates are quite different depending on the occupation, physical demand and employer-provided accommodations. Vocational therapists and social workers are beginning to assist survivorship care plans in some countries with the provision of RTW counseling.

14.3.4. *Functional decline and comorbidities*

Not every survivor has a smooth recovery pathway; many of them undergo long-term or permanent impaired functioning. The latter is particularly relevant to elderly and comorbid patients with diabetes, cardiovascular disease or arthritis. The issue of frailty is becoming a pressing problem in geriatric oncology, because treatment and diagnosis of cancer hastens aging.

A 2023 cohort study on lung cancer survivors found that pre-treatment frailty and low baseline physical activity levels predicted poor post-treatment mobility and independence. These results suggest that baseline risk stratification, coupled with predictive modeling, could identify patients needing more intensive post-treatment support (Lv et al. 2022).

14.4. Predictive analytics and machine learning approaches

The integration of clinical oncology, biostatistics and AI forms the basis of predictive analytics in cancer rehabilitation. Predictive models, drawing upon historical and current data, make estimations about the recovery pathways and timelines of functions in different organs and systems. This section focuses on the key predictive models using statistics and machine learning (ML) to estimate RTF timelines of cancer survivors and discusses their strengths and weaknesses.

14.4.1. *Traditional statistical approaches*

For years, oncology research has used predictive models that include linear regression, logistic regression and Cox proportional hazards models. While these models offer predictability, clarity and interpretability, they are still fundamental in that they assess the effects of variables (e.g. age, stage, treatment, etc.) on outcome(s) that are binary events or time-to-event (Teshale et al. 2024). For example, logistic regression was used to estimate the probability of work return in colorectal cancer survivors for one-year post diagnosis using clinical and demographic variables. Cox models have also been used for assessing time to regain post-therapy basal functions. Such models are sound, clear and can result in a useful answer and also have the problem of being linear, proportional and unable to identify nonlinear or high-dimensional interactions among variables.

14.4.2. *Emergence of ML in RTF prediction*

As a branch of AI, ML offers a more adaptable approach for large and heterogeneous datasets. Machine learning can identify and analyze hidden patterns in data and handle missing or incomplete information in fundamentally different ways compared to traditional analytical approaches. It can manage complicated nonlinear associations and interactions without the need for a researcher to devise a hypothesis a priori. Some of the more common ML techniques for predicting recovery include:

– *Random forest (RF)*: these are ensemble-based techniques that construct a set of trees and average to enhance predictions. RFs are able to manage heterogeneous data. They have been used to predict recovery during the post-acute rehabilitation for cognitive and physical recovery (e.g. predicting delay of recovery).

– *Gradient boosting machines (GBMs), especially XGBoost*: these algorithms focus on predicting the residuals and have been used to predict the outcomes of interest to a very high degree of precision. In the oncology setting, for example, they predict the length of stay and readmission rates.

– *Support vector machines (SVMs)*: SVM classifies complex functional outcomes (e.g. recovered vs. not recovered) based on an extensive set of features. It is useful but complex, hence requiring greater tuning and less transparency (Campagnini et al. 2022).

– *Neural networks and deep learning*: suited for large datasets of IT-based imaging, sensors and wearables. Deep learning can be used, for example, to monitor post treatment fatigue or inactivity based on data from an accelerometer.

– *K-nearest neighbors (KNNs)*: an uncomplicated yet efficient technique for function status prediction, founded on the resemblance to past patient profiles.

14.4.3. *Time-to-event and survival ML models*

Older methods of survival analysis, such as Cox models, usually fail to identify nonlinear relationships and time-varying effects. By contrast, the more recent ML survival models are able to predict not just if recovery will happen, but when, as well:

– *Random survival forests (RSFs)*: RFs can be modified to handle censored survival data. Such models serve to estimate the time it takes to RTW, regain mobility or resolve fatigue symptoms.

– *DeepSurv*: a deep learning-based extension of the Cox model that can incorporate high-dimensional inputs such as genomic data, imaging and longitudinal PROs.

– *Multi-task logistic regression (MTLR)*: this simultaneously estimates multiple survival probabilities over time, offering a more granular prediction of recovery likelihood across time points (Suresh et al. 2022).

The functional timelines have been modeled using these approaches, taking into account that recovery is a complex process that takes weeks or months, not a binary event.

Model type	**Input data sources**	**Functional domain predicted**	**Strengths**	**Limitations**
Logistic regression/ Cox models	Demographic and clinical data (age, tumor stage and comorbidities)	Probability and timing of return to work/physical recovery	Simple, interpretable; suitable for small datasets	Assumes linearity; limited for nonlinear; high-dimensional data
Random forest (RF)	EHR; wearable metrics; PROs	Physical function; fatigue recovery	Handles missing data and mixed variables; high accuracy	Limited interpretability; prone to bias if unbalanced data
Gradient boosting machines (XGBoost, LightGBM, etc.)	EHR; lab biomarkers; activity metrics	Multidimensional recovery timelines (physical, cognitive, etc.)	Excellent predictive accuracy; adaptable to different outcomes	Requires hyperparameter tuning; potential overfitting
Support vector machine (SVM)	Wearable sensor data; PROs	Classification of delayed vs. normal recovery	Works well with small; high-dimensional datasets	Low interpretability; tuning kernel parameters is complex
Neural networks/deep learning (CNN, LSTM and DeepSurv)	Longitudinal wearable and sensor data	Time-to-event recovery, cognitive restoration, fatigue	Captures temporal dependencies; powerful for continuous data	Black-box nature; data-hungry

Random survival forest (RSF)	Survival data; clinical covariates	Time to regain baseline function	Models censored survival outcomes; captures nonlinear effects	Interpretation complexity
Federated learning frameworks	Multicenter datasets without data sharing	Cross-population generalizable RTF prediction	Enhances privacy; allows collaboration	Computationally intensive; limited implementation experience

Table 14.1. *Machine learning and predictive analytics models for forecasting return-to-function (RTF) in cancer survivors*

14.4.4. *Interpretable ML for clinical use*

Complex models, while potentially more accurate, tend to be opaquer – this is a major barrier for clinical implementation. As a result, recent work has shifted towards interpretable ML, which aims to provide a balance between explainability and predictive power.

– *SHAP (SHapley Additive exPlanations)*: SHAP represents a single framework for explaining individual predictions of a model by quantifying the value added (or subtracted) by each feature. SHAP can determine, for instance, that in predicting poor functional recovery, age, fatigue scores and chemotherapy dose intensity were the major contributing elements.

– *LIME (Local Interpretable Model-agnostic Explanations)*: this offers local explanations for black-box models, aiding clinicians in understanding specific patient-level predictions.

– *Rule-based ML models*: algorithms such as decision trees and rule fit models create transparent decision pathways that can be validated against clinical intuition.

14.4.5. *Data challenges and feature engineering*

Effective prediction hinges on data quality, feature selection and model training:

– *Missing data*: incomplete documentation or loss to follow-up is common in oncology datasets. Addressing these gaps can be performed through techniques such as imputation or probabilistic modeling.

– *Feature engineering*: creating informative features – such as rate of symptom improvement, baseline activity levels, or comorbidity scores – is crucial to improving model performance (Alsaleh et al. 2023).

– *Temporal sequencing*: integrating time-stamped data such as symptom progression or wearable device inputs requires careful structuring, particularly in recurrent or episodic cancers.

– *Multimodal inputs*: combining EHRs, lab results, imaging, genomic data and PROMs increases model robustness but also demands sophisticated data integration techniques.

14.5. Integrating patient-generated and wearable data

In oncology rehabilitation, as in other fields of medicine, the use of patient-generated health data (PGHD) along with data from predictive analytics tied to wearable devices, presents new opportunities to deliver more predictive and personalized rehabilitation.

The use of predictive analytics with clinical data from electronic health records/health information systems structured data remains disconnected from real-time changes in the patients' functional abilities, symptom distress, patient compliance with rehabilitation and other activities prescribed by the healthcare team. PGHD and wearable sensors provide true field data and valuable data streams integrated in personalized RTF predictive analytics.

14.5.1. *What constitutes PGHD?*

Patients generate, document and collect data outside of healthcare meetings, and this data is called PGHD. In regard to cancer rehab, this most commonly entails:

– *PROs*: pain, fatigue, emotional distress, physical functioning and quality of life are assessed regularly through mobile apps, websites and telephone interviews.

– *Symptom diaries*: records that patients keep to note and report treatment side effects or exercise to track progress.

– *Documentation of nutrition and sleep*: eating and sleeping patterns, which are tracked either through logs or sensors.

Such data give a more detailed picture of recovery as compared to the periodical clinical snapshots that are normally taken.

14.5.2. *Role of wearable devices in function monitoring*

Fitness trackers, smartwatches and biosensors as wearable technologies provide constant and objective measurements of movement, activity and physiological reactions.

The key measures in cancer survivorship are:

– *Gait analysis and step count*: helpful in the quantification of mobility and endurance. Recovery of functions with delays and complications may be an indicator of declines.

– *HRV*: associated with autonomic function and stress resilience. Lower HRV can reflect fatigue or overexertion.

– *Sleep metrics*: disrupted sleep is linked to delayed healing, fatigue and reduced overall function.

– *Posture and balance tracking*: particularly important in elderly or neurologically affected survivors to assess fall risk and mobility recovery.

The metrics, upon long-term monitoring, provide a dynamic image of functional capacity and allow the clinicians to time the interventions customized to the particular person.

14.5.3. *Predictive modeling with wearable data*

Recent research has shown that wearable-derived data in ML models is a worthwhile possibility:

– In a cohort of breast cancer survivors, accelerometer data were used to predict fatigue and adherence to exercise regimens. Wearable data-based models are always more effective as compared to those only depending on clinical data or self-reported data.

– In lung cancer survivors, continuous tracking of physical activity via smartwatches allows for early detection of functional decline, predicting hospital readmission and guiding tele-rehabilitation efforts (Kumar et al. 2026b).

– Pediatric cancer survivorship research has also leveraged wearables to monitor neurocognitive development and recovery, highlighting the utility across age groups.

Wearable data can significantly enhance the accuracy of the model when it is combined with other clinical and PRO data. Such combined inputs allow dynamic prediction-estimates of risks are regularly updated as new information arrives.

14.5.4. *Mobile health platforms and engagement tools*

Mobile health platforms serve as platforms for gathering, incorporating and visualizing PGHD and wearable information:

– *interactive dashboards* that allow patients and clinicians to track progress towards recovery milestones;

– *automated alerts* that notify care teams when a patient's metrics deviate from expected recovery patterns;

– *gamification and coaching features* to encourage adherence to rehabilitation programs and promote physical activity;

– *support integration* with electronic health records to develop an integrated picture of the health of the patient.

These technologies have demonstrated the potential of enhancing activities, enhancing self-efficacy and promoting functional recovery.

14.5.5. *Challenges and considerations*

Although promising, incorporating patient-generated information and wearables into predictive analytics faces a number of challenges, including:

– *Data standardization*: the accuracy of the device varies, and data is in different formats, as well as proprietary algorithms; hence, they are hard to compare and integrate.

– *Compliance and data accuracy*: wearables and data entries are not always used by many patients, so there are gaps and this aspect may affect predictions.

– *Digital literacy and access*: older adults or socioeconomically disadvantaged persons can be unfamiliar with or have no access to the required technology.

– *Privacy and data protection*: constant collection presents ethical issues of informed consent, information sharing and access by third parties.

To overcome these constraints, there would be the need to look into the collaboration of oncologists, data scientists, engineers, ethicists and patients to develop equitable, secure and clinically integrated systems (Brown et al. 2022).

14.6. Broader AI and predictive techniques in oncology

Oncological predictive analytics and AI go well beyond predictions of RTF. The algorithms that are employed to predict the schedule of functional recovery are variations of larger procedures that have so far been used in diagnosing cancer, its prognosis, treatment planning and survivorship. It is within this context that refining, validating and integrating RTF tools need to be understood to integrate them into the oncology ecosystem.

14.6.1. *Prognosis and survival prediction models*

Survival analysis has been revolutionized using AI in cancer care. ML models are now able to estimate the overall survival, progression-free survival, disease-specific mortality and the response to treatment. Random survival forests, gradient-boosted survival models and deep-learning time-to-event models are all techniques that can be used directly on predicting functional milestones such as returning to work or overcoming fatigue (Yin et al. 2024).

There is an increasing amount of literature employing AI to forecast survival in certain cancers with remarkable accuracy. As an example, the models that predict survival of colorectal and lung cancer patients, based on clinical and imaging data, have AUCs greater than 0.85. Such survival models can usually be used to make dual-outcome predictions, where the risk of mortality is predicted with a functional recovery outcome achieved at the same time, particularly in weaker or elderly groups.

14.6.2. *Length of stay and readmission prediction*

In cancer surgery and inpatient oncology, the major indicators are the length of stay and the 30-day readmission rates in the hospital. Predictive models, which use data including tumor type, treatment type, pre-operative risk factors and post-operative complications, are useful in the optimization of care paths and discharge plans.

As an example, AI models forecast LOS with the mean absolute error of 9.7 days in the case of colorectal cancer. ML applications predicting recovery in gastric cancer applications use variables such as nutrition, surgical operation duration and inflammation armors to predict recovery (Zhou et al. 2023). Similarities between these approaches and return-to-work models include both being dynamic in quantifying recovery and being individualized in providing care.

14.6.3. *Radiomics and imaging-based predictions*

Radiomics obtains numerous quantitative characteristics of medical images, including computed tomography (CT), magnetic reasoning imaging (MRI), and positron emission tomography (PET) images. Such radiomic signatures are able to predict the aggressiveness of tumors, the probability of metastasis and the treatment outcome. Recently, radiomics has also been applied to forecast functional outcome (find imaging biomarkers associated with post-operative complications or functional deterioration).

Physical recovery can hardly be predicted through imaging, although it may aid in the prediction of functional loss due to the location of the tumor (e.g. on the spinal cord), tissue damage during treatment or organ deformity. With the inclusion of the radiomics of predictive analytics, we can approximate long-term disability risk as a preoperative estimate.

14.6.4. *Knowledge-informed ML*

Black-box AI models tend to be not interpretable and general. To solve this, knowledge-aware ML combines expert medical knowledge, for example, cancer biology or rehabilitation science and data-driven pattern recognition.

As an example, in models where treatment toxicity is predicted, clinicians incorporate expert-curated rules in the pipeline, which makes the model more understandable and trustworthy. The same principle may be applied to the predictions of returns to work: rehabilitation guidelines, frailty scores or published data on RTW can serve as priors that bind the outputs of a model and ensure that they remain clinically and medically relevant (Oliveira et al. 2016).

Knowledge-based models also work better in different healthcare contexts, particularly in cases involving limited or diverse training data. They assist decision-makers in not just visualizing the prediction but also the rationale behind the prediction, which is vital for clinical trust and ethical application.

14.6.5. *Real-world applications and decision support systems*

Prediction systems based on AI are integrated into electronic health records and clinical decision support systems. The latter platforms are able to notify clinicians that a cancer survivor is at a high risk of unsuccessful recovery or advise custom rehabilitation, depending on the anticipated requirements.

These tools are pilot tested by some of the institutions to direct early discharge, refer patients to physical therapy or initiate psychosocial assistance. Others develop AI-based dashboards to monitor the recovery progress of each patient over time, modifying care services on received observations or information (Callejas Carrión and Flynn 2023).

In practice, little is adopted in reality, even though it is promised. The majority of models are academic lab-based, which are not externally validated or regulated and clauses are reluctant due to concerns about the transparency of the algorithms used, liability and data control.

14.7. Future directions in predictive analytics for cancer survivorship

Predictive analytics is becoming increasingly sophisticated, and its use on cancer survivorship, particularly in relation to predicting RTF schedules, will become even more extensive, precise and available. The new methods of computation, improved data infrastructure, wearable gadgets and individualized rehab plans will fuel this advancement (Rane et al. 2024). The methods below are likely to influence the field in the next decade.

14.7.1. *Multimodal data integration*

Predictive models that will be created in the future will integrate a great variety of data to provide a more comprehensive view of the health of a survivor. These include:

– *EHRs*: treatment dates, comorbidities, lab results and medication use.

– *PROs*: fatigue, pain, quality of life and emotional status.

– *Wearables and remote sensors*: step count, HRV, sleep patterns and gait speed.

– *Imaging and radiomics*: tumor location and treatment-related structural changes.

– *Genomics and biomarkers*: indicators of inflammation, healing capacity or treatment response.

Fusing these data types using AI methods such as deep learning and ensemble modeling can improve prediction accuracy and adaptability across patient subgroups (Radford et al. 2024). These models have the noteworthy ability of updating predictions in real time whenever new information is presented during the recovery journey.

14.7.2. *Real-time and adaptive prediction models*

The use of real-time adaptive learning systems is becoming a substitution for static models that are trained once and not used again. These tools update predictions and are based on:

– new symptoms or PRO entries;

– modifications in the level of physical activities;

– improvement in planned treatments (e.g. radiation therapy);

– new complications or readmission.

These systems have been shown to be more adaptable to dynamic cancer recovery and provide a kind of living roadmap to patients and clinicians that changes at each survivorship phase. Practically, it implies that the rehabilitation plans, estimates of the returns to work and support referrals can be constantly narrowed down to become correct and more personalized (Gudur et al. 2024).

14.7.3. *Federated and privacy-preserving learning*

Federated learning enables predictive models to be trained on decentralized data to solve issues such as dataprivacy and institutionalsilos without accessing raw data (Myakala et al. 2024). In this model, the algorithms are distributed to every institution or device and they are trained locally; also, the model parameters and underlying data are brought together in a single central location.

This has significant consequences in the survivorship research in which small samples comprising rare types of cancer are distributed across numerous locations. Federated learning allows collaborative learning without violating confidentiality and increases the representativeness of training information. This has been further enhanced by combining it with different privacy and secure multiparty computation to further improve data security (Diaz-Asper et al. 2024).

14.7.4. *Explainable and human-centered AI*

Predictive analytics need to be incorporated into clinical care by making the models explainable, actionable and patient-focused:

– underlining the most important aspects that promote a prediction (e.g. fatigue score, hemoglobin level);

– the ability to experiment with hypothetical scenarios where people can say "what if?";

– supplying graphs of recovery curves.

In addition, human-centered design presupposes a collaborative design of tools with survivors, oncologists, physiotherapists and caregivers. This gives the presentation of predictions in clear and accessible language, put ethically, and aligned with the specific objectives – walking unassisted, resuming driving or returning to work (Yadav et al. 2024).

14.8. Conclusion

Predictive analytics can alter the way we help the survivors of cancer to resume normal functioning. Advanced models can predict the recovery of each patient by combining large, diverse data, which are electronic health records, wearable devices, imaging and PROs. This enables clinicians to independently and objectively decide when to initiate rehab and the intensity of rehab and the ways to employ it, ultimately enhancing physical functioning, quality of life and the distribution of resources. This chapter brings into focus some areas of predictive models' development and what is already being used. These are musculoskeletal, cardiopulmonary and neurological post-cancer rehab. Recent models are becoming more dynamic, incorporating ML approaches that change dynamically with changes in the condition of a patient. Oncology and rehab teams already benefit from such tools such as risk-stratification dashboards and individual recovery schedules that can help them to customize care to the needs of individual survivors. However, these technological developments are associated with considerable obstacles. The issue of data privacy, algorithmic bias and limited generalizability remains a concern. The training data used in many models is not representative and not externally validated and, therefore, cannot be useful in diverse or underserved populations. Besides, the interpretability of models is critical to clinical trust and adoption – providers need to make predictions understandable, justifiable and explain them to the patients in a transparent, ethical manner.

14.9. References

Adeniran, I.A., Efunniyi, C.P., Osundare, O.S., Abhulimen, A.O. (2024). Data-driven decision-making in healthcare: Improving patient outcomes through predictive modeling. *International Journal of Scholarly Research in Multidisciplinary Studies*, 05(01), 059–067.

Alsaleh, M.M., Allery, F., Choi, J.W., Hama, T., McQuillin, A., Wu, H., Thygesen, J.H. (2023). Prediction of disease comorbidity using explainable artificial intelligence and machine learning techniques: A systematic review. *International Journal of Medical Informatics*, 175, 105088.

Balaji, L., Farrar, M.A., D'Silva, A.M., Kariyawasam, D.S. (2023). Decision-making and challenges within the evolving treatment algorithm in spinal muscular atrophy: A clinical perspective. *Expert Review of Neurotherapeutics*, 23(7), 571–586.

Brown, S.A., Sparapani, R., Osinski, K., Zhang, J., Blessing, J., Cheng, F., Hamid, A., Berman, G., Lee, K., BagheriMohamadiPour, M. et al. (2022). Establishing an interdisciplinary research team for cardio-oncology artificial intelligence informatics precision and health equity. *American Heart Journal Plus: Cardiology Research and Practice*, 13, 100094.

Callejas Carrión, Z. and Flynn, T. (2023). Cancer survivorship and AI for well-being white paper: A European research projects collaborative perspective. White Paper, A European research projects collaborative perspective [Online]. Available at: https://hdl.handle.net/10481/81842.

Campagnini, S., Arienti, C., Patrini, M., Liuzzi, P., Mannini, A., Carrozza, M.C. (2022). Machine learning methods for functional recovery prediction and prognosis in post-stroke rehabilitation: A systematic review. *Journal of NeuroEngineering and Rehabilitation*, 19(1), 54.

Diaz-Asper, C., Hauglid, M.K., Chandler, C., Cohen, A.S., Foltz, P.W., Elvevåg, B. (2024). A framework for language technologies in behavioral research and clinical applications: Ethical challenges, implications, and solutions. *American Psychologist*, 79(1), 79.

Dinov, I.D. (2016). Methodological challenges and analytic opportunities for modeling and interpreting Big Healthcare Data. *Gigascience*, 5(1), s13742-016.

Feliciano, E.M., Vasan, S., Luo, J., Binder, A.M., Chlebowski, R.T., Quesenberry, C., Banack, H.R., Caan, B.J., Paskett, E.D., Williams, G.R. et al. (2023). Long-term trajectories of physical function decline in women with and without cancer. *JAMA Oncology*, 9(3), 395–403.

Gudur, A., Patil, V., Jain, A., Garg, N. (2024). Hybrid genetic algorithm and deep learning approach for lung nodule detection and classification in chest X-rays. *International Journal of Intelligent Systems Applications in Engineering*, 12(3s), 577–587.

Kumar, A., Batta, P., Ahuja, S., Rathore, P.S. (eds) (2026a). *AI-driven Innovations in Physiotherapy and Oncology 1*. ISTE, London and Wiley, Hoboken, NJ.

Kumar, A., Batta, P., Ahuja, S., Rathore, P.S. (eds) (2026b). *AI-driven Innovations in Physiotherapy and Oncology 2*. ISTE, London and Wiley, Hoboken, NJ.

Li, C., Lin, Y., Xiao, X., Guo, X., Fei, J., Lu, Y., Zhao, J., Zhang, L. (2025). Development and validation of a risk prediction model for kinesiophobia in postoperative lung cancer patients: An interpretable machine learning algorithm study. *Scientific Reports*, 15(1), 19412.

Lv, J., Wu, C., Li, J., Chen, F., He, S., He, Q., Zhou, G., Ma, J., Sun, Y., Wei, D. et al. (2022). Improving on-treatment risk stratification of cancer patients with refined response classification and integration of circulating tumor DNA kinetics. *BMC Medicine*, 20(1), 268.

Myakala, P.K., Jonnalagadda, A.K., Bura, C. (2024). Federated learning and data privacy: A review of challenges and opportunities. *International Journal of Research Publication and Reviews*, 5(12), 1867–1879.

Obeagu, E.I. and Obeagu, G.U. (2024). Predictive models and biomarkers for survival in stage III breast cancer: A review of clinical applications and future directions. *Annals of Medicine and Surgery*, 86(10), 5980–5987.

Oliveira, R., Postolache, G., Moreira, I., Domingos, S., Vicente, S., Pedro, L. (2016). Technology use and preference among physiotherapy patients. In *20th European Congress of Physical and Rehabilitation Medicine*, Estoril Congress Center, Estoril.

Post, K.E. (2019). Understanding patient engagement in breast cancer survivorship care: A national web-based survey. Doctoral Dissertation, Boston College.

Radford, K., Kettlewell, J., das Nair, R., Morriss, R., Holmes, J., Kellezi, B., Timmons, S., Jones, T., Tresidder, H., Andrews, I. et al. (2024). Development of a vocational rehabilitation intervention to support return-to-work and well-being following major trauma: A person-based approach. *BMJ Open*, 14(10), e085724.

Rane, N., Choudhary, S.P., Rane, J. (2024). Ensemble deep learning and machine learning: Applications, opportunities, challenges, and future directions. *Studies in Medical and Health Sciences*, 1(2), 18–41.

Sarfati, D., Koczwara, B., Jackson, C. (2016). The impact of comorbidity on cancer and its treatment. *CA: A Cancer Journal for Clinicians*, 66(4), 337–350.

Shaik, M.A., Choo, P.Y., Tan-Ho, G., Lee, J.C., Ho, A.H. (2024). Recovery needs and psychosocial rehabilitation trajectory of stroke survivors (PReTS): A qualitative systematic review of systematic reviews. *Clinical Rehabilitation*, 38(2), 263–284.

Stubblefield, M.D., Custodio, C.M., Franklin, D.J. (2006). Cardiopulmonary rehabilitation and cancer rehabilitation. 3. Cancer rehabilitation. *Archives of Physical Medicine and Rehabilitation*, 87(3), 65–71.

Suresh, K., Severn, C., Ghosh, D. (2022). Survival prediction models: An introduction to discrete-time modeling. *BMC Medical Research Methodology*, 22(1), 207.

Teshale, A.B., Htun, H.L., Vered, M., Owen, A.J., Freak-Poli, R. (2024). A systematic review of artificial intelligence models for Time-to-Event outcome applied in cardiovascular disease risk prediction. *Journal of Medical Systems*, 48(1), 68.

Wood, J.P., Connelly, D.M., Maly, M.R. (2010). 'Getting back to real living': A qualitative study of the process of community reintegration after stroke. *Clinical Rehabilitation*, 24(11), 1045–1056.

Yadav, T., Varadharajulu, G., Gudur, R. (2024). Multimodal therapeutic exercises and their role in mitigating chemotherapy-induced peripheral neuropathy: A longitudinal study on patient outcomes. *Cureus*, 16(11), e74214. doi: 10.7759/cureus.74214.

Yang, Y., Liang, J., Yu, B., Chen, Y., Ren, J.S., Shi, B. (2024). Latency correction for event-guided deblurring and frame interpolation. In *Proceedings of the IEEE/CVF Conference on Computer Vision and Pattern Recognition.*

Yin, D., Li, Q., Karimi, A., Mitsakos, N., El Mahbes, R., Cox, J. (2024). Integration of AI and domain expert knowledge to improve ESP efficiency and performance. In *SPE Middle East Artificial Lift Conference and Exhibition*, D011S003R001.

Zaree, A., Dev, S., Khan, I.Y., Arain, M., Rasool, S., Rana, M.A., Kanwal, K., Bhagat, R., Prachi, F.N., Puri, P. et al. (2023). Cardiac rehabilitation in the modern era: Optimizing recovery and reducing recurrence. *Cureus*, 15(9), e46006. doi: 10.7759/cureus.46006.

Zhou, C.M., Wang, Y., Yang, J.J., Zhu, Y. (2023). Predicting postoperative gastric cancer prognosis based on inflammatory factors and machine learning technology. *BMC Medical Informatics and Decision Making*, 23(1), 53.

15

AI-Enabled Monitoring of Neuromuscular Recovery in Cancer Rehabilitation

Survivors of more aggressive strategies of cancer treatment – surgery, chemotherapy or radiation therapy – can experience lingering effects, particularly neuromuscular deficits. Weakness, neuropathy and balance dysfunction are significant contributors to impaired mobility and the overall functioning of survivors. This makes the monitoring of neuromuscular recovery crucial for the formulation of individualized rehabilitation strategies. The effectiveness of rehabilitation can influence therapeutic outcomes and prolonged quality of life can be negatively or positively impacted. Although traditional assessment methods have merit, they are inherently episodic and subjective, and, more critically, they lack the sensitivity to identify small or initial shifts in the patient's motor control. There is significant potential in artificial intelligence (AI) for healthcare, particularly in continuous, automated and fine-resolution evaluation. AI-powered tools, particularly wearables, machine learning and computer vision systems, facilitate remote and real-time monitoring of recovery, muscle engagement and movement patterns. This allows rehabilitation to be more flexible and responsive to patients whose functional status is changing.

This chapter concentrates on cancer rehabilitation, particularly the monitoring of neuromuscular systems, which has AI-infused technologies integrated into the architecture and systems for clinical application. Enhanced monitoring of functional superficial and deep neuromuscular systems and remote rehabilitation have more precision, improved speed of care and rapid response healthcare. These include issues such as the quality of the data used, how understandable the models are, biases present in the algorithms and the ethics of patient privacy, fairness and bias contained in the algorithms.

15.1. Introduction

Cancer treatments can save people's lives and also have consequences that can limit a person's body and affect their quality of life. Neuromuscular problems such

Chapter written by Rashmi GUDUR and Abhay KASHETWAR.

as weakness, CIPN, balance and mobility problems, joint and muscle tightness and difficulty with fine and gross motor skills may occur. Neuropathic sequelae after chemotherapy and nerve-related after surgical trauma due to the disruption of muscle and nerve connections, and radiation, may cause fibrosis that tightens muscle and connective tissue and coordinated movement disorder (Nori et al. 2020). Prioritizing the rehab challenges is vital to improving the quality of life after cancer.

In cancer care, recovery assessments relying on CPT and goniometry, muscle testing, APM and scaled PROMs have undoubtedly dominated the practice, although their adoption looks more like a rationalization of the process (Voet et al. 2024). It is crucial to understand the rigor of motor recovery and the flattened, vertical loss and cancer recovery paradox. It creates the need for more purposeful, continuous and accurate communication regarding the loss and recovery of motor function and the need for AI to fill that gap.

The use of artificial intelligence (AI) can pivot integrated machine learning (ML) and data development algorithms to provide usable information on a patient's motor function, consolidating and simplifying disparate and complex multimodal datasets.

AI technology integrated with wearable sensors (inertial measurement units (IMUs), electromyographic and surface electromyographic (sEMG) sensors), computer vision (CV) and biomechanical simulations have the ability to follow and assess movement, muscle activation, gait and posture control, and asymmetries and control posture, all in real-time and at a very high time resolution. These advanced systems can perform remote real-time assessment of neuromuscular recovery and recuperation, and often long before clinical assessment and evaluation, these systems can identify early signs of deterioration or plateauing of active recovery (Shen 2024).

Deterioration and recovery can happen in real time, which allows for the recovery to be far more prompt, targeted and goal-directed to specific rehabilitation task interventions that can include modifications to the prescribed exercises, the implementation of specific balance or strength training and even the integration of multiple disciplines to support the recovery (Shubert 2011). The recovery monitoring algorithms in AI technology also allow for more personalized and specific adjustments to be made to rehabilitation strategies, focusing on the response to the treatment as well as existing comorbidities, which is the basis of value-based care.

This chapter discusses the application of AI monitoring systems in cancer rehabilitation and the fundamental technologies that enabled these systems, specifically sensor arrays, data processing systems and decision support algorithms. The primary applications addressed include remote supervision of rehabilitation, automated adaptive rehabilitation and the tracking of functional rehabilitation outcomes.

In the long run, smart supervision during rehabilitation can enhance results, promote autonomy and elevate international norms of care for rehabilitated cancer patients.

15.2. Neuromuscular impairments in cancer survivors

Aseptic necrosis is the destruction of bone tissue due to a lack of blood supply, which can occur after a fracture, bone surgery, or dislocation, or accompany systemic diseases such as sickle cell anemia, hematopoietic malignancies or after the use of high-dose corticosteroids. Bone tissue and stem cells are extremely sensitive to hypoxia, leading to rapid tissue death and necrosis (Zajączkowska et al. 2019). Distal bone segments are more susceptible to necrosis due to their distance from the main blood supply.

If the body has finite physiological resources to sustain a certain region, the extra bone tissue may endure necrosis due to the lack of compensation resources. Extensive necrosis may lead to BS or vascular necrosis, which can cause a range of motion issues, stiffness, chronic pain and disability. Nerve tissue stepping from the CNS and other regions can suffer from injury, leading to the loss of the affected region.

The above factors are related to flexor and extensor muscles suffering from tissue hypoxia and necrosis. Hypoxia and necrosis of tissue within muscles cause pain and restrict range of motion (Pattanshetty and Rao 2021). For main extensor and flexor muscles of the body, the loss of 10%–15% of range of motion is present, leading to altered and abnormal limb gait, which can cause the limb to appear as if it is stroking.

Increased generalized muscle weakness, decreased stability of posture and poorly coordinated and regulated motor skills can arise from cancer-related fatigue, deconditioning and cancer cachexia, which is a syndrome with identifiable waste of muscles and exhaustion and can also include generalized weakness (Masri 2023). These factors can inhibit easy movement and increase the risk of falling, which can lead to a permanent loss of independence due to restricted movement.

In rehabilitation, standard approaches include the MRC muscle grading scale, functional independence measure (FIM) and timed up and go (TUG), as well as patient-centered instruments such as Functional Assessment of Cancer Therapy: Fatigue (FACT-F) and the Neuropathy Screening tools. These tools are also steeped in subjective reporting, inter-rater dependency, topicality and relative unwillingness to address the issue and hence do not speak to the complete subject. Continuous, objective and sensitive tools are needed to address the gaps in the assessment of dynamic and static neuromuscular status (Harris et al. 2022).

CV and AI-enabled tools of predictive analytics, along with passive wearable sensors, can provide analytics on movement ease, joint angles, posture control and other specific metrics of muscle activation during gait and other higher-level motor activities. These predictive analytics can indicate and assist in the refinement of rehabilitation and other clinical tools to implement seamless transitions and interventional pathways.

The technology systems we have discussed could significantly aid in the expansion of support services within cancer rehabilitation units, such as surveillance and patient assistance tools and evaluation of rehabilitation progress analytics.

15.3. AI technologies for neuromuscular monitoring

15.3.1. *Wearable sensor-based systems*

For rehabilitation of cancer survivors, AI tools start using wearable sensor systems. These systems include sEMG sensors, pressure insoles and IMUs to assess the biomechanics of muscle functions in the range of seconds. IMU accelerometers and gyroscopes capture the limbs' and trunk's movements, while the electromyography (EMG) sEMG sensors capture the timing and amplitude of the muscle action to assess the movement relationship in the neuromuscular systems (Malesevic et al. 2023).

Pressure insoles aid in the phase and asymmetry analysis of the gait by providing a likely set of contacts of the feet. Capturing the essence of this highly intricate and varied dataset involves supervised ML systems. Supervised models, such as support vector machines (SVMs) and random forests, have succeeded in classifying gait patterns, identifying alterations associated with chemotherapy-induced peripheral neuropathy and differentiating movement patterns in various treatment phases.

These models incorporate hand-crafted features such as stride variability, stance-to-swing ratios and timing of peak muscle activation.

As previously mentioned, advanced methods in deep learning, especially convolutional neural networks (CNNs) and recurrent neural networks (RNNs), analyze complex sensor data and automate feature extraction within the data stream. Since these algorithms can supervise users in real time without the need for custom-designed features, they become more scalable and generalizable.

The power of wearable technology comes from its portability and the ecological validity it provides. Unlike assessments in clinical settings, wearable technology enables assessments during unrestricted everyday activities.

15.3.2. *CV and pose estimation*

The advent of AI has enabled CV to assess human activities using RGB, depth and infrared cameras, thus enabling motion capturing without the use of markers. These systems use pose estimation structure algorithms to assess the anatomical positions of the body and the changes in body positions across time to generate movement of body parts (kinematic) and the angles and symmetry of the muscle (dynamic) for the neuromuscular system. Models such as OpenPose, HRNet and Convolutional Pose Machines (CPMs) are core to pose estimation systems.

These pioneering architectures employ deep learning techniques and use multistage CNNs to extrapolate body joint positions, even in cases of occlusion or challenging lighting conditions (Samkari et al. 2023).

Extracted joint trajectories can now be evaluated with ML models to determine the level of stability, control, sway, balance asymmetries of the body and posture control of survivors from certain cancers recovering from musculoskeletal impairments or surgeries. In the case of telerehabilitation, CV systems permit the monitoring of patients in a remote setting.

AI-enabled video systems analyze the data in real time and measure and quantify the step width, gait speed, ability to turn and overall fluidity of movement, so that clinicians can determine the level of rehabilitation the patient has achieved without the in-person need for the patient. This ability to operate from a distance is in great need for patients residing in rural or lower socioeconomic areas.

15.3.3. *Multimodal data fusion*

The use of several sensors or combined data from wearables, EMG, CV or even patient feedback can provide a more in-depth and holistic, well-rounded perspective on neuromuscular recovery. Multimodal data fusion attempts to capture the strengths of each sensor to mitigate the shortcomings that a single one cannot fulfill. Fusion architectures, including multistream CNNs or transformers, are capable of learning and processing datasets from temporally aligned multistreamed sources. These nets can capture cross-domain relations, for instance, associating electromyography with movement kinematics through CV or IMUs.

The precision of diagnostics and the accuracy of monitoring are further enhanced by the richer neuromuscular profiles obtained. AI-powered predictive analytics can use comprehensive longitudinal multimodal datasets to forecast recovery trajectories, classify patients at risk of delayed motor recovery and recommend tailored recovery strategies (Ali 2022). These predictive insights can help make changes to a predetermined course of action to avoid permanent disability or a decline in functionality.

15.3.4. *Edge versus cloud processing*

AI monitoring systems can be divided between edge and cloud processing. Both methods provide unique benefits. Edge AI operates by completing calculations on the physical device or phone. This method has low latency and provides real-time feedback, such as posture or walking corrections, without the use of the Internet. Edge computing also improves patient privacy, an important consideration in managing sensitive medical information (Hartmann et al. 2022). High-capacity processing, such as deep learning model training and large population datasets, is reserved for cloud-based AI. Decentralized computing allows for the constant updating of models, supporting cross-patient learning and helping clinicians spot major common rehabilitation outcome patterns. A hybrid edge–cloud system is often optimal, combining the best aspects of both methods and ensuring immediacy, scalability and security.

Table 15.1 shows the summary of AI modalities and computational frameworks employed for monitoring neuromuscular recovery in cancer rehabilitation, highlighting their data inputs, analytical techniques, clinical roles and implementation challenges.

Technology/modality	Primary data sources	Core AI techniques	Clinical objectives	Key advantages	Major limitations/challenges
Wearable sensor systems (sEMG, IMUs, Pressure Insoles, etc.)	Electromyographic signals; acceleration; gyroscope; plantar pressure	Supervised ML (SVM, RF); deep learning (CNN, RNN)	Real-time assessment of gait; muscle activation; postural stability	Portable, objective and enables home use and continuous monitoring	Sensor drift; placement variability; limited contextual awareness
Computer vision and pose estimation systems	RGB/Depth/IR video; body joint trajectories	CNN-based pose estimation (OpenPose, HRNet, CPM, etc.); motion analytics	Kinematic/dynamic analysis; posture and balance tracking; telerehabilitation	Non-contact assessment; scalable via smartphone/webcams	Lighting/occlusion sensitivity; privacy concerns; high computation
Multimodal data fusion platforms	Combined EMG, IMU, CV and patient feedback data	Multistream CNNs; transformers; sensor-fusion networks	Holistic neuromuscular recovery profiling; prediction of motor improvement	Cross-domain synergy; richer insights; robust diagnostics	Data alignment complexity; high training data requirements

Technology/modality	Primary data sources	Core AI techniques	Clinical objectives	Key advantages	Major limitations/challenges
Edge AI systems	On-device data streams (wearables, smartphones, etc.)	Lightweight ML inference; federated learning	Real-time feedback (e.g. gait correction) with privacy preservation	Low latency; offline capability; improved data security	Limited computational resources; small model capacity
Cloud AI systems	Centralized; large-scale datasets	Deep learning (DL) training; predictive analytics	Population-level trend modeling; longitudinal recovery analysis	Scalable, high compute power; multipatient learning	Requires stable Internet; higher privacy risks
Hybrid edge–cloud architectures	Distributed wearable + cloud datasets	Model synchronization; edge–cloud collaboration frameworks	Integrative analytics balancing immediacy and scalability	Combines strengths of both approaches	Integration and maintenance complexity

Table 15.1. *Overview of AI technologies for monitoring neuromuscular recovery in cancer rehabilitation*

15.4. Clinical applications in cancer rehabilitation

15.4.1. *Baseline quantification and personalized goal setting*

With the initial stage of any cancer rehab exercise, there is the need to define a set of quantifiable objectives. These objectives span across muscle strength, gait, peripheral neuropathy, postural stability and other attributes. For AI to accurately assess the conditions of a patient, AI-powered assessment tools, which include wearable sensors, CV and other technologies, need to be employed. AI is able to replace emotions with data and is able to set the standard, which is reproducible. Its normative dataset is age, sex and cancer-type-informed estimates of muscle strength, postural stability, gait asymmetry and peripheral neuropathy. Rubrics of conventional medicine are much more subjective. AI contextualizes a patient's functional status and finds unique deficits based on age, sex and cancer type (Ganti 2025). With such a set baseline, achievable rehab goals and thresholds are set, which are unique to the patient recovered from breast cancer with dissection and ascendant lymphedema. There are AI estimates of their mobility, which are above the post-surgery AI normative curve. Such estimates enhance clinical decision-making and patient actions that achieve the minimally defined targets. These are more easily acceptable goals, which are AI-driven.

15.4.2. *Monitoring recovery and adapting rehabilitation*

While patients are in different phases of cancer treatments and are counted as survivors, monitoring them becomes highly significant. Also, tracking the rate of recovery becomes more important, especially in the case of neuromuscular recovery.

AI processes can receive and analyze sensor and vision data in real time and can sometimes note changes in motor performance that indicate improvement, stagnation or regression. Changes in performance most often happen undetected between appointments between a clinician and patient. The changes in performance can be credited to a lack of changes in rehabilitation plans. For example, if a patient is experiencing degeneration of balance due to chemo and has been diagnosed with chemo-induced neuropathy, the system can clinically suggest a change in the ability of the patient to walk independently due to the assistive augmented self-training (APTA Oncology 2025).

In addition to this, the changes in balance can clinically be analyzed to receive a prescription due to the more than necessary stimulation experienced. This patient model not only helps in achieving the goals of rehab but also prevents adverse complications.

15.4.3. *Remote telerehabilitation and home monitoring*

In cancer rehabilitation, immunosuppression, general fatigue, and limited access to care centers often necessitate remote telerehabilitation and home monitoring, thereby reducing the frequency of in-person appointments. Telerehabilitation has AI with the capability of guiding the aide on the use of the wearable devices. The devices include smart insoles, units of measurements and sensors for electromyography, which can process data and stream this for the doctor to analyze.

To assess the quality of movements, identify improper exercise and provide instant corrective feedback, CV systems can access smartphones or home cameras (Mukund 2025). They provide a new way for care systems to continue operating in dislocated or restricted physical environments. In addition, systems can converse with telehealth platforms to support virtual consultations, check-ins and inter-professional teamwork. In the telehealth systems, accessibility and continuity of care are provided with the required surveillance and protection of data needed.

15.4.4. *Complication risk stratification*

AI techniques can analyze longitudinally collected patient data, including performance, treatment history and other pertinent information and can stratify patients and evaluate risks of complex outcomes. For example, predictive models can identify patients at high risk for progression of chemotherapy-induced peripheral neuropathy, impaired postural control and/or slow musculoskeletal recovery (Müller et al. 2021). Such stratification allows clinicians to provide timely and more focused interventions that may include balance training, prescribing assistive devices, modifying medications or making referrals to other specialists. AI also helps in predicting risks associated with polypharmacy, high treatment intensity and compromised motor functions, which allows for safer care planning. AI can raise alerts for risks related to immobility, muscle disuse and deconditioning and decreased rehabilitation effort. This allows the rehabilitation team to more intensively address the focus of therapy in a more streamlined manner using the intensity of the devices.

15.4.5. *Outcomes research and quality improvement*

AI-powered platforms improve the support of personalized care. They also facilitate large-scale research. Using fully anonymized and aggregated data from different hospitals, researchers can evaluate the relative effectiveness of various rehab protocols, cancer types or demographic factors and create usable clinical guidelines (Stout et al. 2021). In addition, with the increasing scope of research, these datasets also aid in training AI models to enhance their accuracy and reduce bias, thus improving the generalization of the models.

Real-world usage patterns, adherence trends and functional trajectories assist rehabilitation providers in strategizing areas for enhancement, setting improvement goals and ultimately improving the service quality of cancer survivorship care.

15.5. Challenges and considerations

15.5.1. *Data quality, labeling and generalizability*

AI systems that monitor systems used for neuromuscular recovery in cancer rehabilitation charge the model with data and teach it, which constrains the model's accuracy with the quality, consistency and representativeness of the data used. One of the primary problem areas is the data heterogeneity across different systems. Patient data can be captured with the assistance of a multitude of wearable sensors, smartphone interfaces or CV systems in different settings, including in the clinic or during home rehabilitation (Toh et al. 2023).

These factors, which include hardware specifications such as sampling rates and sensor placements, environmental conditions such as lighting and individual patient factors such as body type, movement patterns, cancer type and side effects of cancer treatment, introduce inconsistencies or noise, which limit the model's reliability and generalization across populations. Also, the problem of mislabeled or unmatched data represents a critical bottleneck. While there is a plethora of datasets for general motor function or gait analysis in healthy populations, there is a glaring lack of annotated datasets that capture the unique neuromuscular impairments of cancer survivors.

Neuropathy caused by chemotherapy and surgery, along with radiation fibrosis, surgical trauma and exhaustion syndromes, may lead to the need for domain-specific labels, which are costly, tedious and highly manual. Besides, labeling those domains with gold standards the way they are meant to restricts scalability because it almost always requires expert clinicians or therapists. Still, the potential for bias and underrepresentation remains. It is possible for datasets to represent the younger, urban, tech-savvy populations almost exclusively and neglect older adults, underrepresented groups, those in rural areas and patients with lesser-known cancers. This results in bias in the underlying model and inequitable outcomes in AI-enabled rehabilitation. It, therefore, becomes imperative to include geographic, demographic and clinical diversity in datasets to promote equity and the underlying model's robustness.

15.5.2. *AI model understanding and clinician's confidence*

Direct trust and reliability are vital for the incorporation of AI tools in clinical rehabilitation. Clinicians and therapists are usually hesitant to use AI "black boxes", whereby the system derives some conclusions and offers recommendations and fails to drill down to justify how those conclusions were reached. Lack of transparency in vital healthcare choices, in case AI suggestions run counter to the clinician's decision, raises the refusal to accept and also raises apprehensions of responsibility, which inhibits assimilation of new advanced ideas.

As AI becomes more common, there is a need to use more explainable techniques such as explainable artificial intelligence (XAI). Neural networks have attention maps, feature attribution has Shapley values, and CV uses saliency overlays. These tools allow clinicians to know how certain features and movements affect predictions (Jin et al. 2022).

For example, when a model predicts a patient is likely to deteriorate in gait, visualizations showing changes in stride length, or angle asymmetry in joints, provide feedback. It is also worth noting that interpretability helps human and AI clinicians in making decisions. Unrestricted models enable rehabilitation clinicians to validate, negate or add to the algorithm, creating collaboration, not replacement. Research in this area of interpretability should be a design focus rather than an appendage, particularly when AI is used with vulnerable populations, such as cancer survivors.

15.5.3. *Privacy, security and data sharing*

The sensitive nature of patient data such as videos, wearable sensors and EMRs comes with significant privacy and security threats. Ethics and reputation are compromised when a person who has not been given permission accesses the information, or it is deliberately misused, and a data breach occurs. The developers need to implement strong data safety measures such as end-to-end encryption, differential privacy, data anonymizing and secure cloud storage.

Edge computing has the potential to strengthen these protections. By being able to process data at the "edge" of the network, for example, on smartphones and edge-enabled wearables, sensitive data can be protected because it would not have to be sent to centralized servers, improving access, latency and real-time feedback (Putra et al. 2024). This is important in low-resource settings, for example, where Internet access is patchy.

15.5.4. *Regulatory and reimbursement barriers*

The regulatory context of AI-enabled tools for cancer rehabilitation is still multifaceted and developing. AI-enabled tools that undertake the monitoring of a patient's condition or guide them in some clinical decisions are likely to fall within the category of Software as a Medical Device (SaMD), which in turn is vetted by the U.S. Food and Drug Administration (USAFDA) and Europe's CE marking along with other regulatory bodies. These bodies still require a perimeter of ultra-stringent and controlled clinical trials, which is often a time burden and arduous. The overstated effort is still worth funding and developing the tech for. Subsequently, regulatory bodies are still slower than the tech itself to innovate the reimbursement policies.

Many healthcare facilities are still hesitant to adopt remote or AI physiotherapy due to the lack of clinical reimbursement codes (Ho et al. 2023). These healthcare organizations may lack the incentive to invest in AI healthcare solutions. These concerns must be the focus of advocacy and policy reform, along with addressing barriers to value-based healthcare.

15.5.5. *Integration into clinical workflow*

The efficacy of AI monitoring systems primarily depends on their ease of integration into clinical workflows. Standalone tools that need the user to unlearn processes or retrain extensively will almost always be met with resistance by therapists and other clinicians. Integration with existing EMRs is crucial to minimizing documentation and facilitating longitudinal patient tracking. The tools must also be user-friendly. AI tools need to be aligned with rehabilitation practices (Lee et al. 2025). Therapeutic dashboards should focus on the primary objectives of patient management, while reactive AI feedback should be goal-oriented, and the suggestions laminated. Automated systems should be complemented with structured transformative training programs aimed at improving the workforce and redefining the approach to patient management.

15.5.6. *Equity and access*

Access to AI in cancer rehabilitation is made through the enduring barrier of equality. Consider as an example the geographical periphery of the low- and middle-income economies or rural areas within these countries. These areas are likely to lack the underlying infrastructure needed to support advanced AI solutions, such as high-speed Internet and cloud services, as well as the expensive wearable devices. These gaps for the purpose of AI solutions need to be low. These gaps also need to be easy to use, friendly to low-bandwidth situations and more importantly available on devices that are routinely used, such as smartphones. Real-world applications of AI and affordable units that use a combination of low-cost sensors can dramatically improve the accessibility of AI-driven rehabilitation (Goyal and Fiorini 2025). Also, the need for inclusive design, which accounts for the range of languages, accessibility barriers, low literacy and high digital use skills, ensures that no patient will be left behind.

15.6. Future research directions

Incorporating AI within healthcare has numerous applications with the potential to AI-enable neuromuscular rehabilitation for cancer survivors within the advancing healthcare systems. Doing so will require the management and implementation of numerous strategic and technical approaches, which will then unlock the true potential of AI systems. This subsection highlights the most relevant directions for the future, which are considered to be of greatest potential, as well as the most requirements.

15.6.1. *Constructing large, various and longitudinal datasets*

The absence of large-scale diverse and longitudinal datasets is a critical obstacle in constructing AI systems designed for cancer rehabilitation. The majority of data that is used tends to be produced within homogeneous populations as well as short-term interventions, thus making it more difficult to generalize and be clinically robust (Wilhelm and Grossman 2010).

The cancer survivor population has yet to be captured as a dataset from numerous angles, such as motion files, videos, EMG and their own feedback. This includes, but is not limited to, their age, ethnicity, cancer type, cancer treatment and other skills. Multiorganization collaborations can help fill these data gaps (Hedges et al. 2021). The hospitals, rehabilitation facilities and academic institutions need to strategically align their data collection, data tagging and collection methods to collaborate on data partnerships.

Utilizing standardized data constructs (such as HL7 Fast Healthcare Interoperability Resources (FHIR)) and other interchangeable data formats will also improve data gaps. Also, it is critical to understand and define the recovery pathways and improve therapy regressions, determine the intervention needed and teach AI to track and analyze intervention data months and years after treatment.

15.6.2. *Transfer learning and self-supervised learning*

In relation to cancer rehabilitation, acquiring comprehensive labeled databases is a key challenge. Traditional methods can no longer be the only techniques employed. Transfer learning techniques allow the adjustment of pre-trained models from other domains (such as general gait analysis, tracking elderly mobility or the kinematics of sports) to be customized on smaller datasets specific to cancer, which drastically minimizes the data needed for the functions of the model to be operational. Self-supervised learning (SSL) is another effective method.

Predictive modeling techniques can help automate feature extraction, even when there is no labeled data. SSL unlocks rich feature learning through pose ordering or other surrogate labeling tasks. Motion analysis and biosignals have proven successful in generalized CV tasks and can be readily applied in rehabilitation settings where there is no ground truth but abundant raw sensor and video data. With SSL and domain-specific fine-tuning, it is possible to develop robust AI systems with little manual data labeling and annotation (Patil et al. 2024).

15.6.3. *Personalized and XAI systems*

Predictive AI systems require more than just customized predictive analysis to be acceptable for deployment in the clinic. XAI allows for steps in the analysis and decision process to be tracked, which is important for clinician and patient rationale when providing asymmetries in gait, compensatory movements or progression assessments (Sawangjai et al. 2019). AI should classify walking patterns but go further to explain "instability is detected during the mid-stance phase due to insufficient hip abduction strength". This adds explainability to the AI, which is equally important to personalization. AI mechanisms should integrate data to emphasize the individual, baseline target and goal, underlying conditions and personal preferences for the patient. Proactive patient player systems that modify the intensity of exercises or fatigue patterns to remind patients independent of clinician supervision improve engagement and adherence to the program.

With the use of wearable technologies and behavioral patterns, AI is able to design custom rehab programs with improved results.

15.6.4. *Clinical trials and effectiveness studies*

While several pilot studies and feasibility trials have demonstrated the technical feasibility of AI cancer rehabilitation monitoring, a strong clinical proof is still lacking. There is a gap of comprehensive large-scale randomized controlled trials (RCTs) focused on the efficacy, safety and cost-effectiveness of AI versus standard practice in care. These studies should address the short- and long-term outcomes of the interventions in relation to the improvements in physical function, quality of life, adherence, menopausal status and overall healthcare utilization. Such trials would also ideally assess effectiveness in diverse patient populations and cancer types. Implementation studies are needed to evaluate AI systems' effectiveness in real-world practice. System usability, clinician training necessary and integration with electronic health records (EHRs) should also be addressed (Kumar et al. 2026a).

15.6.5. *Policy and reimbursement advocacy*

Without institutional and financial backing, the most sophisticated AI systems will have difficulty scaling. Structural advocacy is imperative to ensure that AI cancer rehabilitation monitoring is recognized as a reimbursable service.

Stakeholder engagement with healthcare payers, regulatory bodies and associations should focus on developing guidelines, coverage policies and standardized coding for rehabilitation services delivered remotely and digitally (Kumar et al. 2026b). Demonstrating cost-effectiveness remains one of the strongest selling points. Research should focus on the clinical and economic values, such as the decrease of rehospitalizations, as well as the reduction of the cost and volume of in-person visits to the healthcare system, and the overall improved productivity of the workforce. These policies also need to include components related to ethical use, as well as the security and privacy of the sensitive health information of especially vulnerable members of the population.

15.6.6. *Emerging technological innovations*

Several of the portable technologies ought to improve the breadth of functionality and ease of use of AI-powered rehabilitation platforms, including, but not limited to:

– *Micro-portable sensors*: the integration of soft and flexible electronics with low-cost sensors have enabled the construction of portable, low-cost and gentle skin-mounted sensors that may be embedded in clothing, smart textiles and even patches (Richmond et al. 2017). These sensors can monitor and measure movements such as joint angles, muscle activity and balance in real time without disrupting the person's movement and without elaborate setup procedures.

– *Augmented and virtual reality (AR and VR) systems*: standalone, these technologies can often provide engaging, personalized and, at times, gamified rehabilitation. When coupled with AI, AR and VR systems, they can provide immersive and interactive rehabilitation, in real-time, as they track and monitor user movements with sensors and provide guided feedback for movements to enhance engagement (Rauniyar et al. 2023).

– *Federated learning*: federated learning addresses varying institutional privacy concerns and limitations on sharing patient data by training AI models without centralizing patient data (Gudur et al. 2024). Each site advances a shared model using its private data and only exchanges model weights – not the data. This approach facilitates both cross-border and cross-institutional collaborations while maintaining the confidentiality of the data.

15.7. Conclusion

AI is changing how we approach neuromuscular rehabilitation for cancer survivors by introducing real-time, customizable evaluations of their motor skills. Tailored recovery tracking is made possible through the latest advancements in multifunctional recovery monitoring tech, including sensors, CV and wearables. ML and adaptive analytics provide timely, tailored and real-time functional progress monitoring, no matter how remote or clinical the setting, which is vital for cancer survivors enduring fatigue and neuromuscular disorders. There are seven documented challenges to be addressed to aid recovery for patients and survivors of cancer. These include the explainability gap of AI, the overreach of Big Tech, the inter-patient population data heterogeneity, the clinical workflow bottleneck and ethical constraints. The inequitable distribution of new digital health technology and the resultant health inequities must be counteracted, which calls for the collection of more inclusive and well-defined datasets, transparently defining the AI models to be deployed and closing the explainability gap. AI approaches should undergo proper strategizing, validation and detailed validation with RCTs. For scaling up, the integration of cost-efficient advanced technology clinical mentors, clinically scalable interfaces, coupled with appropriate policies and reimbursement frameworks, will be crucial. In the integration of everyday medical practice with AI, clinicians, data scientists, engineers and policy-makers will need to work together. With these changes, AI-enabled neuromuscular monitoring will be fully incorporated into rehabilitation to greatly increase the functionality and quality of life of cancer survivors worldwide.

15.8. References

Ali, H. (2022). AI in neurodegenerative disease research: Early detection, cognitive decline prediction, and brain imaging biomarker identification. *International Journal of Engineering Technology Research & Management*, 6(10), 71.

APTA Oncology (2025). Combined sections meeting (CSM) 2025 poster and platform presentations. *Rehabilitation*, 43(1), 38–66.

Ganti, V.K. (2025). *Beyond the Stethoscope: How Artificial Intelligence is Redefining Diagnosis, Treatment, and Patient Care in the 21st Century*. Deep Science Publishing, New Delhi.

Goyal, S. and Fiorini, L. (2025). Practical implementation and integration of AI in mental healthcare. In *Adversarial Deep Generative Techniques for Early Diagnosis of Neurological Conditions and Mental Health Practises: Theoretical Insights with Practical Applications*, Kumar, A., Ortiz-Rodriguez, F., De Vasconcelos, J.B. et al. (eds). Springer, Cham, 373–415.

Gudur, R.A., Bhosale, S.J., Gudur, A.K., Kale, S.R., More, A.L., Datkhile, K.D. (2024). The effect of CYP2C19* 2 (rs4244285) and CYP17 (rs743572) SNPs on adriamycin and Paclitaxel based chemotherapy outcomes in breast cancer patients. *Asian Pacific Journal of Cancer Prevention*, 25(6), 1977.

Harris, E.J., Khoo, I.H., Demircan, E. (2022). A survey of human gait-based artificial intelligence applications. *Frontiers in Robotics and AI*, 8, 749274.

Hartmann, M., Hashmi, U.S., Imran, A. (2022). Edge computing in smart health care systems: Review, challenges, and research directions. *Transactions on Emerging Telecommunications Technologies*, 33(3), e3710.

Hedges, J.R., Soliman, K.F., Southerland, W.M., D'amour, G., Fernández-Repollet, E., Khan, S.A., Kumar, D., Shikuma, C.M., Rivers, B.M., Yates, C.C. et al. (2021). Strengthening and sustaining inter-institutional research collaborations and partnerships. *International Journal of Environmental Research and Public Health*, 18(5), 2727.

Ho, D., Sapanel, Y., Blasiak, A. (2023). *Medicine without Meds: Transforming Patient Care with Digital Therapies*. JHU Press, Baltimore.

Jin, D., Sergeeva, E., Weng, W.H., Chauhan, G., Szolovits, P. (2022). Explainable deep learning in healthcare: A methodological survey from an attribution view. *WIREs Mechanisms of Disease*, 14(3), e1548.

Kumar, A., Batta, P., Ahuja, S., Rathore, P.S. (eds) (2026a). *AI-driven Innovations in Physiotherapy and Oncology 1*. ISTE Ltd, London and Wiley, Hoboken, NJ.

Kumar, A., Batta, P., Ahuja, S., Rathore, P.S. (eds) (2026b). *AI-driven Innovations in Physiotherapy and Oncology 2*. ISTE Ltd, London and Wiley, Hoboken, NJ.

Lee, M., Kim, K., Shin, Y., Lee, Y., Kim, T.J. (2025). Advancements in electronic medical records for clinical trials: Enhancing data management and research efficiency. *Cancers*, 17(9), 1552.

Malesevic, N., Svensson, I., Hägglund, G., Antfolk, C. (2023). An integrated approach for real-time monitoring of knee dynamics with IMUs and multichannel EMG. *Sensors*, 23(21), 8955.

Masri, S. (2023). A systemic view of cancer cachexia: Consequences of altered metabolic networks. *Journal of Cachexia, Sarcopenia and Muscle*, 14, 3–59.

Mukund, A. (2025). Wearable technology and sensor data in assistive systems: Enhancing rehabilitation through predictive analytics. In *Predictive Algorithms for Rehabilitation and Assistive Systems*, Korupalli, V.R.K., Ranasinghe, P., Feldman, D. (eds). IGI Global, Hershey, PA, 253–290.

Müller, J., Kreutz, C., Ringhof, S., Koeppel, M., Kleindienst, N., Sam, G., Schneeweiss, A., Wiskemann, J., Weiler, M. (2021). Chemotherapy-induced peripheral neuropathy: Longitudinal analysis of predictors for postural control. *Scientific Reports*, 11(1), 2398.

Nori, P., Kline-Quiroz, C., Stubblefield, M.D. (2020). Cancer rehabilitation: Acute and chronic issues, nerve injury, radiation sequelae, surgical and chemo-related, part 2. *Medical Clinics*, 104(2), 251–262.

Patil, D.S., Yadav, T., Gudur, A. (2024). To compare the effect of swallowing and non-swallowing exercises on dysphagia in post operative oral cancer individuals: A comparative study. *Asian Pacific Journal of Cancer Prevention*, 25(8), 2945.

Pattanshetty, R. and Rao, M.S. (2021). Cancer-related fibrosis: Prevention or treatment?: A descriptive review. *Journal of Dr. YSR University of Health Sciences*, 10(4), 222–228.

Putra, K.T., Arrayyan, A.Z., Hayati, N., Damarjati, C., Bakar, A., Chen, H.C. (2024). A review on the application of internet of medical things in wearable personal health monitoring: A cloud-edge artificial intelligence approach. *IEEE Access*, 12, 21437–21452.

Rauniyar, A., Hagos, D.H., Jha, D., Håkegård, J.E., Bagci, U., Rawat, D.B., Vlassov, V. (2023). Federated learning for medical applications: A taxonomy, current trends, challenges, and future research directions. *IEEE Internet of Things Journal*, 11(5), 7374–7398.

Richmond, T., Peterson, C., Cason, J., Billings, M., Terrell, E.A., Lee, A.C., Towey, M., Parmanto, B., Saptono, A., Cohn, E.R. et al. (2017). American Telemedicine Association's principles for delivering telerehabilitation services. *International Journal of Telerehabilitation*, 9(2), 63.

Samkari, E., Arif, M., Alghamdi, M., Al Ghamdi, M.A. (2023). Human pose estimation using deep learning: A systematic literature review. *Machine Learning and Knowledge Extraction*, 5(4), 1612–1659.

Sawangjai, P., Hompoonsup, S., Leelaarporn, P., Kongwudhikunakorn, S., Wilaiprasitporn, T. (2019). Consumer grade EEG measuring sensors as research tools: A review. *IEEE Sensors Journal*, 20(8), 3996–4024.

Shen, D. (2024). AI based fatigue assessment system for athletes using smart textile wearables. PhD Thesis, Université de Lille/Donghua University, Shanghai.

Shubert, T.E. (2011). Evidence-based exercise prescription for balance and falls prevention: A current review of the literature. *Journal of Geriatric Physical Therapy*, 34(3), 100–108.

Stout, N.L., Santa Mina, D., Lyons, K.D., Robb, K., Silver, J.K. (2021). A systematic review of rehabilitation and exercise recommendations in oncology guidelines. *CA: A Cancer Journal for Clinicians*, 71(2), 149–175.

Toh, S.F., Fong, K.N., Gonzalez, P.C., Tang, Y.M. (2023). Application of home-based wearable technologies in physical rehabilitation for stroke: A scoping review. *IEEE Transactions on Neural Systems and Rehabilitation Engineering*, 31, 1614–1623.

Voet, N., Pater, R., Garmendia, J., Sistiaga, A., Labayru, G., Gallais, B., de Groot, I., Muslemani, S., Gagnon, C., Graham, C. (2024). Patient-reported outcome measures in neuromuscular diseases: A scoping review. *Journal of Neuromuscular Diseases*, 11(3), 567–577.

Wilhelm, F.H. and Grossman, P. (2010). Emotions beyond the laboratory: Theoretical fundaments, study design, and analytic strategies for advanced ambulatory assessment. *Biological Psychology*, 84(3), 552–569.

Zajączkowska, R., Kocot-Kępska, M., Leppert, W., Wrzosek, A., Mika, J., Wordliczek, J. (2019). Mechanisms of chemotherapy-induced peripheral neuropathy. *International Journal of Molecular Sciences*, 20(6), 1451.

16

Automated Motion Capture Systems for Oncology Physiotherapy Using AI

Due to myopathy induced by surgery and steroids, deconditioning following surgery and chemotherapy-induced peripheral neuropathy, often combined with radiation fibrosis, treatment results in declined functionality, especially with regard to moving around. Restorative care specialists are trained in physiotherapy, but care brought to motion capture is not instantaneous or subjective and is often confined to the attention of the therapist. The rapid advancement of AI in motion capture tools (MoCap) allows for a revolutionary, non-biased way of capturing data, while also tailoring rehabilitation treatments and care. This chapter is mainly focused on the newest advancements in the oncology physiotherapy field involving physiotherapy and artificial intelligence. It goes through the timeline of the optic systems used with markers up to the more recent computer vision systems with markers and sensors. The focus is on the hospital and outpatient systems of advanced AI techniques for the pose estimation and gait analysis, activity recognition and anomaly detection. The aim of using sensors, RGB and depth cameras and algorithms is to facilitate real-time tracking and evaluation of body motion in order to measure the accuracy of physiotherapy treatment and assess the quality of movement, which in turn aids in the evaluation of improvement in functionality from a declined state. In contrast, patients whose treatment AI algorithms were trained on datasets with varying components can also be dissociated. Both responsive and non-responsive treatment therapies are used in order to implement flexible strategies to modify the treatment. The absence of standardized pre-validated models in the domain is also of concern, in addition to the variety of data and the integrity of patient analytics. In the end, we outline emerging domains such as federated learning, edge computing for home rehabilitation and explainable AI placed within clinical decision support. AI-powered motion capture systems could provide equitable access to continuous and detailed physiotherapy evaluations for cancer survivors.

Chapter written by Anand GUDUR and Mrudula NIMBARTE.

16.1. Introduction

16.1.1. *Background*

The quality of life a person has ultimately depends on the level of movement and physical activity they are able to perform. In some branches of medicine, this factor is particularly important. In the case of advanced-stage cancer, the patient's physical ability tends to decline even more. The decline is due to several interconnected factors, including fatigue, chemicals, limited mobility and even the physical impacts of surgery. Therefore, we would argue that strengthening the tissues, along with coordination and independent body movements, should be a priority of supportive care physiotherapy. Therapy is reliant on the assumption of real-time physical assessments by qualified physiotherapists who work on set schedules. They are able to perform this verification by observing the individual's movement patterns and participation in functional movement assessments, including the Timed Up and Go (TUG) test, the six-minute walk test (6MWT) and muscle strength assessments. While the physical assessments are useful to the clinician, they are, more often than not, inadequate in capturing the full magnitude of a movement associated with a given functional level. This is not a complete approach in oncology as cancer-related fatigue often manifests as a "fully devoid action", where the patient feels incapable of movement even if their muscle markers are stable. Physiotherapists have the difficult task of distinguishing between a patient's actual physical capacity and their subjective perception of decline. When Patients do not show false positives (FP) or overt symptoms, it can lead to a "slip of monitoring" where early interventions are missed. Without multilayer attention the subtle decline in a patient's mobility might be overlooked until it becomes a permanent disability. (Maffoni and Magnani 2025).

16.1.2. *Rise of motion capture in physiotherapy*

The application of motion capture technology (MoCap) in documentation is seen in a wide range of fields, with the most pronounced within the context of sports and animation. MoCap systems offer a more sophisticated framework of level recording in physiotherapy, because the parameters of movement exposed to a slip of monitoring, such as degree of joints, limb motion patterns and even the total extent of displacement within a given time period use simplistic mechano metrics.

Automation and artificial intelligence (AI) progression have widened the applicability and lowered the price and accessibility of motion capture systems used in clinical settings. Marker-based optic systems are now becoming obsolete and being replaced by inertial measurement units (IMUs), smartphones, depth cameras and even standard videos, which are cheaper systems for extracting 3D motion and pose data (Galasso 2024). With AI technologies integrated, these systems are able to detect clinical movement abnormalities, differentiate between them and generate automated feedback to patients and clinicians.

16.1.3. *Relevance in oncology*

Cancer survivors experience profound and complex changes in their physical conditions. With the neuromuscular system impacted by numerous surgeries and even chemotherapy and long-term medications, signs of deconditioning may go overlooked by the patient and their clinicians due to irregular appointments. Those who live in more remote areas lack access to the proper physiotherapy for their condition. AI automation systems, which are driven by automated motion capture technology, serve as an exciting prospect in remote feedback solutions and telerehabilitation. Such systems in oncology rehabilitation, which needs precise and flexible systems, are able to provide patient-centered solutions for clinicians to address common patient concerns (Hincks et al. 2019).

16.2. Overview of motion capture technologies in physiotherapy

A MoCap system is a technology that records, quantifies and analyzes human movement. In physiotherapy oncology, these systems provide objective, quantitative and continuous records of an individual's range of motion. This facilitates functional decline and treatment planning. These systems have evolved from laboratory-based monitoring systems moored, to real-time and portable systems that can be used in both clinics and homes (Meng et al. 2012). This chapter explains the different types of MoCap systems and their use in rehabilitation for cancer patients.

16.2.1. *Marker-based optical MoCap*

For a long time, marker-based optical systems have been the gold standard in biomedicine. These systems have multiple infrared cameras and use markers placed on specific anatomic points to capture three-dimensional motion. Their accuracy is to about a sub-millimeter.

Benefits:

– high precision and temporal resolution;

– joint angle and movement path accuracy;

– provides the gold standard of validation for gait and posture analysis.

Drawbacks:

– expensive and needs highly specialized laboratory environments;

– not practical for home or outpatient care;

– needs specialized and trained personnel for setup.

16.2.1.1. *Relevance to oncology*

While useful for detailed assessments in research or tertiary care connecters, these systems are not practical for longitudinal monitoring in most oncology rehabilitation settings (Stout et al. 2019). But they are helpful in validation for portable systems.

16.2.2. *Markerless capture systems*

Within the field of computer vision and deep learning, the use of markerless motion capture systems with RGB or RGB-D cameras (e.g. Sketch, Intel RealSense, Microsoft Kinect, standard webcams) has become popular. These systems compute the body's critical positions and joints from a video sequence.

Key technologies:

– real-time human pose estimation has algorithms such as OpenPose, MediaPipe, PoseNet and HRNet;

– single or multiple 2D views pose estimation using depth sensors and 3D Optical Systems.

Advantages:

– scalable, unobtrusive and low cost;

– remotely accessed and utilized on telehealth applications;

– real-world settings are available with low patient prep.

Limitations:

– lighting, clothing and occlusion greatly impact accuracy;

– patients with abnormal posture or complex movement are not as reliable;

– the camera's field of view and calibration are critical.

16.2.2.1. *Relevance to oncology*

The use of markerless systems with built-in AI analysis can automate movement evaluations without physical patient visits, enabling continuous assessment for patients confined to their homes (Herath et al. 2024).

16.2.3. *Wearable sensor systems*

Wearable motion capture systems use IMUs, accelerometers, gyroscopes and magnetometers to measure body orientation, acceleration and movement dynamics.

Common devices:

– consumer wearables such as Apple Watches, Fitbits and smartphones equipped with sensors.

Benefits:

– can be utilized anywhere outside of laboratory, for example, normal day-to-day activities;

– offers long-term monitoring and high-frequency sampling;

– resistant to lighting conditions and other types of obstruction and occlusion.

Drawbacks:

– calibration and correct placement of sensors is crucial;

– with time, data can deviate and be influenced by soft tissue artifacts and tissue movements;

– lack of visual cues can hinder a full understanding of the data.

16.2.3.1. *Relevance to oncology*

Wearables assist in cancer care by tracking fatigue, gait speed and exercise adherence. IMUs also unobtrusively give quantitative estimates for chemotherapy-induced neuropathy, post-operative recovery and fall risk assessments (Beauchamp et al. 2020).

16.2.4. *Hybrid systems*

Some new systems integrate visualization and wearable sensors to utilize both types together.

For instance, a patient could wear IMUs and simultaneously be filmed with an RGB-D camera that improves the overall accuracy and completeness of the captured movements.

Benefits:

– adds the spatial precision of the camera to the motion capture capabilities of the wearables;

– better reliability and compensates for data loss through redundancy.

Drawbacks:

– complex synchronization and data merging, as well as increased computation and energy usage.

16.2.4.1. *Relevance to oncology*

Hybrid systems have a clinical trial or distinct evaluation focus predicting complexities in post-surgery recoveries and rehabilitation intensity adjustments for high-risk patients (Cascella 2024).

16.2.5. *Smartphone-based capture*

Smartphones have embedded accelerometers and other components that make them an affordable and readily available motion capture device. Mobile apps have step counting and other activities that are classified as simple assessments such as the TUG test or upper limb motion assessments (Madhushri 2017).

Advantages:

– cost-effective and user friendly;

– compatibility with telemedicine and other standard medical record systems;

– no instruction or formal training is needed to operate the system.

Limitations:

– less precise than dedicated systems;

– concerns include battery performance, privacy of the information collected and differences in devices used.

16.2.5.1. *Relevance to oncology*

Remote mobile assessment and monitoring of physiotherapy is extremely valuable in countries with few resources and during periods of recovery at home, and this is supported by the common ownership of smartphones.

System type	Core technology	Advantages	Limitations	Relevance in oncology physiotherapy
Marker-based optical systems	Infrared cameras with reflective markers placed on anatomical landmarks	High precision and temporal resolution; accurate joint-angle tracking; gold-standard validation reference	Expensive; requires controlled lab environment; not suitable for home or outpatient use	Ideal for validating newer portable systems and detailed motion analysis in tertiary rehabilitation centers
Markerless vision systems (RGB/ RGB-D)	AI-driven computer vision algorithms (OpenPose, MediaPipe, HRNet, etc.) using 2D/3D cameras	Low-cost, contact-free; scalable for remote use; minimal setup	Sensitive to lighting, clothing and occlusion; less accurate for abnormal postures	Enables telerehabilitation and continuous remote assessment of cancer patients unable to attend clinics
Wearable sensor systems (IMU-based)	Inertial measurement units with accelerometers, gyroscopes and magnetometers	Long-term monitoring; usable in natural environments; not affected by lighting	Requires calibration; prone to drift and soft-tissue artifacts; limited visual context	Tracks gait, fatigue and neuropathy; quantifies day-to-day mobility and adherence to physiotherapy
Hybrid systems	Integration of vision (RGB-D) and wearable sensors for multimodal data capture	Combines spatial accuracy and temporal reliability; redundancy minimizes data loss	Complex synchronization and data processing; higher computational demand	Provides comprehensive assessments for complex post-surgical rehabilitation cases
Smartphone-based systems	Built-in accelerometers and cameras with app-based data analytics	Affordable, widely available, easy to use and supports telemedicine	Less precise; device variability; privacy concerns	Facilitates remote monitoring and home-based physiotherapy in resource-limited settings

Table 16.1. *Comparison of motion capture systems used in oncology physiotherapy*

16.3. AI techniques in motion data processing

AI assists in analyzing motion capture data, offering insightful interpretations. In oncology physiotherapy, AI plays a significant role in automating detection, classification and quantification of movement patterns to identify a functional decline, treatment side effects or a decline in therapy response early on (Davids et al. 2022). This section discusses different AI algorithms to analyze motion data in various capture forms and their applications in cancer rehabilitation.

16.3.1. *Preprocessing and feature extraction*

Motion data captured needs to be cleaned, organized and simplified before it can be used in machine learning (ML) algorithm models.

Common first steps taken in processing are as follows:

– smoothing noise, which can include using certain filters for IMU signals such as Kalman or Butterworth;

– merging data from different sensors, for example, using both the accelerometer and gyroscope to calculate orientation;

– normalizing data to account for differences in users' body sizes and the positions of the cameras;

– dividing continuous motion into discrete components or tasks, such as cycles of walking and sitting or walking strides.

The types of features captured include the following:

– gait parameters;

– kinematics of different joints;

– dissimilar movements in a certain time span;

– scores designed for certain tasks.

16.3.2. *Pose estimation and skeleton tracking*

With the vision technology in use, the first and most essential piece of AI is doing the pose estimation. This is the process of identifying certain body parts (such as elbows, knees or hips) and framing them into a skeletal structure in every video frame.

Some well known AI models include:

– OpenPose which does 2D skeletal creation for many users with the use of part affinity fields;

– MediaPipe Pose, a 2D and 3D mobile application done in real-time with a 3D pose feature;

– HRNet, a network with high precision for joint positioning;

– VIBE, video-based 3D pose estimator through adversarial learning.

These models and many others utilize deep convolutional neural networks, which are trained on sizable datasets such as a COCO, MPII or Human3.6M (Pavllo et al. 2019). Some techniques use smoothing or recurrent layers to improve stability over time across frames.

Application in oncology:

– quantify abnormal joint ranges during upper limb rehab post-mastectomy;

– identify and classify abnormal gaits due to chemotherapy-induced neuropathies;

– evaluate patient compliance to prescribed physiotherapy through webcams.

16.3.3. *Gait analysis and classification*

AI-based tools can automatically analyze and evaluate gait patterns by interpreting series of motion data. This is important in oncology since a lot of cancer patients have gait issues due to fatigue, neuropathies or post-surgery recovery.

Common ML models:

– *Support vector machines*: useful for high-dimensional non-sample datasets.

– *Random forests and XGBoost*: nonlinear relationships and interacting features.

– *K-nearest neighbors*: basic classifications of gait types.

– *Deep learning frameworks*: patterns in time series data.

Features include:

– IMU metrics enhanced by visual recordable gait parameters and lower limb movements;

– kinematics derived from vision systems: knee flexion, footlift, trunk oscillation, etc.;

– gait indices such as the gait deviation index (GDI) or gait variability index (GVI).

Potential applications:

– identify patients at high risk of falling;

– separate patients based on response to therapy;

– evaluate post-treatment balance difficulties after radiation or surgery for brain tumors.

16.3.4. *Activity recognition systems that automate recognition of activities*

AR systems utilize movement data to identify the specific forms of actions of a person (mobility such as walking, standing, sitting or performing any form of physical activity). They are crucial to understanding the patient's behavior away from the clinic as well as for monitoring compliance with scheduled physiotherapy appointments (Nascimento et al. 2020).

Approaches include:

– windowed feature extraction + classical ML (SVM, Decision Trees, etc.);

– deep convolutional neural network (CNN)/recurrent neural network (RNN) on raw sensor data in an end-to-end fashion;

– hybrid of spatial (CNN) and temporal (long short-term memory (LSTM)/Gated Recurrent Unit (GRU)) features.

Real-world application:

– monitoring compliance to a schedule of daily stretching exercises;

– post-discharge consideration of sedentary behavior;

– automatic prompting after sessions of chemotherapy to mitigate prolonged sitting or standing.

16.3.5. *Anomaly detection and personalized modeling*

Within the cancer rehabilitation domain, less focus is placed on differences from the individual's baseline compared to the population metrics. Anomaly detection and unsupervised learning can identify small functional alterations, particularly in the initial stages of decline.

Some common strategies are as follows:

– *Abnormal motion autoencoders*: an autoencoder neural network built to memorize and learn normal motion patterns; abnormality is signified by high reconstruction error.

– *One-class SVMs*: encapsulating "normal" data and creating a boundary around "normal" data;

– *Pattern recognition (e.g. k-means, DBSCAN, etc.)*: finding subgroups or outlying patients.

The use of longitudinal data to identify deviations from the baseline of a patient increases the sensitivity to change in ML models (Tayob and Feng 2022).

Use cases:

– identify and report patient CIPN-associated gait changes before the onset of symptoms;

– informs clinicians about therapy stagnation or deterioration;

– facilitates the planning of goal achievement and progress.

16.3.6. *Closed-loop systems and real-time feedback*

AI systems can offer complete supportive rehabilitation through closed loops. Machine learning models can assist in real-time supportive feedback during movements, optimization and adherence.

Example applications:

– provide real-time arm movement during leg pair gait training;

– audio re-instruction during gait training in a non-stance phase;

– real-time performance assessment in gamified physiotherapy applications.

Required capabilities:

– real-time inference (low latency);

– on the device model deployment (privacy and autonomy);

– continuous data streaming.

16.3.7. *Multimodal and multitask*

Some sophisticated systems use multimodal learning frameworks to combine motion data with other inputs such as heart rate, electromyography (EMG) and patient outcomes.

Benefits:

– greater accuracy through the use of augmenting signals;

– support joint learning of multiple tasks (e.g. predicting fatigue and decline in gait);

– provide comprehensive patient monitoring beyond mobility (Jeong et al. 2022).

Example AI models:

– multimodal transformer models;

– graph neural networks (GNNs) connecting joint and sensor models;

– use of attention in feature interpretation across channels.

16.4. Clinical applications in oncology rehabilitation

Oncological rehabilitation increasingly incorporates AI-powered motion capture systems. These systems assist in tracking recovery trajectories, therapy-induced decreases and customizing physiotherapy approaches.

16.4.1. *Monitoring chemotherapy*

Chemotherapy drugs such as taxanes, vinca alkaloids, and platinum-containing compounds can cause peripheral neuropathy, which may impair gait, balance, and fine motor control. (Laforgia et al. 2021). These consequences can be dose-dependent and may become disabling, and interventions to restore mobility and overall quality of life may offer little to no promise.

AI-enabled motion capture systems have the following aims:

– proactive identification of minor shifts in gait using IMU or camera systems with ML classifiers;

– periodic assessments of change in deterioration or improvement;

– identifying risk and classifying gait patterns against neuropathy severity scales (e.g. CTCAE, total neuropathy score, etc.).

Example: wearable sensor studies with SVM classifiers have shown the ability to capture the first signs of CIPN well ahead of traditional clinical detection, leading to timely dose modification or neuroprotective strategies.

16.4.2. *Evaluating recovery of functions post-cancer interventions*

Surgery for cancer, such as thoracotomies, mastectomies and abdominal or limb-sparing orthopedic reconstructions, is commonly associated with restrictions in mobility. Some milestones on recovery can be objectively verified as mastered, as well as deviations from the anticipated recovery roadmap.

Some uses are as follows:

– improving ultra limb mobility retention following breast surgery recovery using pose tracking;

– assessment of sit-to-stand "postures" after abdominal surgery to track recovery;

– symmetry of limb loading in patients after onco-orthopedic operations.

Advantage:

All patients can be evaluated from a distance that enables real-time adjustment of the rehabilitation procedure based on the measurements of the movement and posture algorithms (Nascimento et al. 2020).

16.4.3. *Managing and detecting lymphedema*

Range of motion and functioning of a limb can be greatly hindered by lymphedema, in particular after the removal of the lymph nodes, as well as post-radiation therapy.

How motion tracking and AI can assist in this:

– assessing asymmetrical arm movement in breast cancer survivors;

– identifying "self-protective" movement patterns in the earlier stages of lymphedema;

– secondary algorithm tracking of upper limb exercises for real-time correction monitoring.

Clinical benefit:

– Progressive changes and initial stages of irreversible fibrosis can be ruled out. Limb functionality can be improved for the long-term.

16.4.4. *Exercising and telerehabilitation*

During the pandemic, the compliance levels to home-based exercise regimens were proven to be an indicator of exercise adherence. The latter was the focal point of analytics done on physiotherapy (Sassi et al. 2024).

Advances:

– remote frameworks for monitoring and evaluating therapeutic exercises accessed via smartphones and webcams;

– engaging platforms with embedded video analysis for motivational reinforcement;

– alerts for breakdowns in the set range of motion or repetitions.

Example: rehabilitation platforms incorporating MediaPipe and LSTM models have achieved over 90% accuracy in classifying physiotherapy movements with real-time corrective feedback (Jones et al. 2020).

16.4.5. *Fatigue and frailty assessment*

One of the most debilitating symptoms of cancer and its treatment at any stage is cancer-related fatigue (CRF). There is still no way to objectively quantify it.

AI-powered approaches:

– Algorithms designed for activity recognition can determine the levels of sedentary behavior and quantify energy expenditure using IMU data (Yammouri and Ait Lahcen 2024).

– Patterns and variability of gait as well as postural sway may be indicators of fatigue or frailty.

– Models for unsupervised anomaly detection can identify changes in activity levels with self-reported fatigue that may be described as borderline fatigue.

Relevance: assists in disentangling the psychological factors from physical fatigue to better customize exercise programs.

16.5. Validation, performance metrics and outcomes

It is clear that AI motion capture systems in clinical oncology physiotherapy require validation against multiple datasets. This involves analysis of the systems' technical performance, clinical relevance and patient impact. This section explains the methods used to evaluate these systems and the metrics defined to prove their reliability, safety and utility.

16.5.1. *Technical validation of motion capture systems*

Systems using motion capture technology must prove that metrics such as accuracy, human motion capture under different conditions, repeatability and robustness are in equilibrium (Van der Kruk and Reijne 2018).

Benchmarking against gold standards:

– Marker-less and worn systems are often validated against Vicon and other fixed-marker optical systems.

– Joint angles, step length, angular velocity, etc. are evaluated using:

- mean absolute error (MAE);
- root mean square error (RMSE);
- Bland–Altman agreement analysis;
- intraclass correlation coefficients (ICCs).

System validation:

– IMUs and other motion capture systems are validated against derived kinematics in lab-controlled settings.

– Vision systems are validated against annotated datasets (e.g. COCO, HumanEva, etc.) for accuracy in frame-wise pose estimation.

16.5.2. *Model validation and evaluation metrics*

Performance of AI systems that analyze motion data is evaluated in the context of using multiple ML methods. The choice of metrics is bound to the task type:

For classification tasks (e.g. recognizing an activity or classifying a gait):

– *Accuracy*: the overall correct predictions divided by the total predictions.

– *Sensitivity (recall)*: the true positive rate of the positive class; critical in identifying a decline.

– *Specificity*: the true negative rate of the negative class; critical to avoid false detection of a positive.

– *Precision*: the percentage of positives predicted that are true positives.

– *F1 score*: the mean of precision and of recall.

– *Confusion matrix*: to display the results of a classifier and its performance on various classes.

For regression tasks (e.g. forecasting gait speed or ROM):

– mean squared error (MSE) and MAE;

– Pearson/Spearman correlation with clinical assessments;

– R^2 (coefficient of determination).

For time-to-event or longitudinal predictions:

– C-index (concordance index) for predicting survival or decline;

– time-dependent AUC.

16.5.3. *Clinical validation and utility*

Apart from technical accuracy, clinical validation also checks if the system enhances decision-making, outcomes or the efficiency of workflows.

Types of validity:

– *Content validity*: Are the metrics captured relevant to the rehabilitation of cancer patients?

– *Construct validity*: Do the outputs of the system correlate to known functional scales (Eastern Cooperative Oncology Group Performance (ECOG) status, Karnofsky, TUG, etc.)?

– *Criterion validity*: Does the system correlate with the gold-standard assessments of a physical therapist or other existing tests?

Clinical trial outcomes AI-powered MoCap systems have shown:

– Recognizing the decline of gait due to CIPN has been detected earlier than with other scales.

– Improvement of adherence and engagement in home-based rehab via real-time feedback may be enhanced by 30–50%.

– TUG or 6MWT tests remote assessment with 85–95% accuracy as compared to in-clinic evaluations (Hwang et al. 2022).

16.5.4. *Usability and acceptance of patients*

It remains that acceptability from users, particularly oncology patients, who tend to be older, fatigued or immunocompromised, may impact the overall success of AI-driven MoCap systems (Suhr and Keese 2025).

User-centered validation processes entail:

– system usability scale (SUS);

– NASA task workload index (mental, physical and temporal demands);

– open-ended feedback from participants and focus groups.

16.6. Challenges and ethical considerations

While the prospects of AI-driven motion capture systems and their potential for revolutionizing oncology physiotherapy practice are immense, there are problems with their clinical application. These are multifaceted, technical, clinical, ethical and systemic. This section outlines the most crucial problems and the issues that must be dealt with for their safe, fair and equitable application.

16.6.1. *Data quality and variability*

Motion capture technology faces challenges around capturing uniform, high-quality data, user movement integration and fluidity across varying users and conditions. Some of the challenges are:

– *Sensor placement symmetry*: users of any wearables should be instructed on how to put wearables on, body escape, limb capture and layering and how to wear sensors.

– *Environmental concerns*: background and foreground movement, illumination and overlapping controls of lighting and cameras, and vision-controlled video systems.

– *Psychological and physical factors*: outfits, limb proportions, body and posture, motion and the changes in body movement that a person usually perceives as automatism.

This principle is essential to AI model performance, which hinders the progress of generalization across patients or situations without extensive preprocessing and normalization techniques (Maleki et al. 2022).

16.6.2. *Small and diverse collections of datasets*

There is a lack of sufficient quality annotated datasets to train AI systems with, and cancer datasets tend to be the hardest to come across.

The issues with this include:

– Sample sizes are usually small in a domain specialized to oncology medicine.

– Class imbalance, for example, there are more instances of normative functions and less of severely dysfunctional gaits.

– The diversity in types of cancer, treatment schedules and other illnesses makes creating a model that is easy to use more difficult.

Consequently, many AI models are the subject of overfitting and diminished external validity outside the initial populations of the study (Starke 2022).

16.6.3. *Scientific transparency and reliance*

Having much of the eHealth biometric predictive model function without any rationale in terms of clinical care, and particularly in fields such as oncology where decisions regarding care are particularly complex, this is concerning (Mane and Yadav 2025).

Diagnosticians wonder:

– Which behavior or combination of behaviors triggered the system?

– How certain is the system of its prediction?

– Am I allowed to block or micro-modulate the model?

These questions require the application of some version of XAI, for example, saliency maps for pose identification or feature attribution for certain time series. XAI does reinforce model trust and user acceptance in more diverse fields where the gap is greatest in the outcomes (Ho 2023; Afridi and Khan 2024).

16.7. Future directions and research opportunities

With advancements in AI systems concerning motion capture, several new pathways have emerged concerning their efficacy, expansion and clinical employment in the field of physiotherapy for oncology patients.

16.7.1. *Custom tailored and dynamic models*

Deficits of oncology patients are often judged at the population level, which is the direct opposite of how oncology patients are assessed, as there is an excessive degree of variability in the field (Mollura et al. 2020).

Important Strategies:

– *Individualized goal setting*: early identification of clinical symptoms and/or subtle mobility enacted changes (e.g. walking orthodox CIPN) is made possible by ML frameworks in baselines with motion data generated over an extended duration.

– *Adaptive learning*: constant model updates as new data streaming becomes available, in the hope that it triggers changes in the static mental model, which parallels the changes in constructs representing recovery and decline (Rockne et al. 2019).

– *Transfer learning*: models are proven to work with very large sets of data, but model performance is severely lacking because it is tuned with very little, often clinical-grade data, and can perform better with unique rehabilitation requirements stemming from rare cancer subtypes.

16.7.2. *Clinical implication*

Adaptive systems are expected to have the greatest impact on rehabilitation processes by enabling dynamic therapy adjustments as patients transition from rest to exertion, in contrast to static, non-responsive therapeutic systems. (Kumar et al. 2026a).

16.7.3. *Real-time, on the edge computing and motion analysis*

Reactive analysis of recorded motion is the guiding principle for interactive rehabilitation and can be considered the primary requisite for rehabilitation programs delivered asynchronously.

Trends:

– *On-device AI*: performing on-device processing on smartphones, wearables and edge sensors (NVIDIA Jetson, for example).

– *Low-latency inference*: provides instantaneous feedback (e.g. posture correction, gait alerts).

– *Energy-efficient models*: model pruning, quantization and knowledge distillation to decrease power and memory usage (Kumar et al. 2026b).

– *Impact*: removes reliance on cloud connectivity, improves data privacy and allows use in resource-poor or rural settings.

16.7.4. *Federated and privacy-preserving learning*

Collaborative model training across institutions can increase dataset diversity without violating any patient privacy (Kulbacki et al. 2024).

Federated learning (FL):

– Supports model training in multiple sites/devices without sharing raw data (decentralized model training).

– Increases generalizability across cancer types, regions and demographics.

– Adheres to the Health Insurance Portability and Accountability Act (HIPAA) and General Data Protection Regulation (GDPR) laws while enabling the development of large-scale AI.

Privacy enhancements:

– Privacy for individuals so that their data cannot be reverse-engineered.

– Homomorphic encryption protects data while operations are performed on it.

16.8. Conclusion

Innovations in cancer rehabilitation have created an immediate need for improved scalable and tailored individualized approaches for assessing and tracking patient functionality. In the dynamic and complex arena of oncology physiotherapy, AI-automated motion capture systems can seamlessly fulfill these needs. These systems' ability to capture continuous and uninterrupted high resolution of patient movement allows clinicians to more accurately determine the early stages of functional decline, monitor therapy results and customize therapy more than ever before. Motion capture with AI expands this potential further. Advanced algorithms in time-series ML, such as LSTMs and convolutional networks, can detect subtle shifts in decline that are easily missed in traditional assessments. This aspect is extremely important in oncology patients. They frequently undergo cycles of functional upgrades and downgrades due to the toxic effects of chemotherapy, recovering from surgery and fluctuating disease progression. Although most current applications are still in the research and early clinical stages, substantial technical, clinical and ethical obstacles still need to be addressed for effective, wider deployment. These include the lack of model interpretability, privacy concerns and the need to address data quality from various devices, along with clinical transparency of the decision-making process using federated and privacy-preserving learning systems. In addition, the domain requires oncology-specific datasets, agreement among stakeholders on evaluation metrics and implementation science frameworks to guarantee equitable distribution across various health architectures.

16.9. References

Afridi, A. and Khan, S.N. (2024). Digital transformation in healthcare rehabilitation: A narrative review. *Journal of Process Management and New Technologies*, 12(1–2), 16–30.

Beauchamp, U.L., Pappot, H., Holländer-Mieritz, C. (2020). The use of wearables in clinical trials during cancer treatment: Systematic review. *JMIR mHealth and uHealth*, 8(11), e22006.

Cascella, M. (2024). *Exploring AI in Pain Research and Management*. Springer, Cham.

Davids, J., Lidströmer, N., Ashrafian, H. (2022). Artificial intelligence for physiotherapy and rehabilitation. In *Artificial Intelligence in Medicine*, Lidströmer, N. and Ashrafian, H. (eds). Springer, Cham, 1789–1807.

Galasso, S. (2024). Development of a machine learning-based marker-less gait analysis system for clinical applications. PhD Thesis, Università degli studi di Cassino e del Lazio Meridionale, Cassino.

Herath, H.M., Jayasekara, A.G., Madhusanka, B.A. (2024). *Vision Attentive Technology: AI Approaches for Functional Mobility Assessments in Elderly Healthcare*. CRC Press, Boca Raton.

Hincks, C., Jones, C., Reynolds, L., Robertson, N. (eds) (2019). *ANZCA Bulletin*, 28(3). Australian and New Zealand College of Anaesthetists, Melbourne.

Ho, A. (2023). *Live like Nobody is Watching: Relational Autonomy in the Age of Artificial Intelligence Health Monitoring*. Oxford University Press, New York.

Hwang, R., Fan, T., Bowe, R., Louis, M., Bertram, M., Morris, N.R., Adsett, J. (2022). Home-based and remote functional exercise testing in cardiac conditions, during the COVID-19 pandemic and beyond: A systematic review. *Physiotherapy*, 115, 27–35.

Jeong, H., Jeong, Y.W., Park, Y., Kim, K., Park, J., Kang, D.R. (2022). Applications of deep learning methods in digital biomarker research using noninvasive sensing data. *Digital Health*, 8, 20552076221136642.

Jones, J.M., Fitch, M., Bongard, J., Maganti, M., Gupta, A., D'Agostino, N., Korenblum, C. (2020). The needs and experiences of post-treatment adolescent and young adult cancer survivors. *Journal of Clinical Medicine*, 9(5), 1444.

Kulbacki, M., Segen, J., Kulbacki, M., Chaczko, Z., Bożejko, W., Paz Suárez Araujo, C., Klempous, R., Bonikowski, M. (2024). Advancements and applications of medical human digital twin technology in cerebral palsy diagnosis, therapy, and rehabilitation. In *International Conference on Computer Aided Systems Theory*. Springer, Cham, 125–139.

Kumar, A., Batta, P., Ahuja, S., Rathore, P.S. (eds) (2026a). *AI-driven Innovations in Physiotherapy and Oncology 1*. ISTE Ltd, London and Wiley, Hoboken.

Kumar, A., Batta, P., Ahuja, S., Rathore, P.S. (eds) (2026b). *AI-driven Innovations in Physiotherapy and Oncology 2*. ISTE Ltd, London and Wiley, Hoboken.

Laforgia, M., Laface, C., Calabro, C., Ferraiuolo, S., Ungaro, V., Tricarico, D., Gadaleta, C.D., Nardulli, P., Ranieri, G. (2021). Peripheral neuropathy under oncologic therapies: A literature review on pathogenetic mechanisms. *International Journal of Molecular Sciences*, 22(4), 1980.

Madhushri, P. (2017). A model based framework for mobility assessment of older adults using wearable systems. PhD Thesis, The University of Alabama, Huntsville.

Maffoni, M. and Magnani, A. (2025). Subjective cognitive decline. In *The Palgrave Encyclopedia of Disability*, Bennett, G. and Goodall, E. (eds). Springer, Cham, 1–14.

Maleki, F., Ovens, K., Gupta, R., Reinhold, C., Spatz, A., Forghani, R. (2022). Generalizability of machine learning models: Quantitative evaluation of three methodological pitfalls. *Radiology: Artificial Intelligence*, 5(1), e220028.

Mane, A.S. and Yadav, T. (2025). The short-term efficacy of mulligan traction straight leg raise on low back pain associated with hamstring tightness in young adults. *Cureus*, 17(3).

Meng, X.L., Zhang, Z.Q., Sun, S.Y., Wu, J.K., Wong, W.C. (2012). Biomechanical model-based displacement estimation in micro-sensor motion capture. *Measurement Science and Technology*, 23(5), 055101.

Mollura, D.J., Culp, M.P., Pollack, E., Battino, G., Scheel, J.R., Mango, V.L., Elahi, A., Schweitzer, A., Dako, F. (2020). Artificial intelligence in low-and middle-income countries: Innovating global health radiology. *Radiology*, 297(3), 513–520.

Nascimento, L.M., Bonfati, L.V., Freitas, M.L., Mendes Junior, J.J., Siqueira, H.V., Stevan Jr, S.L. (2020). Sensors and systems for physical rehabilitation and health monitoring–A review. *Sensors*, 20(15), 4063.

Pavllo, D., Feichtenhofer, C., Grangier, D., Auli, M. (2019). 3d human pose estimation in video with temporal convolutions and semi-supervised training. In *Proceedings of the IEEE/CVF Conference on Computer Vision and Pattern Recognition*, 7753–7762.

Rockne, R.C., Hawkins-Daarud, A., Swanson, K.R., Sluka, J.P., Glazier, J.A., Macklin, P., Hormuth, D.A., Jarrett, A.M., Lima, E.A., Oden, J.T. et al. (2019). The 2019 mathematical oncology roadmap. *Physical Biology*, 16(4), 041005.

Sassi, M., Villa Corta, M., Pisani, M.G., Nicodemi, G., Schena, E., Pecchia, L., Longo, U.G. (2024). Advanced home-based shoulder rehabilitation: A systematic review of remote monitoring devices and their therapeutic efficacy. *Sensors*, 24(9), 2936.

Starke, G.M. (2022). Trusting black-box algorithms? Ethical challenges for biomedical machine learning. PhD Thesis, University of Basel, Basel.

Stout, N.L., Sleight, A., Pfeiffer, D., Galantino, M.L., deSouza, B. (2019). Promoting assessment and management of function through navigation: Opportunities to bridge oncology and rehabilitation systems of care. *Supportive Care in Cancer*, 27(12), 4497–4505.

Suhr, M. and Keese, M. (2025). The role of virtual physical therapy in the management of musculoskeletal patients: Current practices and future implications. *Current Reviews in Musculoskeletal Medicine*, 18(8), 289–301. doi: 10.1007/s12178-025-09964-9.

Tayob, N. and Feng, Z. (2022). Personalized statistical learning algorithms to improve the early detection of cancer using longitudinal biomarkers. *Cancer Biomarkers*, 33(2), 199–210.

Van der Kruk, E. and Reijne, M.M. (2018). Accuracy of human motion capture systems for sport applications; state-of-the-art review. *European Journal of Sport Science*, 18(6), 806–819.

Yammouri, G. and Ait Lahcen, A. (2024). AI-reinforced wearable sensors and intelligent point-of-care tests. *Journal of Personalized Medicine*, 14(11), 1088.

17

Machine Learning to Forecast Rehabilitation Needs After Oncological Surgery

With the shifts in survivorship rates, the focus of care in oncology has shifted to the quality of life after surgery. Recovery from surgery involves more than the physical aspects; the psychological components must also be addressed. Within the context of the cancer type as well as the psychosocial factors, post-surgical comorbidities can mean the rehabilitation needs to cover more. Planning and assessment of post-surgical care can be vague and broad in scope and therefore more costly in terms of resources. Recent advancements in machine learning (ML) can enhance understanding of the complex clinical and behavioral aspects of patients and assist rehabilitation specialists.

This chapter summarizes the literature on the use of ML to predict post-oncological surgery rehabilitation needs. It describes the datasets that include electronic health record (EHR), imaging, biosensors and patient-reported outcomes (PROs). It evaluates the performance of various supervised and unsupervised ML techniques, including decision trees, support vector machines, applied neural nets, ensemble methods and deep learning. It also underscores the clinical concern of these models for timely identification of complex patients, defining rehabilitation resource models and resource-optimized rehabilitation.

This chapter also explores biases in the literature around data privacy and model obfuscation. The end of this chapter suggests a potential integration of ML into clinical diagnostics and decision-supporting system, while focusing on the incorporation of interdisciplinary teamwork and ML model user-centered validation and design. Proactively anticipating the needs of a patient in recovery after a surgical operation could result in the saving of financial resources, the enhancement of the recovery process and the positive outcome of even the cancer patients.

Chapter written by Trupti YADAV and Abhay KASHETWAR.

17.1. Introduction

The World Health Organization (WHO) continues to consider cancer an important health concern, as it is one of the top causes of morbidity and mortality. There is, however, an increasing trend in cancer survivorship. This is due to the advances cancer medicine has made in precision diagnostics, early cancer detection, surgical techniques, chemotherapy, radiotherapy, immunotherapy and targeted medication therapies. There has also been a change in approach toward cancer survivors, as there is an increasing focus on the quality of life and survivorship due to the growth in life expectancy. This is especially the case in life-extending surgical procedures (MacDonald et al. 2021).

"Onco surgery", an important branch in surgical oncology, especially defined for the purpose of cancer treatments, falls into this category of life-extending surgeries. Like in many other procedures, surgery in onco surgery also has palliative and curative approaches (Pinto et al. 2016). The difference is that oncological surgery is integrated into the treatment as the primary approach, or the primary approach is integrated into the other onco modalities. Yet, the surgery in these cases is often accompanied by other difficult challenges. This is especially the case for the challenges associated with postoperative care, which deeply affect the quality of life in the physical, emotional and social domains. There is a growing consensus that in such cases, steps such as rehabilitation and physiotherapy can make an important difference.

Rehabilitation is recognized as an important factor in addressing recovery but is still difficult and not done optimally. Each of the patients differs as a result of the type of cancer, the complexity of the surgery, additional diseases, age, overall health and psychological dimensions. Most rehabilitation approaches are reactive and apply a "one-size-fits-all" surgery. Most clinicians depend on subjective assessment as well as patients' self-reports on a functional assessment done at set timelines, post-discharge. Recovery is a gradual process, and these steps do not take the complexity of each individual case into account and therefore leads to an inefficient allocation of resources (Liu et al. 2012). As a result, some patients are over-rehabilitated, while others do not have sufficient timely rehabilitation, which increases the length of stay, readmission and possibility of disability. Machine learning (ML) aims to address the intricacy and variability of rehabilitation needs within the healthcare setting with the aid of artificial intelligence (AI) technology.

ML is capable of assessing and predicting patterns, which will help formulate focused, personalized plans for rehabilitative care. Healthcare does not require the fine-tuned adaptability of traditional systems and rule-based systems; hence, the reason why ML is best suited for this domain. With respect to rehabilitation of patients who have undergone cancer surgeries, ML has the ability to foresee the need for intensive rehabilitation, the timing and type of rehabilitation workouts and the recovery path. Numerous predictive models can be built based on data from electronic health records (EHRs), surgical and imaging reports, wearable devices, demographic data and outcomes. ML can estimate the probability of surgical patients experiencing delayed recovery and fatigue post-rehabilitation as well – for instance, patients who underwent surgery on the abdomen and patients who underwent head and neck cancer surgeries. The focus of this chapter is to explore systems for predicting rehabilitation needs for patients who have undergone oncological surgeries. It outlines the different data types visioned for model development, the prevalent ML paradigms such as decision trees, support vector machines (SVMs), neural networks and the post-training model testing approaches (Almadani et al. 2025).

In addition, it integrates the use of ML in the workflows of clinical practices for rehabilitation planning in real-world applications and pilot studies. It is this purposeful recalling of activities in the past and thinking of the consequences that allows the authors to capture the essence of the domain while improving on the areas that need further investigation and greater integration of clinical practice. Incorporating ML into rehabilitation planning, as global cancer care systems aspire to deliver care that is more efficient, equitable and personalized, is an important advancement (Jacobs et al. 2024).

17.2. Rehabilitation needs after oncological surgery

17.2.1. *Types of rehabilitation*

Rehabilitation after an oncological surgery is a complex process that aims to support the patients in achieving the best physical, mental and emotional status possible. This is done through a variety of processes. Each process is designed to address the unique issues and challenges a patient experiences due to the treatments they undergo.

– *Physical rehabilitation*: this component of the strategic plan aims to recover the mobility of the patient, their physical and mental strength, as well as their functional independence. Therapeutic surgery, immobility, chemotherapy and radiotherapy treatments can all lead to muscular atrophy, joint stiffness, fatigue and a greatly restricted range of motion (Tsuji 2022). Holistic physiotherapy works by setting specialized single-patient exercises that aim to train the patient on improving their cardiovascular fitness, correcting their postural and retraining neuromotor skills that were lost. During the post-operative care period after a patient has undergone breast cancer surgery, physiotherapy and functional exercises to retrain the shoulder girdle range of motion may be vital. Gynecological procedures may require pelvic floor physiotherapy and functional retraining.

– *Occupational therapy (OT)*: this domain of rehabilitation is dedicated to helping patients with self-care activities such as getting dressed, bathing, preparing a meal and functional work. Patients who have undergone surgeries that bring about the disuse of a limb and have their dexterity and energy levels altered greatly benefit from the practice of OT. Reintegration into the patient's normal life is quite often accompanied by the use of specialized equipment, energy-saving methods and alteration of the patient's social environment. After surgery to the upper limb, patients can be actively assisted by therapists to gain the necessary control of their hands to do the buttons of their clothes, as well as utensils.

– *Speech and swallowing therapy*: after surgery, patients who develop head and neck cancers often struggle with the activities of speech and swallowing. Speech-language pathologists assist them with the rehabilitation of safe swallowing and speech articulation (Starmer et al. 2017). Therapy can include strengthening the compensatory strategies, the vocal cords and alternative means of communication with devices for patients who have lost vocal structures. Early intervention is essential to avoid aspiration and access to the nutrients.

– *Psychosocial rehabilitation*: psychosocial rehabilitation includes support groups, counseling, cognitive behavior therapy and other mental health techniques to help patients recover from the anxiety, depression and cognitive issues related to chemo brain, as well as to assist them in overcoming other issues such as recurrence fears, resilience loss and mental health. These patients, who additionally face changes in physical appearance and social loneliness, are in dire need of emotional support.

– *Nutrition support*: nutrition rehabilitation aims to aid healing post-operation by providing sufficient calories and proteins to replace energy. Patients who experience surgical intervention on the head and neck, pancreas and gastrointestinal tract are more likely to experience malnutrition because of the loss of anatomy and difficulties with swallowing. In these cases, integrated enteral nutrition, custom meal plans, adjusted textures and diet additions are used.

Including integrated rehabilitation improves quality of life and health in long-term survivors from various aspects.

17.2.2. *Needs for rehabilitation and its determinants*

Taking into account different clinical, demographic and psychosocial parameters, formulating rehabilitation plans for each individual following oncological surgery entails complexity and uniqueness. These parameters serve to clarify which rehabilitation processes and techniques should be used during recovery.

– *Cancer type and stage*: the type and location of cancer have significant relevance to how much rehabilitation a patient will need. For example, someone with bone cancer will require intensive orthopedic rehabilitation and patients with cancer in the nerve tissues will need neurorehabilitation. In general, the higher the grade of cancer, the more active symptom management and supportive rehabilitation care the patient will need.

– *Surgical complexity and its complications*: the type of surgery performed and whether its approach is open or closed, and whether critical structures in the body get removed, determine functional healing and recovery. Complications such as infections, nerve damage and the body losing the capability to close wounds will, more likely than not, extend rehabilitation time as well as make the care more complex (Hakkarainen et al. 2014).

– *Pre-surgical functional status*: this refers to what a patient has in the area of movement, strength and independence. These will determine the success a patient will attain from rehabilitation processes. Patients with functional reserve are likely to recover more quickly than patients with disabilities. They will need more intensive and aggressive rehabilitation during recovery.

– *Age, sex and comorbidities*: as a rule, older individuals will recover much slower than younger individuals, and this is largely due to the natural decline in body physiology. Recovery will also be slower in the presence of these prevailing chronic conditions: diabetes, heart disease, pulmonary disease and the like, due to greater than usual fatigue.

– *Support systems and psychosocial factors*: the presence of rehabilitative and community caregivers is a valuable resource for rehabilitation and recovery. The psychosocial attributes of resilience correlate with therapy participation, while active health literacy largely drives therapy participation. Individuals with robust supportive networks tend to recover more functionally (Meraz et al. 2023).

– *Side effects of treatment and pain*: chronic pain, fatigue from chemotherapy and other palliative care practices are major barriers to participation in rehabilitation at all levels. They must be addressed to improve the rehabilitative approach, patient motivation and overall quality of life. The imposition of all these and attempting to balance them creates a model where the affected domains increase in complexity, and the classical pathways of rehabilitation cease to be viable.

The complex systems perspective navigates advanced self-organization for learning. The development of scaled pattern recognition soars along self-reinforcing positive feedback loops, demonstrating the multitude of complex patterns and relationships to provide complex, data-driven predictive rehabilitation patterns. This predictive power enables rehabilitation resource allocation to be more focused and effective (Adeyinka et al. 2022).

17.3. ML in healthcare

17.3.1. *Definition and scope*

Each subfield of AI has a different purpose and focus. ML concentrates on creating algorithms which learn and formulate rational conclusions and/or predictions automatically. Such algorithms improve their predictions and decisions with more data and exposure overtime (Helm et al. 2020). ML is rapidly growing in importance in healthcare because of large quantities of datasets such as clinical, genomic, imaging and sensor data prevalent in contemporary medicine. It is used in the early detection and diagnosis of a disease, estimating prognoses, categorizing and classifying images, stratifying patients by risk, personalizing treatment and improving healthcare operations. ML can also predict the need for rehabilitation, prescribe therapy and monitor student progress in real time.

17.3.2. *Types of ML models*

ML methodologies can be decreased to four core pillars with separate functions in healthcare, which include:

– *Supervised learning*: this is the process of teaching algorithms using labeled records. Each dataset contains the input and the desired output. The algorithms learn to associate which results belong with which inputs and are capable of classifying and executing regression functions. Supervised learning can also be used in rehabilitation to identify patients who will need intensive post-operative rehabilitation based on preoperative and perioperative data:

- *Examples*: linear regression, logistic regression, decision trees, random forest, SVMs and artificial neural networks.

– *Unsupervised learning*: these models work with unlabeled data to uncover hidden patterns or groupings within the dataset. This approach is valuable for discovering subgroups of patients with similar rehabilitation needs or recovery trajectories that may not be evident through traditional analysis:

- *Examples*: K-means clustering, hierarchical clustering and principal component analysis (PCA).

– *Reinforcement learning (RL)*: RL is a subset of ML where the agent learns a series of actions to get to very specific goals. In rehabilitation, RL can optimize therapy routines by adjusting therapy intensity and personalizing socially over time.

– *Deep learning*: deep learning is the most specialized type of ML, which focuses on networks that have multiple hidden layers. To provide the most accurate analysis, deep learning is especially recommended for advanced imaging such as magnetic resonance imaging (MRI) data, motion data capture or readings collected from motion sensors and other portable devices (Gandi et al. 2025). Deep learning still bears promises, especially in macroscope ML, where it can automate the assessment and prediction of functional outcomes in cancer rehabilitation such as complex outcomes.

Each ML method described enhances the capability of efficient rehabilitation planning and precision physiotherapy for cancer survivors.

17.4. Data sources for forecasting rehabilitation needs

A range of datasets is necessary in order to anticipate rehabilitative needs after cancer surgery. Personalized predictive modeling is supported by myriad data types. The primary data sources are EHRs, medical imaging, wearable technology, patient self-report and novel biomolecular data.

17.4.1. *EHRs*

EHRs significantly contribute to the accuracy achieved in predictive rehabilitation analytics. These records contain structured information such as the International Classification of Diseases (ICD) diagnostic codes, Current Procedural Terminology (CPT) procedure codes, lab test results and prescriptions. In addition to structured data, EHRs contain narrative texts, operative and discharge summaries and other unstructured data. The data describe the type and stage of the cancer, other concurrent illnesses, the therapies, some of the surgical details and most of the surgical data, all of which define the rehabilitation potential (Pompili 2019). To perform quantitative analysis, unstructured electronic records will need to be analyzed using the Natural Language Processing (NLP) to capture clinically relevant portions in clinician's notes, reports and summaries. Over time, EHR data illustrate the trends related to the recovery stages, functional baselines and the incidences of readmissions or complications.

17.4.2. *Radiology data*

Rehabilitation may be impacted by structural changes, complications and functional impairments. These are assessed by post-operative MRI and other images, such as computed tomography (CT) and X-rays. For instance, imaging can identify post-surgical edema, joint instability, muscle atrophy or nerve damage (Preuss et al. 2022). Using Radiomics, along with deep learning techniques, we can extract imaging features that are not visually discernable, possibly providing sensitive markers for tissue integrity or the progression of pathology that other observers might miss. These features can be used for predictive modeling aimed at flagging patients with increased risk of prolonged recovery or persistent functional limitations.

17.4.3. *Wearable sensors*

Wearable sensors can continuously and objectively track patients' activities, physiological metrics and behavioral activities. Accelerometers, gyroscopes, pulse oximeters and smartwatches can monitor patients' gait speed, static and dynamic balance, daily step count, heart, heart rate variability and sleep quality metrics (Vijayan et al. 2021). This data is useful to assess the functional status of cancer survivors beyond the confines of the hospital. Over long periods, data captured by wearables can demonstrate trends in recovery, predict episodes of fatigue or decline and track the effectiveness of the rehabilitation program. Incorporating this data into predictive ML models augments their granularity in terms of time and individualization.

17.4.4. *Patient-reported outcome measures*

Patient-reported outcome measures (PROMs) illustrate the individual's perspective regarding pain and fatigue, depression, emotional status and functional status barriers along the recovery pathway. Response to recovery in the overall emotional state and the degree of rest needed is located within approaches such as the EORTC QLQ-C30 or even the more modern Patient-Reported Outcomes Measurement Information System (PROMIS) system. PROMs are outcomes that provide information on the patient's perspective and the psychological factors that might affect motivation, adherence to rehabilitation and recovery (Patchick et al. 2015). This information can be valuable when added to predictive models in forecasting the rehabilitative resources needed. Incorporating PROMs improves the models' sensitivity and specificity.

17.4.5. *Genomic and biomolecular data*

Genomic and biomolecular data are progressively more recognized in the era of precision medicine as potential predictors of rehabilitation outcomes. Genomic data such as the presence of some variants may affect the ability to repair the tissue, the inflammatory response or the development of cognitive and neuromuscular deficits such as those seen after chemotherapy. Systemic stress, inflammatory response, poor nutritional condition or some underlying pathology can be suggested using high levels of certain cytokines, proteomic and metabolomic signatures, among others. This is an emerging area, and incorporating omics data into rehabilitation forecasting may help to provide specific patients that are at molecular risk for complications, slow recovery or persistent dysfunction (Saceleanu et al. 2023).

17.5. ML models for predicting rehabilitation needs

17.5.1. *Decision trees and random forests*

Using a series of feature values that create a tree-like structure, decision trees are simple models that are straightforward to explain and easy to interpret. Decision trees are capable of modeling simple if-then statements, including the processes of figuring out if a patient after surgery might need physiotherapy with the help of post-surgical mobility scores or clinical indicators such as comorbidities (Rivlin et al. 2023). Some of the most important positive aspects of decision trees are the ability to work with numerical and categorical variables, the presence of missing values and the almost complete lack of preprocessing. Random forests advance decision trees by assembling an average of many trees, increasing the accuracy of predictions and diminishing the chances of overfitting. Random forests are an effective method to deal with noise and a high degree of variability, particularly in healthcare datasets. Random forests can be used in rehabilitation forecasting to predict a patient's discharge disposition, recovery time and need for home-based or inpatient rehabilitation services.

17.5.2. *SVMs*

SVMs are classified as powerful models used to perform classification, as in the case where there is a hyperplane that best divides the data points into different classes. They work well in high-dimensional spaces and in situations wherein the relationships among the variables are nonlinear. In the case of rehabilitation after post-oncological surgery, SVMs can be used in the classification of risk groups for patients using preoperative data, intraoperative data and data from the early recovery period. Kernel functions (e.g. radial basis functions) help in the defining of SVMs by complex relationships that may be intractable to linear models.

17.5.3. *Logistic regression*

Logistic regression continues to enjoy popularity in the prediction of clinical outcomes because the method is simple and easy to work with. The model predicts the probability of a binary event – in this case, the likelihood for a patient to be classified as needing intensive rehabilitation – by a linear function of the values of the independent variables. Logistic regression is appreciated by practicing clinicians because it is able to generate odds ratios that demonstrate the level of influence of various predictors. Although it is a basic model, logistic regression frequently performs well under conditions where the data is arranged well and linear and independence assumptions are not greatly contravened (Njenga 2018).

17.5.4. *Neural networks and deep learning*

Deep learning and neural network models, in particular, are greatly dominant in working with intricate and large datasets – medical imaging, CT and MRI scans, and also time-series sensor data, such as data obtained from wearables on gait and movement. These models contain several distinct layers of connected nodes (or "neurons"), which are capable of approximating and fine-tuning a dataset and statically "learning" from the information provided. The ability of deep learning to unearth intricate details in imaging or motion data with a predictive functional relationship to the functional outcomes leads to nuanced analysis and precision in forecasting rehabilitation requirements. Such neural networks can be too intricate and lacking in clarity, which is why they are not widely adopted in the medical field, unless equipped with explainability tools.

17.5.5. *Gradient boosting machines (e.g. XGBoost, LightGBM, etc.)*

The gradient boosting machines (GBMs) with XGBoost and LightGBM are versatile ensemble methods, with each new tree in a sequence correcting the mistakes made by the previous one, and decision trees being constructed one after the other. They have a reputation of being the best in terms of performance on organized data, which is why they are commonly used in healthcare-related competitions.

Within oncological rehabilitation, GBMs predict various outcomes such as length of stay, complication risk and optimal rehabilitation modality. Hence, GBMs are suitable for use as part of clinical decision support systems (Tewarie et al. 2021). Table 17.1 shows an overview of key ML models applied in forecasting rehabilitation requirements following oncological surgery, including input data types, predictive focus, advantages and current implementation challenges.

Machine learning model	Input data types commonly used	Primary predictive tasks	Strengths	Limitations/challenges
Decision trees	EHR, demographic and perioperative variables	Classifying patients by rehabilitation intensity; predicting discharge readiness	Simple to interpret; handles mixed data types; minimal preprocessing	Can overfit with small datasets; limited generalization
Random forests	Combined clinical and functional datasets	Forecasting recovery duration; estimating readmission risk	High accuracy; robust to noise; handles missing data	Less interpretable; high computational load
Support vector machines (SVMs)	Preoperative, intraoperative and early recovery data	Classification of risk groups and rehabilitation needs	Performs well on small and high-dimensional datasets	Requires feature scaling; kernel choice affects results
Logistic regression	Clinical indicators, demographics and PROMs	Predicting binary rehabilitation outcomes (e.g. intensive vs. standard rehab)	Easy to interpret; widely accepted in clinical contexts	Limited for nonlinear relationships; assumes feature independence
Neural networks/deep learning	Imaging (CT/MRI); wearable sensor data; speech and motion data	Predicting functional recovery, fatigue and impairment severity	Captures complex nonlinear relationships; supports multimodal inputs	Requires large labeled datasets; lacks interpretability ("black box")
Gradient boosting machines (XGBoost, LightGBM, etc.)	Structured EHR and surgical data	Predicting length of stay, complication probability and optimal rehab pathway	High predictive accuracy; strong performance on tabular data	Computationally intensive; sensitive to hyperparameters

Table 17.1. *Comparison of machine learning models used to predict rehabilitation needs after oncological surgery*

17.6. Applications in oncology rehabilitation forecasting

17.6.1. *Colorectal surgery*

Surgery to treat colorectal cancer comes with a set of challenges. These may include reduced mobility, loss of bowel control and fatigue. ML models have started to analyze EHRs (operating notes, hospital stay, lab and preoperative status) and PROMs – items to investigate perceived functional limitations (Campbell et al. 2022). ML algorithms can identify patients who are at the greatest risk of not recovering or recovering with functional deficits. Those patients can be highlighted early so that the appropriate amount of rehabilitative physiotherapy can be administered. This greatly improves the potential for the long-term outcomes and reduces the risk of hospital readmissions.

17.6.2. *Breast cancer surgery*

Many patients experience complications following a mastectomy or even a lumpectomy. These include chronic pain, lymphedema of the arm and a reduced range of movement. ML models have been developed for a variety of datasets that include demographics, genetic risk factors and lymph node radiation. As a means to assist clinicians in prescribing the most effective therapies, these models can assess the chances of a patient developing lymphedema or shoulder dysfunction and suggest early physical or compression therapy, aside from strength training exercises (Davies et al. 2020). In addition to enhancing rehabilitation plans, ML helps improve the life and disability of breast cancer survivors, which is a desirable outcome.

17.6.3. *Head and neck cancer*

Surgical treatment of head and neck cancer is associated with several complications, particularly with the functioning of the speech and the management of the airway. Clinical prognostication about the functional decline of head and neck cancer patients remains limited.

There is a new head of deep learning model development, which uses multimodal data, such as CT and MRI, biomechanical sensors and audio of the patients' speech. With these algorithms, clinicians should be able to classify patients with varying degrees of dysphagia, aphonia or aspiration into different risk groups and start treatment before their speech and swallowing progress worsen. Less suffering is associated with the functioning of the body, and patients do not have to rely much on complex feeding systems (Donohue 2021).

17.6.4. *Thoracic and abdominal cancer*

Patients with thoracic (lung cancer) or abdominal (pancreatic, gastric and liver cancers) cancers undergoing surgical treatment face a particular risk of pulmonary complications, along with the bread-and-butter complications of surgery and nutritional deficits. Random forests and GBMs are examples of ensemble ML techniques that have been used to evaluate and estimate the perioperative risk of patients using respiratory function test results, nutritional scores and the amount of blood lost during surgery. These models enable prompt referral to pulmonary rehabilitation and dietetics, facilitating enhanced recovery and a decreased risk of complications, including pneumonia and malnutrition.

17.7. Model validation and evaluation

17.7.1. *Metrics*

In the particular case of rehabilitation needing predictions post-oncological surgery, the performance of the ML models has to be evaluated using metrics that have clinical meaning. In this case, the following metrics are applicable:

– *Accuracy*: this is the ratio of the total number of predicted outcomes that are correctly predicted (i.e. correctly predicting which patients will and which patients will not need rehabilitation and attributing the appropriate outcome) to the total number of predicted outcomes. It is important to note that within unbalanced datasets, one of the most dominant classes can skew the results of the predicted outcomes so that accuracy is not always the most insightful metric to use.

– *Precision and recall*: these metrics are important metrics because of the datasets being unbalanced. With precision, patients labeled positive predictions for rehabilitation whom the model has incorrectly classified as not needing rehabilitation are very small (Gudur et al. 2023). Recall, which is also known as sensitivity, quantifies the number of patients classified as positives who indeed need rehabilitation and are captured as positives by the model. These metrics assist in assessing the model from clinical points of view, which is very important in practice.

– *F1 score*: this is particularly useful in the clinical cases of numerous false positives and false negatives. It is the precision and recall of the system in question.

– *AUC-ROC (area under the receiver operating characteristic curve)*: this metric assesses how well the model differentiates among classes at various points, checking the model's threshold-based discrimination capabilities. It excels because, unlike many other measurements, it is continuous, thereby enabling a more precise assessment than a simple accuracy score. In classification contexts, particularly in medicine, a higher AUC is associated with a better-performing model.

17.7.2. *Validation methods*

Validation is critical to confirming that the ML models have not simply learned the training data but are also able to extend their knowledge to different patient groups:

– *Cross validation*: in k-fold cross-validation, the data is split into k subsets, and the model is trained and tested k times, each time using a different fold as the hold-out set. This approach reduces the likelihood of overfitting and offers a more accurate prediction of model performance.

– *Train-test split*: the data is then divided into training and test sets, a method termed as train-test split. This technique is helpful during the first few stages of performance evaluation, but on its own is not very effective, especially for smaller sets of data.

– *External validation*: testing the model on data from other hospitals or geographically distinct areas improves the model's applicability and trustworthiness for real-world clinical use (Owusu-Adjei et al. 2023).

17.7.3. *Challenges*

Although ML has the potential for predicting the needs for rehabilitation, a number of challenges still remain:

– *Small sample sizes*: due to the limited availability of patient data, many oncology rehabilitation studies lack depth, which inhibits model training and validation.

– *Gaps in or disordered data*: completed and free-text portions in electronic health documents can make data manipulation and feature mining challenging.

– *Identifying and examining indexical needs*: ground truth determination can be problematic owing to the highly individualized and subjective nature of rehabilitation (Singh et al. 2022).

– *Outcome measures not uniformly defined*: the absence of uniformly defined measures reported across different institutions also restricts model comparison and generalization.

17.8. Future directions and research opportunities

17.8.1. *Personalized rehabilitation pathways*

ML assists in estimating the type, timing and amount of therapy required for a specific patient after oncological surgery. It goes beyond predicting the need for rehabilitation. The type, timing and amount of therapy required for a specific patient after oncological surgery. By integrating various clinical, demographic and other functional variables, ML models are able to tailor rehabilitation plans based on the unique needs of every patient. This assists in early physical, occupational or speech therapy as warranted and increases the adherence and outcomes of therapy (Kumar et al. 2026a).

17.8.2. *Longitudinal modeling*

Recovery as a process is ever-changing and dynamic. Approaches to longitudinal temporal modeling such as Long Short-Term Memory (LSTM) networks, on the contrary, permit dynamic changes to archival data and permit the use of time-series data. For instance, we can effortlessly use data from EHRs, clinical data or data from wearables (Kumar et al. 2026b). These models are able to observe the progression of health status and adjust health predictions as different real-time data becomes available. The contemporary changes to the prediction models permit the rehabilitation strategies to be modified in real time, enabling functional decline.

17.8.3. *Federated learning*

To enhance model usability and tackle data privacy issues, federated learning, similar to other ML applications, allows institutions to build ML models collaboratively without exchanging detailed patient data. Each institution performs local model updating, and only model updates are transmitted to a communal server (Fong et al. 2019). This method provides access to a range of multicenter datasets, augmenting model strength while maintaining compliance with complicated, multidimensional data privacy laws.

17.8.4. *Fusion with robotics and virtual reality*

ML predictive models can directly connect with advanced robotic-assisted therapy and VR (Gudur et al. 2024). These systems can dynamically adjust to a patient's predicted progress and treatment response during therapy, improving the engagement and responsiveness of the rehabilitation experience.

17.8.5. *Cost–benefit analysis*

More research is needed on the impact of ML on rehabilitation forecasting on overall expenditures, particularly regarding complication containment, recovery time and long-term outcome improvements (Feinstein et al. 2024). Such analyses are critical to the implementation of these strategies in clinical practice.

17.9. Conclusion

ML can estimate the post-surgical rehabilitation workload after an oncological procedure, facilitating rapid, data-based and individualized care. Assisted by technology rather than relying on guesswork or rote checklists, ML analyzes all data types and sizes, including structured and unstructured data, EHRs, imaging and other data, wearables, and patient-reported data. ML algorithms are developed to determine and predict the phases, extent and type of rehabilitation, thus extending clinician care and rehabilitation while augmenting clinician functionality. Timelines are crucial, and patient care is improved. However, the integration of ML into the workflow remains unaddressed; interlinked representative data of high-quality, diverse and unrepresented population biases, and accurate and explainable ML models for end-users are only a few patterns to integrate. Issues such as unlinked electronic health systems, clinician trust and engrained clinical routines are other existing problems. To put this in a legible form, very rapid systems are needed. The research also has a lot of ethical implications to consider. Patient data privacy, consent and rights of vertical access or explanation, along with the more abstract implications of stratified data and consent, are all ethically complex. Including professionals from different fields increases the chances of successfully implementing projects, which is why data scientists and ethicists are included. The model will ensure fairness through ongoing validation across different groups and various healthcare contexts. With a focus on the patient and genuine ethical creativity, ML has the potential to become a core component of precision rehabilitation in oncology care, helping to prioritize individualized recovery paths and mitigate inequities associated with disparities in post-surgical care outcomes.

17.10. References

Adeyinka, A., Lamina, Y., Tawo, O., Adeyeye, Y., Minkah, A. (2022). Machine learning efforts that enhance personalized patient care and chronic disease management. *International Journal of Scientific Research in Science and Technology*, 9(2), 647–671.

Almadani, B., Kaisar, H., Thoker, I.R., Aliyu, F. (2025). A systematic survey of distributed decision support systems in healthcare. *Systems*, 13(3), 157.

Campbell, R., Ju, A., King, M.T., Rutherford, C. (2022). Perceived benefits and limitations of using patient-reported outcome measures in clinical practice with individual patients: A systematic review of qualitative studies. *Quality of Life Research*, 31(6), 1597–1620.

Davies, C., Levenhagen, K., Ryans, K., Perdomo, M., Gilchrist, L. (2020). Interventions for breast cancer–related lymphedema: Clinical practice guideline from the academy of oncologic physical therapy of APTA. *Physical Therapy*, 100(7), 1163–1179.

Donohue, C.A. (2021). Proactive dysphagia management of patients with neurodegenerative diseases: Early identification and intervention. Thesis, University of Pittsburgh, Pittsburgh.

Feinstein, M., Katz, D., Demaria, S., Hofer, I.S. (2024). Remote monitoring and artificial intelligence: Outlook for 2050. *Anesthesia & Analgesia*, 138(2), 350–357.

Fong, J., Ocampo, R., Tavakoli, M. (2019). Intelligent robotics and immersive displays for enhancing haptic interaction in physical rehabilitation environments. In *Haptic Interfaces for Accessibility, Health, and Enhanced Quality of Life*, McDaniel, T. and Panchanathan, S. (eds). Springer, Cham, 265–297.

Gandi, R., Suseela, G., Singh, S., Kumar, V.K., Sudhakara, M. (2025). Data-driven rehabilitation using machine learning approaches for adaptive recovery and treatment. In *Predictive Algorithms for Rehabilitation and Assistive Systems*, Korupalli, V.R.K., Ranasinghe, P., Feldman, D. (eds). IGI Global, Hershey, PA, 31–66.

Gudur, A., Patil, V., Gopal, L., Thapliya, S. (2023). Multi-objective optimization for breast cancer risk prediction models with particle swarm optimization. *International Journal of Intelligent Systems and Applications in Engineering*, 12, 531–554.

Gudur, R.A., Bhosale, S.J., Gudur, A.K., Datkhile, K.D. (2024). Effect of cyp1a1, cyp1b1 and cyp2c gene polymorphisms on doxorubicin and paclitaxel. *Bioinformation*, 20(10), 1244.

Hakkarainen, T.W., Kopari, N.M., Pham, T.N., Evans, H.L. (2014). Necrotizing soft tissue infections: Review and current concepts in treatment, systems of care, and outcomes. *Current Problems in Surgery*, 51(8), 344–362.

Helm, J.M., Swiergosz, A.M., Haeberle, H.S., Karnuta, J.M., Schaffer, J.L., Krebs, V.E., Spitzer, A.I., Ramkumar, P.N. (2020). Machine learning and artificial intelligence: Definitions, applications, and future directions. *Current Reviews in Musculoskeletal Medicine*, 13(1), 69–76.

Jacobs, F., D'Amico, S., Zazzetti, E., Gaudio, M., Benvenuti, C., Saltalamacchia, G., Gerosa, R., Gentile, D., Lasagna, A., Pedrazzoli, P. et al. (2024). Digital innovations in breast cancer care: Exploring the potential and challenges of digital therapeutics and clinical decision support systems. *Digital Health*, 10, 20552076241288821.

Kumar, A., Batta, P., Ahuja, S., Rathore, P.S. (eds) (2026a). *AI-driven Innovations in Physiotherapy and Oncology 1*. ISTE Ltd, London and Wiley, Hoboken, NJ.

Kumar, A., Batta, P., Ahuja, S., Rathore, P.S. (eds) (2026b). *AI-driven Innovations in Physiotherapy and Oncology 2*. ISTE Ltd, London and Wiley, Hoboken, NJ.

Liu, D., Mitchell, T.R., Lee, T.W., Holtom, B.C., Hinkin, T.R. (2012). When employees are out of step with coworkers: How job satisfaction trajectory and dispersion influence individual-and unit-level voluntary turnover. *Academy of Management Journal*, 55(6), 1360–1380.

MacDonald, C., Theurer, J.A., Doyle, P.C. (2021). "Cured" but not "healed": The application of principles of palliative care to cancer survivorship. *Social Science & Medicine*, 275, 113802.

Meraz, R., McGee, J., Caldwell, E.P., Ke, W., Osteen, K. (2023). The impact of resilience, health literacy, and social support on medication adherence and self-care among adults with heart failure. *Journal of Cardiovascular Nursing*, 38(5), 415–424.

Muzyczka, K.C. (2020). Understanding racial disparities within total knee replacements. Thesis, University of Florida, Gainesville.

Njenga, S.G. (2018). Analysis of hax delta strategic positioning model on performance of mobile telecommunication companies in Kenya. Thesis, Kabarak University.

Owusu-Adjei, M., Ben Hayfron-Acquah, J., Frimpong, T., Abdul-Salaam, G. (2023). Imbalanced class distribution and performance evaluation metrics: A systematic review of prediction accuracy for determining model performance in healthcare systems. *PLOS Digital Health*, 2(11), e0000290.

Patchick, E.L., Horne, M., Woodward-Nutt, K., Vail, A., Bowen, A. (2015). Development of a patient-centred, patient-reported outcome measure (PROM) for post-stroke cognitive rehabilitation: Qualitative interviews with stroke survivors to inform design and content. *Health Expectations*, 18(6), 3213–3224.

Pinto, A., Faiz, O., Davis, R., Almoudaris, A., Vincent, C. (2016). Surgical complications and their impact on patients' psychosocial well-being: A systematic review and meta-analysis. *BMJ Open*, 6(2), e007224.

Pompili, C. (2019). Early stage non-small cell lung cancer's treatment journey: Patients' perspectives. Thesis, University of Leeds, Leeds.

Preuss, K., Thach, N., Liang, X., Baine, M., Chen, J., Zhang, C., Du, H., Yu, H., Lin, C., Hollingsworth, M.A. et al. (2022). Using quantitative imaging for personalized medicine in pancreatic cancer: A review of radiomics and deep learning applications. *Cancers*, 14(7), 1654.

Rivlin, K., Brenner-Levoy, J., Odum, T., Muzyczka, Z., Norris, A., Norris Turner, A., Bessett, D. (2023). Provider mistrust and telemedicine abortion care preferences among patients in Ohio, West Virginia, and Kentucky. *Telemedicine and e-Health*, 29(3), 414–424.

Saceleanu, V.M., Toader, C., Ples, H., Covache-Busuioc, R.A., Costin, H.P., Bratu, B.G., Dumitrascu, D.I., Bordeianu, A., Corlatescu, A.D., Ciurea, A.V. (2023). Integrative approaches in acute ischemic stroke: From symptom recognition to future innovations. *Biomedicines*, 11(10), 2617.

Singh, H., Mhasawade, V., Chunara, R. (2022). Generalizability challenges of mortality risk prediction models: A retrospective analysis on a multi-center database. *PLOS Digital Health*, 1(4), e0000023.

Starmer, H.M., Ayoub, N., Byward, C., Kizner, J., Le, Q., Hara, W., Holsinger, F.C. (2017). The impact of developing a speech and swallow rehab program: Improving patient satisfaction and multidisciplinary care. *The Laryngoscope*, 127(11), 2578–2581.

Tewarie, I.A., Senders, J.T., Kremer, S., Devi, S., Gormley, W.B., Arnaout, O., Smith, T.R., Broekman, M.L. (2021). Survival prediction of glioblastoma patients: Are we there yet? A systematic review of prognostic modeling for glioblastoma and its clinical potential. *Neurosurgical Review*, 44(4), 2047–2057.

Tsuji, T. (2022). Rehabilitation for elderly patients with cancer. *Japanese Journal of Clinical Oncology*, 52(10), 1097–1104.

Vijayan, V., Connolly, J.P., Condell, J., McKelvey, N., Gardiner, P. (2021). Review of wearable devices and data collection considerations for connected health. *Sensors*, 21(16), 5589.

18

AI-Driven Wearable Sensors for Personalized Cancer Recovery Programs

Increased interest is being oriented towards cancer rehabilitation and, in particular, the early detection and treatment of cancer. Survivorship care aims at enhancing the quality of life of people with cancer. But physical problems persist in the form of debilitating effects on many survivors. Neuropathy, fatigue, musculoskeletal pain and gait issues reduce their quality of life and make them dependent on others, thus such patients require special attention. Survivorship care fails in many cases since its conventional model is disjointed. Evaluations are not connected, inefficient and subjective. This chapter discusses the application of artificial intelligence (AI) to the recovery programs of cancer, specifically the influence of wearable sensor technology. Electromyography (EMG) and inertial measurement unit (IMU) wearable sensors and other devices are useful due to their ability to capture muscle movement, motion and vital signs in real-time. This is a complicated technology that provides specific and detailed information on neuromuscular functioning. The author talks about remote rehabilitation monitoring, fatigue monitoring and fall-risk monitoring and assessment. The reviewed AI methods are supervised learning, deep learning and transfer learning. We also point out the obstacles to successful AI application: poor data and diversity, biased algorithms and privacy issues. Federated learning and explainable AI are listed as possible solutions. Next, we consider the future, such as digital twins, multi-logic analytical systems and gamified augmented reality/virtual reality (AR/VR) systems. Introducing AI in cancer rehabilitation software develops more customizable, responsive software to enhance patient engagement and recovery rates and make the process more accessible and equal.

18.1. Introduction

Cancer is associated with complexities unlike any other disease. On top of all the other challenges the disease adds to one's life and the lives of those close to the patient, the emotional and psychological aspects of the disease must not be

Chapter written by Dhairyasheel PATIL and Sanjay BADJATE.

disregarded. Oncology has achieved notable advancements with targeted therapy, precision medicine and immunotherapy to extend survival, but the challenges around the recovery phase following treatment are still not well addressed. Unfortunately, the most significant issues include extreme fatigue, pain, neuropathy and weakness of the musculoskeletal, cardiovascular and pulmonary systems. And, alarmingly, these symptoms may persist for many months or years, even after the patient is in remission (Lustberg et al. 2023).

In order to function post-recovery, for the rehabilitation to keep the complications neutral, and for the recovery to keep quality of life, timely rehabilitation, ongoing monitoring and rest are vital for post-cancer recovery. All too often, cancer recovery programs rely on patient-reported outcomes and spaced clinical evaluations, which do not accurately capture the post-treatment rehabilitation dynamics.

The use of artificial intelligence (AI) and wearable sensors has already advanced healthcare, particularly in the development of personalized recovery programs. AI coupled with wearable sensors provides remote monitoring of behavioral, physiological and biomechanical parameters, among others (Sripathi and Leelavati 2024). These systems are capable of recording minute measures, such as muscle exhaustion, temperature, stress, emotional and respiratory rates, heart-rate variability, and gait and posture asymmetries.

The data processed by AI provide the clinicians with a platform to create unique recovery routes and allow timely interventions. Wearable recovery systems evolved from simple designs, such as pedometers and fitness trackers, to advanced biosensing systems embedded in garments, patches and implants. Many of these devices are integrated into mobile health systems that apply cloud computing, machine learning (ML) and digital twin technologies to monitor recovery in real-time.

AI enables predictive analytics and anomaly detection in feedback systems that shift recovery from a planned proactive approach to a responsive one based on real-time data from the recovering patient (Singam 2023).

Here, the emphasis is on cancer rehabilitation and the AI sensor technologies and their data collection, analysis and interpretation, and application within oncology. The growing benefits from early detection of deterioration, adherence to recovery pathways and the relief to the healthcare system are also highlighted.

The explanation around the shift in cancer rehabilitation describes the systems in place relative to more administrative barriers, such as the precision of sensors, biased algorithms, regulation and privacy of data. With the increased focus on patient-centered models of care, AI-based wearables facilitate and individualize the course of recovery. This chapter expands the intersection of oncology, bioengineering and AI and presents the latest developments with prospects for growth in the next few years (Sebastian and Peter 2022).

18.2. Summary of wearable sensors

In the last 10 years, the sensor technology has changed considerably. Having started as simple pedometers and heart rate monitors, the field has evolved into complex integrated systems capable of monitoring an astonishing range of physiological and biomechanical parameters. Integrated with "real-world" technologies, these sensor systems provide cancer recovery patients with real-time, unobtrusive and pervasive monitoring of their condition (Lai et al. 2025). This chapter outlines the configuration of the wearable sensor systems and discusses their importance in Cancer Rehabilitation.

18.2.1. *Inertial measurement units*

Perhaps the most popular type of sensor technology integrated into wearable systems is that in the Inertial Measurement Units. These devices are usually composed of accelerometers, gyroscopes and magnetometers, making it possible to gauge movement, orientation and position of the system used. In cancer patients, especially those who are recovering from chemotherapy-induced peripheral neuropathy, musculoskeletal surgeries and patients with radiation fibrosis, IMUs are able to measure the level of activity and movement of limbs, balance and even track an individual's gait (Packel 2024).

18.2.2. *Electromyography sensors*

There are devices known as electromyography (EMG) sensors that measure the electrical activity of skeletal muscles as well as their fatigue and coordination. Surface EMG (sEMG) is now available in the form of armbands, clothing or patches that can be used to monitor muscle activity during rehabilitation exercises (Rampichini et al. 2020). This type of muscle activity monitoring is important in mastectomy surgical patients, orthopedic tumor surgical patients or patients where focused neuromuscular rehabilitation is needed. EMG maestros are able to detect abnormal patterns of muscle utilization and tailor the required therapy.

18.2.3. *Pressure sensors and gait insoles*

Smart or advanced gait monitoring insoles contain sensors that measure pressure distribution, foot contact timing and weight-bearing symmetry. Loss of stride and balance is one of the reasons patients of all ages may be prone to falls, which is worsened with the side effects of chemotherapy and skeletal metastasis. Advanced gait analysis that uses insole sensors can identify different grades of mobility and assist in devising preventive measures.

18.2.4. *Physiological biosensors*

Current state-of-the-art devices can monitor and identify various physiological biosignals, including heart rate, respiration, skin temperature, and blood oxygen levels, without direct contact with the skin or invasive interference (Poli 2023). In patients who survived cancer and received cardiotoxic treatments, as well as those who experience cancer-related fatigue, continuous physiological monitoring enables complication detection and more effective exertion control during rehabilitation.

18.2.5. *Multisensor fusion platforms*

Cancer rehabilitation wearables that combine multiple sensor types onto one platform are the most effective. Take, for example, devices that encapsulate IMU, EMG and HR, as well as temperature sensors. These devices offer insights into the physical and emotional states of the patient as well as the progression of the patient's recovery, giving a more comprehensive picture (Nascimento et al. 2020). These systems are designed for Bluetooth compatibility with smartphones or cloud servers, enabling remote monitoring, visualization of stored data and real-time AI analytics on vast returns.

18.3. AI integration: from data to insight

Although wearable sensors produce and collect a huge amount of real-time data, the importance of the data comes from understanding the data's meaning. AI works with these devices and does more than just track data. They also comprehend, learn and act upon complex physiological data. In cancer rehabilitation, where recovery paths are unique and symptoms vary, AI has a very important function: transforming data into something useful for practice has therapeutic implications (Al-Remawi and Aburub 2024). This part examines the most important AI systems for wearable systems and discusses how these systems are useful for personalized cancer recovery programs.

18.3.1. *Predictive and real-time data processing analytics*

Wearables collect and send data on a user's movements, muscle activity, vital signs, as well as other biochemical data. This data needs to be processed in real-time, so lightweight but advanced AI models are needed that can be processed on device (edge AI) or remotely (cloud systems). These models apply ML techniques such as support vector machine (SVM), decision trees and deep neural networks (DNNs) to search for specific data patterns to provide a prediction on the user's current health. Chemo patients require AI models where longitudinal data on gait speed and heart rate variability (HRV) are used to determine the prediction models that anticipate cancer-related fatigue and peripheral neuropathy flare-ups.

Complications and hospitalizations are lessened when steps to address early warnings are taken and when clinicians adjust exercise prescriptions and pharmacological treatments (Hoffmann et al. 2016). Predictive analytics helps clinicians identify potential timeframes to reach functional milestones, such as independent walking or returning to work, which enhances goal setting and care planning.

18.3.2. *Sensor fusion and multivariate analysis*

Recovery from cancer is a complex process that results in changes within the physical, psychological and metabolic domains. To understand this intricate process, AI applies sensor fusion, the integration of data from several wearables to derive a more complete and precise health status. For example, physical activity, neuromuscular activity and physiological strain that are captured from IMUs, EMG and heart rate monitors can be tracked in real-time.

For high-dimensional, time-series data from sensors, deep learning structures such as convolutional neural networks (CNNs), recurrent neural networks (RNNs) and transformer models are used (Jácome-Galarza et al. 2021). Subtle, complex changes that are easily overlooked are the focus of these models and are often missed by more traditional, rule-based systems. Moreover, multiple models' predictions are combined to enhance the robustness of ensemble learning methods.

18.3.3. *Personalized anomaly detection systems*

Recovery from cancer varies from one survivor to the next, depending on the type of cancer, treatment methods, comorbities and lifestyle, which makes everyone's profile particularly unique. Hence, one-size-fits-all models are unlikely to capture significant changes in personal trajectories.

To meet those goals, researchers are working on personalized systems for anomaly detection that learn to flag deviations from baseline patterns of an individual patient that might signal functional or health decline. One such system, UniTS (Unified Temporal and Spatial Anomaly Detection), uses various wearable and ambient sensors to train normal behavior patterns, such as walking, sleeping and using limbs. The system learns from baseline periods of normal (immobility), abnormal (erratic heart rhythm fluctuation, fragmented sleep) and learned baseline periods (baselines of normal values); during learned baseline periods, the system detects and classifies the AI event and flags the occurrence of sudden deviations as abnormal. It does so by giving an LLM-based (large language models) elucidation of the event (Nogaroli 2025). These alerts are sent to the clinicians and caregivers using a mobile interface, thus enabling timely and appropriate reactions. Models of this kind are particularly useful for personalized care of post-operative cancer patients at risk of complications from deep vein thrombosis, infection and lymphedema, whose symptoms can be subtle and non-specific.

18.3.4. *Adaptive feedback and timely interventions*

The use of AI for providing rapid and timely autonomous changes to a system is exciting. It is implemented to alter rehabilitation scheduling and perform other changes in real-time during the therapy process.

Take, for example, the case when a wearable detects a patient's disengagement in strength exercises in the presence of muscle fatigue, doing little or having low motivation. In such an instance, an AI can implement a modified routine, motivational prompts or a routine virtual consultation (Marella 2019). In other, more advanced, cases, AI has the ability to predictively model and, in so doing, recommend personalized interventions. For example, a wearable predictor can determine an increased risk of a fall, due to slow walking and standing or sitting. In such cases, the wearable can suggest the user take a break, inform a caregiver, or even perform low-level guided balance exercises. An example of a wearables loop system. Not only does this improve clinical results but also user engagement by supplying support at the most difficult times.

18.3.5. *Remote clinical integration and patient monitoring*

Remote patient monitoring systems, which use AI, enable the healthcare professional to monitor the patient beyond the walls of a clinic. An AI system synthesizes information from various wearable devices to display recovery metrics, usage patterns and risk indicators in real-time. Extensive research has demonstrated a decline in the rate of hospital readmissions, an increase in the adherence to medication, as well as an increase in life expectancy, especially among cancer patients. More importantly, AI helps the clinicians by filtering and giving only the top priority information (Kanakaprabha et al. 2024).

The use of NLP tools allows for the creation of brief summaries of electronic health records (EHRs) capturing only the most important events or divergences. This enables more efficient decision-making and improved collaboration between members of the multidisciplinary care team.

18.4. Uses in cancer recovery

The use of AI-based wearable sensors is transforming post-cancer rehabilitation from episodic and clinic-based to continuous and patient-centered. This technology facilitates the recovery process by providing appropriate information to the patient in the domains of rehabilitation and psychosocial support.

18.4.1. *Monitoring activity and performance*

Much of the rehabilitation carried out to restore the health of cancer survivors focuses on the exercise component. Cancer survivors lose the opportunity to participate in vigorous exercise due to fatigue, neuropathy, deconditioning and psychological barriers. Smartwatches and activity trackers (e.g. Fitbit, Garmin, Apple Watch) can provide 24-hour monitoring of step count, activity intensity, cadence and energy expenditure. AI can monitor activity patterns and model behaviors to identify periods of baseline deviation, which may indicate physical decline or treatment-related fatigue.

In clinical trials studying oncology patients, disruptions with a reduction in step count were linked with worsened pain, deepened depressive symptoms and a decrease in survivability (Milanti et al. 2021). AI can identify such correlations in realtime, flagging them for clinician evaluation and aid in physical therapy modifications or the addition of supportive care services.

18.4.2. *Frailty and gait assessment*

Frailty and cancer in the elderly population are a major concern because it diminishes the tolerance for treatment, increases the complication risk and worsens the outcome prognostically. There are portable devices with IMUs and pressure sensors that measure gait speed, balance, transitions from sitting to standing and postural sway – all indicators of advanced frailty. AI systems that classify gaits can determine the risk of falls and the presence of limited mobility sooner than a clinician for patients with cancer. For example, a post-chemotherapy recovering oncology patient may show signs of lessened gait symmetry and prolonged double support time, signaling the onset of peripheral neuropathy. The AI systems recognize these patterns using advanced training techniques and modify the rehabilitation program accordingly.

18.4.3. *Assessment of recovery and muscle activation*

Monitoring the active period of complication is very common post-cancer treatment, mainly after surgery, radiotherapy on the limbs or spine and neurotoxic chemotherapy. Wearable sensors that measure muscle electrical activity during movements provide a unique view of muscle activation, sequence coordination and muscle fatigue (Bonnette et al. 2019).

EMG readings capture and model muscle activity in real-time during periods of rest and while performing movements in rehabilitation training. In post-mastectomy cases, functional deficits and pain can elicit compensatory shoulder muscle activity during arm lowering, which can be captured and analyzed using AI-based methods. Predictable patterns can capture these activities. Automated analysis enables the therapist to give specific instructions on biofeedback and exercise in the right biomechanical patterns. Similar tools in AI where EMG is used can predict which muscles are underactive or overactive and aid in the relearning of movements from the limb to the amputated limb during limb-sparing surgeries or spinal tumor removals. These instruments are also capable of identifying other subtle movement disorders that are not easily detectable during physiological examinations.

18.4.4. *Biochemical and inflammatory marker monitoring*

Current-generation biosensors can now conduct real-time biochemical analyses on sweat, interstitial fluid or perspiration on the surface of the skin. In these instances where other biochemical markers may suggest complications, recovering from the procedures is streamlined. One very handy marker is C-reactive protein (CRP), which is continuously monitored. In the instance where there is an overreaction or infection post-surgery (or CT during radiotherapy) to poorly healed tissue or damage, the CRP becomes electrochemical. Elevated levels of these markers may suggest infection during surgery (Clemente-Suárez et al. 2023). AI models are able to capture these periods relative to hyperactive states in temperature, movement and heartbeat. Other markers that can be captured from wearable sensors include glucose, cortisol and saturation of lactate, which are all correlated with exercise, fatigue and metabolic stress.

Thanks to AI, during recovery and metabolic therapies, rest, hydration and nutrition can be suggested to patients, as AI can understand contextual data. Theranostics is a branch of medical science that combines therapeutic and diagnostic procedures. Wearable films with embedded strain sensors can monitor in real-time the shrinkage of tumors by interpreting changes in skin tension over the tumor. This is a leap forward in monitoring the effectiveness of treatment and reduces the burden of multiple diagnostic images.

18.4.5. *Psychosocial and behavioral health monitoring*

Psychological recovery is a vital part of the rehabilitation process, yet its recovery is rarely monitored. Sadness and anxiety, as well as disturbed sleep and emotional strain, are common in this recovery stage and slow down the recovery process and adherence to treatment. Wearable devices with AI can detect emotional changes from heart rate variability, skin conductance, sleep and even the voice used during a conversation. For example, during the sleep period, we talk about the need for rest – a person with low heart rate variability who sleeps at 50% efficiency is exhibiting signs of decreased activity depression.

Certain wearables are paired with AI-enabled conversational interfaces or mental health applications that actively identify distress signals and respond with mindfulness prompts and other CBT techniques or routes to supportive resources (Islam and Bae 2024) AI voice support systems such as My Intelligent Care

Assistant (MICA) can also have health conversations with patients, tracking medication and monitoring cognitive/emotional health through speech-analytical techniques, flagging cognitive or emotional deterents.

18.5. System architectures and practical implementations

To implement AI-based wearable sensors successfully to recover from cancer, designing proper system architectures that ensure smooth data transfer, sound analytics, secure storage and user-friendly interface is needed. Such systems should address different needs of patients, suit clinical workflows and take into account technical aspects, such as power consumption, connectivity and interoperability.

In this section, it is possible to discuss the organization and use of AI-wearable ecosystems in practical healthcare environments and refer to the main design features and practice examples.

18.5.1. *Layered system architecture for cancer recovery*

The majority of wearable health monitoring systems are based on the layered architecture that consists of four fundamental elements:

– *Sensing layer*: it comprises all types of wearables and sensors (e.g. IMUs, EMG, HR monitors, sweat sensors) that measure raw physiological and behavioral information of the patient in real-time.

– *Communication layer*: possible implementation sensor data is transmitted using low-power protocols, including Bluetooth Low Energy (BLE), ZigBee or Wi-Fi to mobile hubs (e.g. smartphones, smartwatches, etc.), which are then connected to the cloud-based projects.

– *Processing layer*: this is where the AI engine is found (it can be at the edge (on the wearable or mobile device) or on cloud A). It does signal preprocessing (noise filtering, data normalization, etc.), feature extraction, machine-learning classification, anomaly detection and personalized feedback generation (Bourechak et al. 2023).

– *Application layer*: this layer will give patients, clinicians and caregivers user interfaces.

This flexible modular architecture is scalable and real-time responsive, all of which are crucial towards personalized cancer rehabilitation programs.

18.5.2. *Conversational AI wearables*

One of the most innovative applications is conversational AI wearables, which is a mixture of health tracking and natural language processing (NLP). One of them is the MICA by The Value Care Group. MICA is a wrist-worn device that has an embedded LLM conversational interface with which it can talk to users in natural language.

Conversational agents, such as MICA, can be used in cancer recovery in several ways:

– *Symptom logging*: patients are able to log in the form of a verbal description of symptoms, pain, emotion and the AI classifies the information and logs it to be reviewed by the clinic.

– *Medication notification*: voice notifications notify the end-user of medication and physiotherapy appointments.

– *Motivational coaching*: AI agents encourage patients to be active and aid mental health by engaging with them in adaptive and understanding dialogue.

These devices will bring more patient engagement, particularly to older adults or even those with limited digital literacy, since these devices have an intuitive, human-like interface (Kumar et al. 2026a).

18.5.3. *Smart garments and integrated sensor systems*

Smart clothes are a form of practical, long-term and passive monitoring. For example, shirts or bras embedded with textile-based EMG sensors, heart rate monitors and respiratory belts allow for the continuous assessment of upper body movement, breathing patterns and muscle activation.

In breast cancer survivors recovering from mastectomy or reconstruction, smart garments can:

– detect compensatory shoulder movement patterns;

– track postural changes;

– identify early signs of lymphedema via skin temperature and circumference monitoring.

AI models process data from these garments to create personalized recovery pathways and provide visual or haptic feedback via connected apps. Table 18.1 summarizes key AI-driven wearable sensor technologies applied in cancer rehabilitation, detailing their monitored parameters, analytical AI methods and primary benefits for personalized recovery programs.

Type of wearable sensor/system	Primary physiological/behavioral parameters monitored	AI techniques used	Clinical applications in cancer recovery	Key benefits
Inertial measurement units (IMUs)	Gait, posture, limb movement, acceleration, etc.	Machine learning (SVM, random forest, etc.), deep learning (CNN)	Detection of balance disorders, gait asymmetry and fall risk post-chemotherapy	Early fall prevention; personalized mobility tracking
Electromyography (EMG) sensors	Muscle activation, coordination and fatigue	Signal processing; neural networks	Rehabilitation post-mastectomy or limb-sparing surgeries; detection of compensatory movements	Precision biofeedback; targeted exercise correction
Pressure sensors and gait insoles	Foot pressure distribution, stride length, weight-bearing symmetry	Predictive Modeling, Anomaly Detection	Monitoring neuropathy-induced gait instability	Objective gait evaluation; fall risk reduction
Physiological biosensors	Heart rate, oxygen saturation, temperature, CRP, glucose and cortisol	AI-driven pattern recognition; edge computing	Monitoring fatigue, infection, inflammation and metabolic stress	Continuous physiological monitoring; early complication alerts

Type of wearable sensor/system	Primary physiological/behavioral parameters monitored	AI techniques used	Clinical applications in cancer recovery	Key benefits
Multi-sensor fusion platforms	Combined motion, EMG and vital signs	Sensor fusion; ensemble learning	Holistic recovery assessment; integrated data visualization	Context-aware insights; improved diagnostic precision
Smart garments (textile sensors)	Muscle movement; respiration; skin temperature	Deep learning (RNN, LSTM, etc.); transfer learning	Monitoring postural changes and lymphedema in breast cancer survivors	Comfortable long-term monitoring; adaptive rehabilitation
Conversational AI wearables (e.g. MICA)	Speech, emotion, activity and symptom reporting	NLP; LLM-based conversational AI	Real-time patient interaction; medication adherence; mental health monitoring	Enhanced patient engagement; emotional support
Point-of-care AI diagnostic platforms	Sweat/interstitial biomarkers, temperature and HRV	Microfluidic data analytics; predictive AI	Early infection and inflammation detection in home-based care	Reduced hospital visits; improved self-management

Table 18.1. *Summary of key AI-driven wearable sensor technologies applied in cancer rehabilitation*

18.5.4. *Point-of-care AI diagnostic platforms*

Several wearable systems now serve as portable diagnostic platforms, capable of conducting point-of-care testing with minimal user intervention. Through the example of wearable biosensors, microfluidics and AI algorithms, the biomarkers in sweat, including glucose, lactate or CRP, can be examined and diagnosed in real-time.

They lessen the reliance on laboratory infrastructure in resource-limited conditions or the post-discharge cancer treatment. AI decodes chemical signals and compares them to clinical parameters and raises alarms or care advice. These channels prove to be particularly applicable in rural oncology practice, where timely follow-ups are usually impeded by transport barriers (Kumar et al. 2026b).

18.5.5. *Cloud-integrated AI platforms and dashboards*

Cancer rehabilitation programs that are scalable need cloud-capable environments, which can consolidate data on thousands of wearables and execute high-level AI algorithms to provide insights on a population scale and at the personal level.

Key components include:

– *Data lake storage*: this is a secure cloud storage that stores time-series sensor data.

– *Analytics engine*: this section offers AI modules assisting in the detection of patterns, stratification of patients based on their risk, and maximization of the rehabilitation plan using longitudinal recovery data.

– *Other system*: these systems include clinician dashboard, risk scores and alerts, which are reported in real-time by patient cohorts and interconnected with EHR systems.

– *Patient app*: an interface designed to provide information on a mobile platform, summaries, track the progress of goals and make suggestions based on the AI analysis.

These types of platforms bear remote patient monitoring (RPM) on a large scale, enhance coordination within a care team and facilitate predictive population health management.

18.6. Benefits and impact

AI-based wearable sensors integration into cancer recovery programs is associated with numerous benefits including clinical, operational, economic and psychosocial ones. These technologies can assist clinicians to be more proactive, personalized and goal-oriented because the systems aid in connecting patients and providers and gathering data in real-time to act more proactively, personalizing treatment and measuring outcomes. In this section, the role of these systems in patients, healthcare systems and public health will be discussed.

18.6.1. *Personalized and adaptive rehabilitation*

Individualized care plans should be among the most innovative features of AI-based wearables. Rather than relying on a universal timeline brought about by the assumption of average recovery rates, for each survivor AI examines the numerical results of the physiological conditions and their general performance.

As an illustration, when the wearables of a patient indicate that they are tired or that the gait stability is reduced, the AI may change the program to prescribe rest and exercises of lower intensity or remind a clinician about a follow-up. This is a feedback loop that enables the program to follow the changes in real-time (Dritsa 2021).

18.6.2. *Early detection and proactive interventions*

The technology of AI is capable of identifying important variations in parameters such as heart-rate variability, muscle activity and temperature. Such changes are common before medical-related complication symptoms. The predictive alerts assist clinicians to intervene early enough so that patients do not worsen at such a stage, leading to emergency care or readmission. The prominent ones are the identification of post-surgery infection by temperature and CRP, detection of fall risk due to minor alterations in gait, and relapse fatigue prognosis and treatment by measuring daily activities and tracking the pattern. Such systems change care into a reactive, then predictive and ultimately proactive model to enhance patient safety.

18.6.3. *Enhanced patient engagement and empowerment*

The wearable devices help survivors to follow the progress, monitor rehabilitation objectives and learn about the behavioral component of their health. Continuous self-monitoring feedback provides many rehabilitation plans with the personal control and independence that they would not have without continuous self-monitoring feedback (Keats et al. 2023).

Devices tend to offer real-time feedback, motivational messages and summaries of visual data through mobile applications. This draws the attention of the users and encourages exercising, following the diet and taking medication. Engagement in care and health behavior observation increases satisfaction and psychological wellbeing – important determinants in survivorship.

18.6.4. *Continuity of care and remote monitoring*

AI wearables aid continuous distance monitoring to allow clinicians to monitor the progress of a patient outside the hospital or rehabilitation facility. This will be especially useful with the patients in rural or underserved regions when face-to-face appointments are challenging to make.

Remote monitoring:

– reduces the need for travel;

– provides clinicians with continuous data between visits;

– supports virtual check-ins based on need, rather than fixed schedules.

This continuity of care improves clinical oversight while reducing costs and patient burden.

18.6.5. *Cost-effectiveness and healthcare efficiency*

Regarding the systems' perspective, AI-based wearables have the potential to lower the cost of healthcare by:

– reducing the number of complications and hospitalization;

– improving recovery times;

– optimizing use of healthcare resources;

– reducing clinician workload through data triage and automation.

Recognizing patients who may be at high risk early on and reducing unwarranted appointments or actions strengthens the principles of value-based care. The systems used in this case are in line with international trends of efficiency and outcome-based reimbursement (Krishna et al. 2024).

18.7. Challenges and future directions

Although promising, AI-enabled wearables in cancer recovery have a long way to go: problems with data quality, bias in algorithms, the lack of interoperability, the uncertainty in regulations and patient privacy. These issues must be resolved in order to achieve mass adoption, safety and equity (Alharbi et al. 2022).

Explainable AI (XAI), federated learning and digital twins are emerging trends that provide opportunities in the future.

This section identifies the existing constraints and the ways in which research and innovation are addressing the same.

18.7.1. *Data quality, standardization and labeling*

Good AI models require big datasets that are well-labeled and of high quality. Nevertheless, the information that is gathered by wearable sensors is often:

– noisy due to motion artifacts or signal drift;

– heterogeneous across device types, sensor placements and manufacturers;

– incomplete, especially when user adherence is inconsistent.

Cancer patients, who may be dealing with fatigue, discomfort or cognitive impairments, may struggle to wear and maintain sensors consistently, impacting the continuity and completeness of data (Nankya et al. 2024).

The labeled datasets of cancer are limited. Unlike orthopedics or cardiovascular domains, rehabilitation in oncology lacks large annotated datasets that pair sensor data with clinical outcomes such as fatigue severity, neuropathy progression or musculoskeletal impairment.

Future directions include:

– developing open-source, cancer-specific datasets with standardized sensor protocols;

– semi-supervised and self-supervised learning techniques to reduce dependence on labeled data;

– encouraging multi-center collaborations to ensure demographic and geographic diversity in datasets.

18.7.2. *Bias, equity and generalizability*

Models of AI trained on populations that are non-representative have the potential to give biased results that could be ineffective when applied to under-represented groups, such as gender, ethnicity, cancer type and socioeconomic status (Abdur Rahman et al. 2019). An example is that gait abnormalities in survivors of breast cancer with lymphedema might not be predicted by models that were trained on prostate cancer patients.

Algorithms' bias may increase inequality within the results and undermine trust in the technology. In addition, the majority of the commercially available wearables are designed and tested in high-income nations and may not be relevant to environmental or cultural variations in movement or rehabilitation objectives.

The strategies that have to be developed in the future have to be aimed at:

– training an inclusive model that reflects the whole range of cancer experiences;

– targeting practices including fair AI implementation models that actively monitor and mitigate bias;

– making contextual validations of various cancer types with varied demographics and care settings.

18.7.3. *Interpretability and clinical trust*

Numerous AI algorithms, in particular, deep learning models, represent black boxes, i.e. highly accurate but with little understanding of how they arrive at the final decision. Interpretability is necessary in cancer rehabilitation, where clinicians are required to justify interventions to the patients and teams.

Clinicians will hardly be satisfied to provide predictions without knowing the reasoning behind them. Indicatively, in a case where a system warns that a patient is at high risk of falls, clinicians should be capable of seeing the features that contributed to this warning, including but not limited to, a shorter stride length, more sway and more fatigue indicators (Kanagarajah 2024).

Solutions include:

– adopting XAI tools that visualize sensor features influencing predictions;

– integrating decision trees or attention maps for transparent model behaviour;

– using natural language generation (NLG) to translate complex predictions into plain-language summaries for both clinicians and patients.

18.7.4. *Regulatory and reimbursement pathways*

The legal intent of AI, as applied to healthcare, is still in development. The majority of AI-based wearables are currently considered wellness devices and cannot be clinically integrated (Patel et al. 2024). These tools need to fulfill the requirements of safety, efficacy and interoperability to be used in cancer rehabilitation programs.

Avoidance of reimbursement models is also a barrier to provider adoption. Unless AI-enhanced rehabilitation technologies are covered by insurance companies or public healthcare systems, they will be accessible to only the richer patients.

Necessary developments include:

– clearer regulatory guidelines from agencies such as the Food and Drug Administration (FDA), European Medicines Agency (EMA) and Central Drugs Standard Control Organisation (CDSCO) for AI and wearable medical devices;

– health technology assessments that evaluate cost-effectiveness and outcomes;

– new reimbursement models that incentivize value-based digital care.

18.7.5. *Future innovations and research directions*

To address current limitations and enhance capabilities, several innovative approaches are gaining traction:

– *Digital twins:* virtual replicas of patients built using wearable data and AI simulations can forecast recovery trajectories, predict adverse events and test interventions in silico before real-world application (Wang et al. 2025).

– *Multimodal AI models*: integrating sensor data with genomics, imaging, EHRs and patient-reported outcomes allows for deeper personalization of recovery strategies.

– *Conversational AI integration*: chatbots and virtual assistants embedded in wearable platforms may guide users, answer questions and support adherence in real-time.

18.8. Conclusion

The development of rehabilitation techniques for cancer patients is just beginning to leverage new technologies. As newly diagnosed cancer patients enter survivorship and rehabilitation, the question of how to help long-term self-management and quality of life functioning presents itself. Wearable technologies with RPM capabilities streamlined and focused for cardiovascular rehabilitation offer an effective solution. Most of the time recovery from cardiovascular interventions is monitored without a clinician, and technology-focused remote monitoring recovery right from discharge is available in rehabilitation. These systems focus and shift to real-time monitoring of recovery. Multifunction wearable devices and AI adapt rehabilitation systems to help cancer patients with prolonged self-managed rehabilitation. AI and wearable devices of this kind aid clinically in remote monitoring and care to support rehabilitation. These devices refine clinical decision-making to evidence clinical outcomes, improving the provision of rehabilitation services to patients in poverty and remote areas. This makes the recovery environment more hospitable for patients. Recovery was once stagnant and distant, in comparison modern recovery is engaged and responsive.

18.9. References

Abdur Rahman, M., Rashid, M.M., Le Kernec, J., Philippe, B., Barnes, S.J., Fioranelli, F., Yang, S., Romain, O., Abbasi, Q.H., Loukas, G. et al. (2019). A secure occupational therapy framework for monitoring cancer patients' quality of life. *Sensors*, 19(23), 5258.

Al-Remawi, M. and Aburub, F. (2024). Clinical applications of AI in post-cancer rehabilitation. In *2024 2nd International Conference on Cyber Resilience (ICCR)*, 26 February, 1–6. IEEE.

Alharbi, A.A., Almalki, M.H., Alhawit, M.E., Alharbi, A.M., Almutairi, G.T., Alhejaili, O.L., Albalaw, M.F., Alkhathami, A.M., Binselm, K.R. (2022). The impact of value-based care models: The role of nurses and administrators in healthcare improvement. *International Journal of Health Sciences*, 6(S10), 2186–2211.

Bonnette, S., DiCesare, C.A., Kiefer, A.W., Riley, M.A., Foss, K.D., Thomas, S., Kitchen, K., Diekfuss, J.A., Myer, G.D. (2019). Injury risk factors integrated into self-guided real-time biofeedback improves high-risk biomechanics. *Journal of Sport Rehabilitation*, 28(8), 831–839.

Bourechak, A., Zedadra, O., Kouahla, M.N., Guerrieri, A., Seridi, H., Fortino, G. (2023). At the confluence of artificial intelligence and edge computing in IoT-based applications: A review and new perspectives. *Sensors*, 23(3), 1639.

Clemente-Suárez, V.J., Martín-Rodríguez, A., Redondo-Flórez, L., López-Mora, C., Yáñez-Sepúlveda, R., Tornero-Aguilera, J.F. (2023). New insights and potential therapeutic interventions in metabolic diseases. *International Journal of Molecular Sciences*, 24(13), 10672.

Dritsa, D. (2021). Spatiotemporal dynamics of urban health: Physiological data driven strategies for enhancing urban health and wellbeing. Thesis, University of Technology Sydney, Sydney.

Hoffmann, T.C., Maher, C.G., Briffa, T., Sherrington, C., Bennell, K., Alison, J., Singh, M.F., Glasziou, P.P. (2016). Prescribing exercise interventions for patients with chronic conditions. *CMAJ*, 188(7), 510–518.

Islam, R. and Bae, S.W. (2024). Revolutionizing mental health support: An innovative affective mobile framework for dynamic, proactive, and context-adaptive conversational agents. arXiv preprint, arXiv:2406.15942.

Jácome-Galarza, L.R., Realpe-Robalino, M.A., Paillacho-Corredores, J., Benavides-Maldonado, J.L. (2021). Time series in sensor data using state-of-the-art deep learning approaches: A systematic literature review. In *Communication, Smart Technologies and Innovation for Society: Proceedings of CITIS*, 28 September, 503–514.

Kanagarajah, A. (2024). AI-driven innovation in healthcare product development: Challenges and ethical implications. Master's Thesis, LUT University, Lappeenranta.

Kanakaprabha, S., Kumar, G.G., Reddy, B.P., Raju, Y.R., Rai, P.C. (2024). Wearable devices and health monitoring: Big data and AI for remote patient care. In *Intelligent Data Analytics for Bioinformatics and Biomedical Systems*, Sharma, N., Cengiz, K., Chatterjee, P. (eds). Scrivener Publishing, Austin.

Keats, M.R., Yu, X., Sweeney Magee, M., Forbes, C.C., Grandy, S.A., Sweeney, E., Dummer, T.J. (2023). Use of wearable activity-monitoring technologies to promote physical activity in cancer survivors: Challenges and opportunities for improved cancer care. *International Journal of Environmental Research and Public Health*, 20(6), 4784.

Kourtis, L.C., Regele, O.B., Wright, J.M., Jones, G.B. (2019). Digital biomarkers for Alzheimer's disease: The mobile/wearable devices opportunity. *NPJ Digital Medicine*, 2(1), 9.

Krishna, S.R., Gudur, A., Jain, S., Deivasigamani, S., Tiwari, M., Venkatesan, K.G. (2024). Deep learning for automatic diagnosis of skin cancer using dermoscopic images. *Journal of Artificial Intelligence and Technology*, 4(2), 114–123.

Kumar, A., Batta, P., Ahuja, S., Rathore, P.S. (eds) (2026a). *AI-driven Innovations in Physiotherapy and Oncology 1*. ISTE Ltd, London and Wiley, Hoboken.

Kumar, A., Batta, P., Ahuja, S., Rathore, P.S. (eds) (2026b). *AI-driven Innovations in Physiotherapy and Oncology 2*. ISTE Ltd, London and Wiley, Hoboken.

Lai, J., Liu, K., Lin, Y., Chen, Z., Ji, X., Wen, J. (2025). Roles and impacts of integrative medical interventions in central nervous system tumor treatment: Multi-technology convergence and the paradigm shift toward functional reconstruction. *CNS Neuroscience & Therapeutics*, 31(7), e70516.

Lustberg, M.B., Kuderer, N.M., Desai, A., Bergerot, C., Lyman, G.H. (2023). Mitigating long-term and delayed adverse events associated with cancer treatment: Implications for survivorship. *Nature Reviews Clinical Oncology*, 20(8), 527–542.

Marella S. (2019). Gamification: A monitoring system for dialysis patients. Master's Thesis, Florida Atlantic University.

Milanti, A., Chan, D.N., Li, C., So, W.K. (2021). Actigraphy-measured rest-activity circadian rhythm disruption in patients with advanced cancer: A scoping review. *Supportive Care in Cancer*, 29(12), 7145–7169.

Nankya, M., Mugisa, A., Usman, Y., Upadhyay, A., Chataut, R. (2024). Security and privacy in E-health systems: A review of AI and machine learning techniques. *IEEE Access*, 12, 148796–148816.

Nascimento, L.M., Bonfati, L.V., Freitas, M.L., Mendes Junior, J.J., Siqueira, H.V., Stevan Jr, S.L. (2020). Sensors and systems for physical rehabilitation and health monitoring–A review. *Sensors*, 20(15), 4063.

Nogaroli, R. (ed.) (2025). Ethical and legal aspects of artificial intelligence (AI) in medical service contracts. In *Medical Liability and Artificial Intelligence: Brazilian and European Legal Approaches*. Springer, Cham, 123–206.

Packel, L. (2024). Oncological diseases and disorders. In *Acute Care Physical Therapy*, Malone, D.J. and Bishop, K.L. (eds). Routledge, New York.

Patel, D.J., Sharma, P., Gudur, R., Banerjee, D., Singh, J. (2024). Prognostic study to detect ailments, including cancer, during solid organ transplantation to prevent postoperative complications using perioperative antibiotics therapy. *Onkologia i Radioterapia*, 18(9), 001–007.

Poli, A. (2023). Measurement and processing of multimodal physiological signals in response to external stimuli by wearable devices and evaluation of parameters influencing data acquisition. PhD Dissertation, Marche Polytechnic University, Ancona.

Rampichini, S., Vieira, T.M., Castiglioni, P., Merati, G. (2020). Complexity analysis of surface electromyography for assessing the myoelectric manifestation of muscle fatigue: A review. *Entropy*, 22(5), 529.

Sebastian, A.M. and Peter, D. (2022). Artificial intelligence in cancer research: Trends, challenges and future directions. *Life*, 12(12), 1991.

Singam, A. (2023). Revolutionizing patient care: A comprehensive review of artificial intelligence applications in anesthesia. *Cureus*, 15(12), e49887. doi: 10.7759/cureus.49887.

Sripathi, M. and Leelavati, T. (2024). The Fourth Industrial Revolution: A paradigm shift in healthcare delivery and management. *Digital Transformation in Healthcare 5.0*, 5(1), 67.

Wang, K., Parrott, N.J., Lavé, T. (2025). Embracing the future of medicine with virtual patients. *Drug Discovery Today*, 30(3), 104322.

19

Computer Vision-Based Range of Motion Analysis in Oncology Physiotherapy

Cancer survivors often experience long-term functional impairments, such as reduced range of motion (ROM), due to surgery, chemotherapy, radiation or prolonged inactivity. Accurate and timely assessment of these impairments is critical for guiding physiotherapy interventions and improving quality of life. Traditional ROM assessment methods – such as goniometry or observational scoring – are limited by subjectivity, variability and the need for in-person clinical supervision. Computer vision (CV)-based approaches offer a promising alternative by enabling automated, non-invasive and markerless motion tracking through standard video input. This chapter explores the current landscape, challenges and opportunities of CV-based ROM analysis in oncology physiotherapy. It examines recent advancements in pose estimation algorithms, machine learning models and sensor fusion techniques that underpin CV tools. The chapter also discusses clinical applications, including post-operative shoulder ROM monitoring in breast cancer patients, gait analysis in chemotherapy-induced neuropathy and home-based rehabilitation systems for survivors. Key limitations, such as pose estimation errors in altered anatomies, lack of oncology-specific datasets and clinical validation gaps, are critically analyzed. Emerging solutions – ranging from personalized movement models to telehealth integration and real-time biofeedback – are also highlighted. Finally, this chapter outlines future directions for research, including model explainability, regulatory alignment and equitable deployment across diverse patient populations. CV has the potential to revolutionize ROM assessment in oncology by enabling scalable, objective and patient-centered physiotherapy monitoring. As technologies mature and clinical integration advances, CV-based systems may become essential tools in modern cancer rehabilitation practice.

19.1. Introduction

Cancer survivors often face long-term musculoskeletal impairments due to surgery, chemotherapy, radiation and hormonal therapies. These impairments can

---

Chapter written by Anand GUDUR and Pankaj THOTE.

significantly limit mobility, joint function and overall quality of life (Lovelace et al. 2019). As oncology care increasingly shifts towards survivorship and rehabilitation, physiotherapy has emerged as a key component in restoring physical function and preventing further disability.

Range of motion (ROM) assessment is a foundational element in physiotherapy, particularly for cancer patients recovering from surgical procedures (e.g. mastectomy, neck dissection, limb salvage surgery), dealing with treatment-induced fibrosis or experiencing chemotherapy-induced peripheral neuropathy (CIPN). Accurate and regular evaluation of joint angles and mobility allows clinicians to tailor exercise protocols, track recovery and monitor for complications such as lymphedema, frozen shoulder or joint contractures (Sadineni 2024). However, traditional ROM assessment methods – such as goniometry or visual inspection – are subjective, time-consuming and require in-person interaction, which may not always be feasible for oncology patients due to fatigue, immunosuppression or geographic limitations.

Computer vision (CV) technologies present a transformative opportunity in this context. CV involves the automated extraction, analysis and understanding of visual information by computers, typically using cameras and artificial intelligence (AI) algorithms. When applied to human motion analysis, CV can capture joint movements, postures and movement patterns without the need for physical markers or wearable sensors – referred to as markerless motion capture (Hellsten et al. 2021). This allows clinicians to assess ROM remotely or in clinics with minimal equipment, increasing access to care and reducing patient burden.

Recent advances in pose estimation models – such as OpenPose, MediaPipe and YOLO-Pose – allow accurate detection of anatomical landmarks and skeletal tracking from simple RGB images or videos (Rana 2024). These tools, combined with machine learning (ML) models, can quantify joint angles in real time and flag abnormal movement patterns, compensations or functional deficits. The growing accuracy, affordability and accessibility of such technologies make them especially promising for use in oncology physiotherapy, where traditional motion labs may be impractical.

Despite significant progress in CV-based movement analysis in general rehabilitation and sports medicine, its application in oncology-specific physiotherapy remains limited. Cancer survivors present unique challenges due to variability in treatment side effects, comorbidities, fatigue and altered biomechanics (Emery et al. 2022). Therefore, systems designed for athletic or post-stroke populations may not directly translate to oncology without careful adaptation and validation.

This chapter aims to provide a comprehensive review of current CV techniques for ROM assessment, their relevance and applicability in oncology physiotherapy, and the specific considerations for deploying these technologies in clinical and home-based settings (Szabo et al. 2023). We also outline a conceptual framework for integrating CV-based ROM analysis into oncology care and discuss limitations, challenges and future directions.

By bridging CV science with oncology rehabilitation, this chapter contributes to the growing field of digital health in cancer survivorship – supporting safer, smarter and more personalized physiotherapy interventions.

19.2. CV techniques in ROM analysis

CV has emerged as a key technology in modern movement science, enabling accurate, markerless tracking of human motion using video-based inputs. In the context of ROM assessment, CV techniques offer clinicians a practical, non-contact alternative to traditional tools such as goniometers or inertial measurement units (IMUs) (Kianifar n.d.). This section outlines the major categories of CV-based ROM analysis techniques, ranging from RGB-D cameras to advanced ML models for pose estimation and real-time motion tracking.

19.2.1. *Markerless motion capture systems*

Traditional motion capture systems – such as Vicon or OptiTrack – use reflective markers and multiple cameras to track joint trajectories with millimeter-level precision. While highly accurate, these systems are costly, require laboratory environments and are unsuitable for outpatient or home-based settings. Markerless motion capture addresses these limitations by using visual input (e.g. RGB or depth cameras) and CV algorithms to infer joint positions without attaching markers to the body (Colyer et al. 2018).

One of the most prominent markerless tools is the Microsoft Kinect system. Kinect uses a combination of infrared (IR) depth sensing and RGB imagery to create a 3D skeleton model of the human body. Early studies have shown that Kinect v2 can capture joint displacements with a root-mean-square error (RMSE) of approximately 3–4 cm and joint angles with RMSE ranging from 10° to 25°, depending on the joint and movement complexity (Ma et al. 2019).

For instance, shoulder and elbow movements, which involve complex 3D rotations, often show higher error margins than knee flexion/extension or hip movements.

Another advantage of Kinect is its built-in SDK for skeletal tracking, which simplifies software development and integration. Clinical applications include gait analysis, balance testing and ROM assessments across orthopedic and neurologic populations.

Some studies have applied Kinect to post-mastectomy shoulder ROM evaluation, showing moderate agreement with clinical goniometers, although limitations remain in fine-grained movement detection.

With the discontinuation of Kinect in mainstream markets, researchers have shifted towards newer depth cameras such as the Intel RealSense and Azure Kinect, offering improved depth resolution, wider field of view and better ambient lighting tolerance.

19.2.2. *RGB camera-based systems and consumer devices*

To reduce hardware dependencies, recent research has focused on using standard RGB cameras – such as webcams or smartphone cameras – for pose estimation. These systems operate without depth information but can still extract 2D or inferred 3D skeletal landmarks using AI algorithms. Their primary advantage lies in their accessibility, low cost and ability to work on consumer-grade devices.

A widely used tool in this category is OpenPose, an open-source pose estimation framework developed by the Carnegie Mellon Perceptual Computing Lab. OpenPose detects anatomical key points (e.g. shoulder, elbow, knee, ankle, etc.) in 2D video frames and tracks them over time.

Although designed for general-purpose human pose recognition, OpenPose has been adapted for clinical applications, such as exercise quality assessment, ROM measurement and movement asymmetry detection (Herath et al. 2024). Its accuracy in joint angle estimation varies depending on the joint and camera positioning but has shown comparable performance to marker-based systems in selected movements (e.g. up to 90% agreement in knee flexion tasks).

Other notable frameworks include:

– *MediaPipe (by Google)*: offers real-time, lightweight models for full-body and hand tracking with minimal hardware requirements.

– *MoveNet*: a fast, accurate model optimized for real-time pose detection on edge devices.

– *YOLO-Pose*: integrates human detection and key point estimation in one model for efficient processing.

These models typically output 2D key points, which can be used to compute planar joint angles (e.g. elbow flexion in sagittal view). For more comprehensive ROM analysis, 3D pose estimation techniques – using multiple camera views or monocular depth inference – can reconstruct joint positions in space and calculate complex angular metrics.

19.2.3. *Deep learning and ML integration*

Beyond raw pose estimation, the integration of ML and deep learning (DL) models has enabled advanced motion interpretation. These algorithms can classify movement quality, detect compensatory strategies and predict deviations from normal movement patterns. In rehabilitation settings, such systems can provide real-time feedback to patients and clinicians, improving adherence and outcomes.

Some common ML models applied in ROM analysis include:

– *Convolutional neural networks (CNNs)*: used for spatial feature extraction from pose sequences or joint coordinate plots.

– *Recurrent neural networks (RNNs) and Long Short-Term Memory (LSTM) networks*: ideal for analyzing time-series data and capturing temporal dependencies in joint movements.

– *Hidden Markov models (HMMs)*: used to model sequential patterns in movement, such as the phases of a reaching task or gait cycle.

– *Dynamic time warping (DTW)*: enables similarity comparison between patient movements and reference patterns, useful for evaluating movement correctness or ROM consistency over time.

For instance, CNN-LSTM models have achieved over 90% accuracy in classifying rehabilitation exercises (e.g. sit-to-stand, shoulder abduction, etc.) and assessing ROM progression in post-stroke patients. Some systems pair pose data with clinical annotations to train supervised models for movement quality grading. Others use unsupervised clustering to detect movement anomalies in oncology or orthopedic populations (Warmerdam et al. 2020).

In the oncology physiotherapy context, DL can help distinguish between limited ROM due to post-surgical restriction versus fatigue or muscle guarding, allowing more nuanced clinical interpretation.

19.2.4. *Real-time analysis and edge computing*

An essential feature of CV-based ROM systems is real-time capability, especially for remote physiotherapy. Real-time processing allows immediate feedback during exercise sessions, improving motivation, compliance and movement accuracy. Recent advances in edge computing and hardware acceleration (e.g. using GPUs or Tensor Processing Units) have enabled CV models to run on smartphones, tablets or small embedded devices such as Raspberry Pi or Nvidia Jetson (Jouini et al. 2024).

For example:

– A telerehabilitation app using MediaPipe can guide a breast cancer survivor through shoulder mobilization exercises, giving real-time visual cues when they deviate from the desired movement arc.

– Real-time joint angle tracking can monitor fatigue over a set of repetitions, prompting rest when compensation or ROM decline is detected.

Such systems also support asynchronous assessment, where patients record their movements and upload videos for offline analysis by clinicians. This hybrid model is particularly useful in low-resource or rural settings.

19.3. Oncology physiotherapy: unique needs and challenges

Oncology physiotherapy plays a crucial role in enhancing quality of life, preserving physical function and mitigating treatment-related impairments in cancer survivors. However, cancer rehabilitation is distinct from other domains such as orthopedic or neurological physiotherapy due to the heterogeneous and often systemic nature of cancer-related impairments. These unique challenges must be considered when applying CV tools for ROM analysis in oncology patients.

Technique	Hardware/tool used	Core features	Advantages	Limitations	Clinical relevance in oncology physiotherapy
Marker-based optical systems	Vicon; OptiTrack (infrared cameras + reflective markers)	3D tracking with sub-millimeter accuracy	High precision; gold-standard validation tool	Expensive; lab-dependent; impractical for clinical use	Used as a reference for validating CV-based and sensorless systems
Markerless depth-sensing systems	Microsoft Kinect, Azure Kinect, Intel RealSense	Infrared + RGB depth mapping for 3D skeletal reconstruction	Real-time data; low cost; moderate accuracy	Sensitive to lighting; limited resolution; discontinued hardware	Enables in-clinic 3D ROM evaluation and patient monitoring
RGB camera-based systems	Standard webcams, smartphone cameras	2D/3D skeletal key point extraction using AI (OpenPose, MediaPipe, YOLO-Pose, etc.)	Lowcost, scalable; contact-free; accessible for telehealth	Limited depth perception; reduced accuracy in occlusion	Ideal for home-based rehabilitation and telephysiotherapy
Deep learning integrated systems	CNN, RNN, LSTM, HMM, DTW frameworks	Automated feature extraction, temporal analysis and movement classification	Detects compensations, quantifies fatigue; supports personalized tracking	Requires training data; computationally intensive	Enables intelligent; adaptive ROM monitoring for oncology patients
Edge computing/hybrid systems	Mobile/embedded AI hardware (Jetson, Raspberry Pi, etc.)	On-device inference with low latency	Real-time feedback; offline functionality	Limited processing power for complex models	Facilitates remote physiotherapy with instant corrective feedback

Table 19.1. *Comparative overview of computer vision techniques for range of motion (ROM) analysis in oncology physiotherapy*

19.3.1. *Clinical context of ROM impairments in oncology*

Cancer and its treatments frequently lead to reductions in joint mobility, muscle strength and coordination. Some of the most common scenarios where ROM assessment is essential include:

– *Post-surgical recovery*: patients who have undergone surgeries such as mastectomy, neck dissection, thoracic surgeries or limb resections often experience joint stiffness and soft tissue contracture. For instance, breast cancer survivors commonly report restricted shoulder flexion and abduction due to surgical scarring and fibrosis, which can significantly affect daily functioning (Levy et al. 2012).

– *Radiation-induced fibrosis*: radiation therapy can cause chronic tissue tightening, reducing ROM in affected areas (e.g. neck, chest wall or upper limbs), often many months post-treatment.

– *CIPN*: neurotoxic agents such as taxanes or platinums may lead to balance deficits, reduced proprioception and altered gait mechanics. ROM assessment in this context helps to detect compensatory strategies that patients may use to maintain mobility.

– *Lymphedema and fluid retention*: swelling in limbs – especially after lymph node removal – can limit ROM and lead to muscle inhibition, often requiring careful monitoring during rehabilitation.

– *Fatigue and deconditioning*: cancer-related fatigue may reduce voluntary ROM not due to structural limitation but due to low-energy availability. Differentiating between voluntary and passive ROM in these cases is essential (Girard 2016).

These diverse impairments demand flexible, precise and individualized assessment strategies: an area where CV can provide valuable assistance.

19.3.2. *ROM assessment challenges specific to oncology*

While CV tools have demonstrated strong performance in general rehabilitation, their deployment in oncology care faces several unique barriers:

a. Patient variability and complexity

Cancer survivors differ widely in age, fitness levels, comorbidities, cancer type and treatment history. For example, an older breast cancer survivor with diabetes and a history of stroke will present differently than a young lymphoma patient undergoing intensive chemotherapy. These variations can affect ROM assessment outcomes and confound interpretation.

CV models must be robust enough to account for:

– asymmetrical postures due to surgical changes;

– temporary or chronic fatigue influencing movement amplitude;

– joint protection or guarded movement due to pain;

– use of assistive devices or orthoses during activity.

Standard CV models trained on healthy individuals or athletes may not generalize well to these scenarios.

b. Movement compensation and subtle impairments

Cancer patients often develop compensatory movement strategies to overcome pain or restriction. For example, a breast cancer patient with limited shoulder abduction may recruit trunk rotation to complete overhead reaching tasks. Without 3D analysis or temporal modeling, such compensation may go unnoticed in simple 2D pose estimation systems.

Thus, CV-based systems must not only measure ROM but also detect abnormal movement patterns, requiring integration with biomechanical models or time-series analysis tools like HMMs or LSTMs (Debnath et al. 2022).

c. Sensitivity to fatigue and treatment timing

ROM performance can fluctuate depending on the timing of chemotherapy or radiation cycles. Fatigue, nausea or neuropathic symptoms may temporarily reduce joint mobility and strength. Therefore, single-session assessments may not reflect a patient's typical capabilities.

To address this, CV systems should ideally support longitudinal monitoring to capture trends and identify true improvements or declines over time.

19.3.3. *Integration challenges in oncology settings*

Beyond patient-level considerations, there are system-level and environmental challenges to implementing CV-based ROM assessments in oncology:

a. Clinical workflow integration

Many oncology rehabilitation programs operate with limited staff, time and equipment. Introducing CV tools must not add excessive burden to therapists or interrupt existing care pathways (Silver et al. 2013). For these systems to be accepted:

– Interfaces must be intuitive and quick to use.

– Setup time (e.g. camera positioning) should be minimal.

– ROM reports should be auto-generated and clinically interpretable.

b. Infection control and immunosuppression

Cancer patients – especially during active treatment – may be immunocompromised and vulnerable to infections. Tools requiring shared physical contact (e.g. goniometers or wearables) may pose a risk. CV's non-contact nature is particularly advantageous here, supporting safer evaluation in both clinics and homes.

c. Accessibility and digital divide

Not all patients may have access to reliable Internet, smartphones or cameras at home. Those from rural or lower socioeconomic backgrounds may be disproportionately excluded unless systems are designed to operate offline, on low-bandwidth networks or via in-clinic kiosks (Kadiri and Owolabi 2020).

19.3.4. *Evidence from related clinical populations*

Although studies specifically focused on oncology are limited, CV-based ROM tools have demonstrated clinical utility in other vulnerable populations, offering valuable lessons:

– In *stroke rehabilitation*: OpenPose and Kinect-based systems have been used to track upper-limb recovery, with high agreement (>90%) compared to therapist ratings. Patients were able to complete assessments at home, with data transmitted to clinicians asynchronously.

– In *orthopedic rehabilitation*: CV systems have measured ROM post-ACL reconstruction, total knee arthroplasty and shoulder repair with mean angular differences of < 10° compared to goniometers.

– In *axial spondyloarthritis*: tools such as the ViMove or smartphone-based apps have successfully captured spinal ROM, with moderate to high correlation to clinical measures ($r^2 = 0.36–0.77$).

These studies validate the feasibility and accuracy of CV-based tools, although oncology-specific validation remains a crucial next step.

19.4. Proposed framework for oncology CV-based ROM analysis

To successfully integrate CV tools into oncology physiotherapy for ROM analysis, a structured and adaptable framework is essential. This framework must accommodate the clinical realities of cancer care, the technological constraints of CV systems and the individualized needs of patients. Below, we propose a comprehensive, step-by-step framework for developing and deploying CV-based ROM assessment solutions in oncology settings.

19.4.1. *Hardware setup and environment*

The choice of hardware directly impacts the accuracy, feasibility and patient experience of CV-based ROM analysis. In oncology physiotherapy, considerations include ease of use, minimal physical burden and safety (Lam 2024).

a. Camera options

– *RGB cameras*: standard webcams or smartphone cameras offer sufficient quality for 2D pose estimation. These are ideal for low-cost, home-based assessments.

– *RGB-D cameras*: devices such as Microsoft Kinect, Azure Kinect or Intel RealSense provide depth data, enabling 3D skeletal reconstruction. They improve ROM accuracy, especially for complex or multiplanar movements.

– *Multicamera systems*: for higher precision, multiple synchronized cameras can capture movement from different angles, reducing occlusion and improving spatial accuracy.

b. Setup recommendations

– Camera positioned at torso height, perpendicular to the movement plane;

– well-lit, non-cluttered background to enhance key point detection;

– clear space for the patient to move without obstruction;

– mounting hardware (tripods, stands, etc.) to ensure stable, repeatable positioning.

c. Patient-centric considerations

– Minimal patient instruction required for setup;

– non-contact systems preferred to reduce infection risk;

– adjustable field of view for seated, standing or bed-bound assessments.

19.4.2. *Pose estimation and skeletal modeling*

Once visual data is captured, the next step is to extract skeletal information through pose estimation algorithms.

a. Model selection

– *OpenPose*: offers full-body key point tracking (COCO or BODY_25 formats), widely used in clinical research.

– *MediaPipe*: optimized for mobile and real-time performance; tracks 33 full-body landmarks.

– *MoveNet/YOLO-Pose:* lightweight models suitable for edge devices with fast inference times.

– *AlphaPose/BlazePose*: alternatives offering higher spatial resolution or robustness under occlusion.

b. Key point tracking

Pose estimation models extract body joint coordinates, typically including:

– upper limb (shoulder, elbow, wrist, etc.);

– lower limb (hip, knee, ankle, etc.);

– spine and torso (neck, mid-back, pelvis, etc.);

– head and facial markers (forehead/neck ROM).

In oncology, tracking upper limb and cervical ROM is especially critical in breast and head/neck cancer survivors.

19.4.3. *ROM computation and kinematic analysis*

With skeletal keypoints available, ROM analysis involves calculating joint angles, trajectories and other motion metrics.

a. Joint angle calculation

– Planar angles (e.g. elbow flexion in sagittal view) are calculated using vector geometry from adjacent joint positions.

– Multiplanar or 3D angles (e.g. shoulder abduction, cervical rotation, etc.) require trigonometric or quaternion-based computations.

b. ROM parameters

– *Maximum angle (flexion/extension)*: peak joint excursion in each direction.

– *Angle range*: difference between max and min joint positions.

– *Movement speed*: angular velocity and acceleration during movement.

– *Symmetry index*: left-right comparison for bilateral joints.

– *Compensation detection*: for example, trunk lean during shoulder abduction.

c. Accuracy improvement strategies

– Apply temporal smoothing (e.g. Savitzky–Golay filters) to reduce noise;

– use multiple trials to account for variability;

– normalize joint positions based on patient body proportions.

19.5. Benefits and opportunities

The integration of CV-based ROM analysis into oncology physiotherapy offers a transformative opportunity to improve patient care, increase accessibility, enhance clinical decision-making and foster a more data-driven approach to cancer rehabilitation (Gandi et al. 2025). While technical challenges remain, the potential benefits span multiple domains: clinical, operational, economic and societal.

19.5.1. *Enhanced accuracy and objectivity in assessment*

Traditional ROM assessments in oncology are typically performed using handheld goniometers or visual estimation by therapists. These methods, while widely used, are subjective and limited by intra- and inter-rater variability. CV systems offer:

– standardized measurement protocols, reducing human bias;

– fine-grained angle detection over time and across multiple joints;

– time-series tracking, allowing continuous monitoring rather than episodic snapshots.

This level of objectivity is especially important in oncology, where small changes in joint function or movement quality can signal significant clinical shifts – such as recovery after radiation-induced fibrosis or early signs of chemotherapy-induced motor deficits (Maltser and Behling-Rosa 2017).

19.5.2. *Remote monitoring and telehealth compatibility*

Cancer patients often have reduced mobility, weakened immunity and limited ability to travel. This makes traditional in-person physiotherapy difficult to maintain consistently. CV-based systems can be integrated into telerehabilitation programs, enabling:

– at-home assessment of ROM through smartphone or webcam;

– synchronous or asynchronous review by clinicians via cloud platforms;

– patient self-assessment, with guided instructions and real-time feedback.

By reducing the need for frequent hospital visits, these systems can support continuity of care without compromising safety, especially important during periods of immunosuppression or public health crises (e.g. Covid-19).

19.5.3. *Scalable and cost-effective care delivery*

Conventional motion capture systems such as Vicon or Qualisys are highly accurate but expensive, space-intensive and limited to research or specialized rehabilitation centers. CV-based alternatives provide:

– low-cost implementations using existing consumer-grade hardware (e.g. smartphones, webcams, etc.);

– software-based scalability, allowing deployment across clinics without major infrastructure;

– minimal training requirements, with intuitive interfaces for clinicians and patients.

This democratization of motion analysis technology enables broader adoption, particularly in resource-limited settings where access to oncology rehabilitation services is often constrained (Barragan-Carrillo et al. 2025).

19.5.4. *Personalized and data-driven rehabilitation*

Cancer patients present with diverse functional baselines and recovery trajectories. CV-based ROM systems generate large volumes of quantitative movement data that can be used to:

– establish individual baselines, making it easier to detect decline or improvement;

– track recovery trends over weeks or months, supporting adaptive therapy plans;

– tailor interventions based on dynamic analysis of joint limitations, movement compensations and functional goals.

This aligns with the broader push towards precision rehabilitation, where treatment is informed not just by diagnosis but also by real-time functional data.

19.6. Limitations and challenges

While CV-based ROM analysis offers many benefits, it also presents several limitations and barriers that must be addressed before routine clinical adoption in oncology physiotherapy (Goncalves Leite Rocco et al. 2024). These challenges span the domains of technical implementation, patient-specific factors, clinical integration, data security and regulatory approval.

19.6.1. *Technical limitations of current CV systems*

Despite recent advances, current CV systems – particularly markerless pose estimation models – still face performance limitations, especially in clinical populations such as oncology patients.

19.6.1.1. *Pose estimation accuracy*

– Many models are trained on healthy, able-bodied datasets such as COCO or MPII, which may not generalize well to cancer survivors with altered postures, limited ROM or use of assistive devices.

– Errors in key point detection (e.g. shoulder, hip) can significantly affect ROM calculations, particularly in joints with subtle or compensatory movements.

– Accuracy may degrade in low-light conditions, with loose clothing or the occlusion of body parts (e.g. arm hidden behind torso).

19.6.1.2. *2D versus 3D limitations*

– Most open-source CV tools offer 2D pose estimation from single-camera input, which cannot accurately capture movements out of plane (e.g. shoulder abduction with trunk lean) (Tian 2023).

– 3D pose estimation requires multiple cameras or depth sensors (e.g. Kinect, RealSense, etc.), increasing system complexity and cost.

– Algorithms for 3D reconstruction from monocular video are improving but remain error-prone in clinical scenarios.

19.6.2. *Patient-related factors in oncology*

Cancer survivors represent a diverse and complex population with unique challenges that may impact the utility of CV-based ROM tools.

19.6.2.1. *Physical heterogeneity*

– Functional impairments vary widely across tumor types (e.g. breast vs. head and neck vs. sarcoma) and treatment regimens.

– Fatigue, pain, neuropathy and balance issues may lead to inconsistent performance during assessments.

– Some survivors may have amputations, stomas, catheters or lymphedema that alter movement or affect body visibility.

19.6.2.2. *Cognitive and emotional barriers*

– Cognitive decline, often related to "chemo brain", may impair patients' ability to follow guided movement instructions accurately.

– Depression or anxiety – common in cancer survivors – may reduce engagement with digital self-assessment tools unless supplemented with support and motivation (Lau et al. 2022).

19.6.3. *Clinical workflow integration*

For widespread adoption, CV-based tools must fit seamlessly into the clinical workflow of oncology rehabilitation teams.

19.6.3.1. *Interoperability with health IT systems*

– Many existing CV platforms are standalone apps or research tools that lack integration with electronic medical records (EMRs), billing systems or clinical dashboards.

– Without integration, therapists must manually transcribe ROM data into reports or assessments, reducing efficiency.

19.6.3.2. *Clinician training and acceptance*

– Physiotherapists may lack familiarity with CV technology, leading to resistance or misuse.

– Clinician education is needed on how to interpret ROM outputs, correct errors and incorporate data into decision-making (Patil et al. 2024).

19.6.3.3. *Validation against gold standards*

– Regulatory and clinical stakeholders require rigorous validation studies comparing CV-based ROM measurements with traditional goniometry or motion lab results.

– Lack of standardization across tools, joint definitions and assessment protocols further complicates validation efforts.

19.7. Future directions and research opportunities

As CV technology continues to mature, its integration into oncology physiotherapy opens numerous avenues for innovation, expansion and clinical transformation.

To fully realize its potential, research must focus not only on improving technical capabilities but also on addressing contextual, human-centered and healthcare system-level challenges (Childs and Cleland 2006). Below, we outline key future directions and research priorities.

19.7.1. *Development of oncology-specific pose estimation models*

Most current pose estimation frameworks – such as OpenPose, BlazePose or MediaPipe – are trained on general population datasets and optimized for able-bodied individuals (Patrício et al. 2020). These models often underperform when applied to patients with postural asymmetries, limited joint mobility or visible surgical alterations.

Key research goals:

– *Curate oncology-specific datasets* including videos of patients undergoing rehabilitation post-mastectomy, laryngectomy or limb-salvage surgery with annotated joint landmarks.

– *Fine-tune existing models* using transfer learning to adapt to reduce ROM, non-standard poses and assistive device usage.

– *Benchmark performance* against traditional physiotherapy tools in clinical studies to evaluate sensitivity, reliability and error margins.

Such efforts can dramatically increase the clinical utility and accuracy of CV tools for this population.

19.7.2. *Integration with telerehabilitation platforms*

The Covid-19 pandemic accelerated the adoption of telehealth, yet remote physiotherapy assessments remain rudimentary, often relying on verbal reports or unsupervised exercise videos (Gudur et al. 2024). CV-based ROM analysis can revolutionize this space by offering quantitative, real-time and clinician-accessible mobility data.

Future opportunities:

– develop CV-enabled telephysiotherapy platforms with real-time visual feedback, automated ROM scoring and clinician alerts;

– embed asynchronous video analysis tools for patients to record exercises and receive automated feedback;

– implement adaptive guidance systems that modify exercise instructions in real time based on detected movement errors.

This can significantly expand access to physiotherapy in rural or resource-limited settings where in-person visits are not feasible (Kumar et al. 2026a).

19.7.3. *Personalized and longitudinal movement modeling*

Every cancer patient's functional baseline, progression and recovery trajectory are unique. Static threshold-based ROM assessments do not account for these individual differences (Kumar et al. 2026b). Future systems should focus on personalized movement modeling to enhance sensitivity and contextual relevance.

Emerging concepts:

– train baseline-aware models that use pre-treatment ROM as a reference to detect deviations over time;

– develop trajectory-based models that capture daily or weekly fluctuations in joint mobility, fatigue or pain;

– combine CV outputs with patient-reported outcomes (PROs) to contextualize motion data with subjective experience.

Personalized modeling enables early detection of subtle functional declines that may precede complications or relapse.

19.7.4. *Multimodal sensor fusion*

While CV offers rich visual data, combining it with data from other modalities can improve robustness and accuracy. Multisensor fusion is particularly valuable in noisy environments or for out-of-plane movements (Renals et al. 2012).

Integrative directions:

– fuse IMU data (e.g. accelerometers, gyroscopes) with CV pose estimates for hybrid ROM metrics;

– incorporate depth sensors (e.g. Intel RealSense, Azure Kinect) for improved 3D joint tracking in compact home setups;

– explore biosignal integration (e.g. electromyography, heart rate variability) to link movement with physiological stress and effort.

Such hybrid approaches can provide a more holistic understanding of patient function during therapy and daily life.

19.8. Conclusion

The integration of CV into ROM assessment marks a significant advancement in oncology physiotherapy, offering objective, scalable and patient-friendly tools for movement evaluation. As cancer survivorship rates improve globally, the need for effective rehabilitation solutions to manage long-term functional impairments has become increasingly urgent. Traditional methods of ROM analysis, while clinically useful, are often limited by subjectivity, time constraints and the necessity of in-person visits – barriers that can be particularly challenging for immunocompromised or fatigued cancer patients. CV offers a non-contact, markerless and automated solution to these limitations. Through innovations in pose estimation algorithms and ML, CV systems can now quantify joint angles, identify functional deficits and monitor rehabilitation progress with high fidelity. These tools are adaptable to both clinic-based and remote environments, facilitating continuity of care across the treatment-to-recovery continuum. Despite the promising potential, current CV applications in oncology remain underdeveloped compared to other fields such as sports medicine or orthopedics. Unique patient characteristics – such as surgical scars, altered biomechanics or fatigue-induced movement patterns – pose new challenges for algorithm development and validation. Ensuring accuracy, usability and clinical trustworthiness will be crucial for broader adoption.

19.9. References

Barragan-Carrillo, R., Asirwa, F.C., Dienstmann, R., Pendhakar, D., Ruiz-Garcia, E. (2025). Global oncology: Tackling disparities and promoting innovations in low- and middle-income countries. *American Society of Clinical Oncology Educational Book*, 45(3), e473930.

Childs, J.D. and Cleland, J.A. (2006). Development and application of clinical prediction rules to improve decision making in physical therapist practice. *Physical Therapy*, 86(1), 122–131.

Colyer, S.L., Evans, M., Cosker, D.P., Salo, A.I. (2018). A review of the evolution of vision-based motion analysis and the integration of advanced computer vision methods towards developing a markerless system. *Sports Medicine-Open*, 4(1), 24.

Debnath, B., O'Brien, M., Yamaguchi, M., Behera, A. (2022). A review of computer vision-based approaches for physical rehabilitation and assessment. *Multimedia Systems*, 28(1), 209–239.

Emery, J., Butow, P., Lai-Kwon, J., Nekhlyudov, L., Rynderman, M., Jefford, M. (2022). Management of common clinical problems experienced by survivors of cancer. *The Lancet*, 399(10334), 1537–1550.

Gandi, R., Suseela, G., Singh, S., Kumar, V.K., Sudhakara, M. (2025). Data-driven rehabilitation using machine learning approaches for adaptive recovery and treatment. In *Predictive Algorithms for Rehabilitation and Assistive Systems*, Korupalli, V.R.K., Ranasinghe, P., Feldman, D. (eds). IGI Global, Hershey, PA.

Girard, D. (2016). Neuromuscular mechanisms contributing to self-reported post-treatment cancer fatigue and the effects of resistance training. PhD Thesis, Charles Sturt University, Sydney.

Goncalves Leite Rocco, P., Reategui-Rivera, C.M., Finkelstein, J. (2024). Telemedicine applications for cancer rehabilitation: Scoping review. *JMIR Cancer*, 10, e56969.

Gudur, A., Patil, V., Jain, A., Garg, N. (2024). Hybrid genetic algorithm and deep learning approach for lung nodule detection and classification in chest X-rays. *International Journal of Intelligent Systems Applications in Engineering*, 12(3s), 577–587.

Hellsten, T., Karlsson, J., Shamsuzzaman, M., Pulkkis, G. (2021). The potential of computer vision-based marker-less human motion analysis for rehabilitation. *Rehabilitation Process and Outcome*, 10, 11795727211022330.

Herath, H.M., Jayasekara, A.G., Madhusanka, B.A. (2024). *Vision Attentive Technology: AI Approaches for Functional Mobility Assessments in Elderly Healthcare*. CRC Press, Boca Raton, FL.

Jouini, O., Sethom, K., Namoun, A., Aljohani, N., Alanazi, M.H., Alanazi, M.N. (2024). A survey of machine learning in edge computing: Techniques, frameworks, applications, issues, and research directions. *Technologies*, 12(6), 81.

Kadiri, C. and Owolabi, B.O. (2020). The new normal of eHealth equity: Transforming chronic disease management amid rural technological constraints and digital divides. *International Journal of Science and Research Archive*, 01(01), 251–269.

Kianifar, R. (n.d.). Automated pose estimation for the assessment of dynamic knee valgus and risk of knee injury during the single leg squat. Doctoral Dissertation, University of Waterloo, ON.

Kumar, A., Batta, P., Ahuja, S., Rathore, P.S. (eds) (2026a). *AI-driven Innovations in Physiotherapy and Oncology 1*. ISTE Ltd, London and Wiley, Hoboken, NJ.

Kumar, A., Batta, P., Ahuja, S., Rathore, P.S. (eds) (2026b). *AI-driven Innovations in Physiotherapy and Oncology 2*. ISTE Ltd, London and Wiley, Hoboken, NJ.

Lam, W.T. (2024). Effects of the application of markerless motion capture (MMC) technology for patients with stroke. PhD Thesis, The Hong Kong Polytechnic University (PolyU), Hong Kong.

Lau, S.C., Bhattacharjya, S., Fong, M.W., Nicol, G.E., Lenze, E.J., Baum, C., Hardi, A., Wong, A.W. (2022). Effectiveness of theory-based digital self-management interventions for improving depression, anxiety, fatigue and self-efficacy in people with neurological disorders: A systematic review and meta-analysis. *Journal of Telemedicine and Telecare*, 28(8), 547–558.

Levy, E.W., Pfalzer, L.A., Danoff, J., Springer, B.A., McGarvey, C., Shieh, C.Y., Morehead-Gee, A., Gerber, L.H., Stout, N.L. (2012). Predictors of functional shoulder recovery at 1 and 12 months after breast cancer surgery. *Breast Cancer Research and Treatment*, 134(1), 315–324.

Lovelace, D.L., McDaniel, L.R., Golden, D. (2019). Long-term effects of breast cancer surgery, treatment, and survivor care. *Journal of Midwifery & Women's Health*, 64(6), 713–724.

Ma, Y., Mithraratne, K., Wilson, N.C., Wang, X., Ma, Y., Zhang, Y. (2019). The validity and reliability of a Kinect v2-based gait analysis system for children with cerebral palsy. *Sensors*, 19(7), 1660.

Maltser, S. and Behling-Rosa, B.A. (2017). Cancer rehabilitation. In *Geriatric Rehabilitation*, Poduri, K.R. (ed.). CRC Press, Boca Raton, FL.

Patil, D.S., Yadav, T., Gudur, A. (2024). To compare the effect of swallowing and non-swallowing exercises on dysphagia in post operative oral cancer individuals: A comparative study. *Asian Pacific Journal of Cancer Prevention: APJCP*, 25(8), 2945.

Patrício, L., Sangiorgi, D., Mahr, D., Čaić, M., Kalantari, S., Sundar, S. (2020). Leveraging service design for healthcare transformation: Toward people-centered, integrated, and technology-enabled healthcare systems. *Journal of Service Management*, 31(5), 889–909.

Rana, M.S. (2024). Leveraging markerless computer vision for comprehensive walking automated gait analysis in rehabilitation. Master's Thesis, Arcada University of Applied Sciences, Helsinki.

Renals, S., Bourlard, H., Carletta, J., Popescu-Belis, A. (2012). Sampling techniques for audio-visual tracking and head pose estimation. In *Multimodal Signal Processing: Human Interactions in Meetings*, Renals, S., Bourlard, H., Carletta J., Popescu-Belis, A. (eds). Cambridge University Press, Cambridge.

Sadineni, H.S. (2024). Orthopedic applications: Advancing physiotherapy in musculoskeletal health. In *Physical Therapy-Towards Evidence-Based Practice*, Nakano, H. (ed.). IntechOpen, London.

Silver, J.K., Baima, J., Mayer, R.S. (2013). Impairment-driven cancer rehabilitation: An essential component of quality care and survivorship. *CA: A Cancer Journal for Clinicians*, 63(5), 295–317.

Szabo, D.A., Neagu, N., Teodorescu, S., Apostu, M., Predescu, C., Pârvu, C., Veres, C. (2023). The role and importance of using sensor-based devices in medical rehabilitation: A literature review on the new therapeutic approaches. *Sensors*, 23(21), 8950.

Tian, I.Y. (2023). Learning clinical body composition metrics from 2D and 3D optical imaging. PhD Thesis, University of Washington, Seattle, WA.

Warmerdam, E., Hausdorff, J.M., Atrsaei, A., Zhou, Y., Mirelman, A., Aminian, K., Espay, A.J., Hansen, C., Evers, L.J., Keller, A. et al. (2020). Long-term unsupervised mobility assessment in movement disorders. *The Lancet Neurology*, 19(5), 462–470.

20

AI-Powered Robotic Assistance for Cancer Patient Physiotherapy

As the rate of survival of the cancer patients grows, there is an urgent need to have proper personalized rehabilitation solutions. Limited movement, muscle weakness, pain, fatigue and even cognitive deficits are common effects for survivors, leading to a significant reduction in their quality of life. This chapter addresses the use of artificial intelligence (AI) powered robotic assistance in improving outcomes in cancer patients' physiotherapy. These new robotic systems, by combining AI and robotics, are able to provide responsive rehabilitation systems that are personalized to patient modalities and real-time performance and provide adaptive AI robotic rehabilitation. This chapter starts with assessing the unique clinical rehabilitation challenges cancer patients face and then detailing the rehabilitation AI technologies and robotic systems that are being developed or in practice. Advantages in robotic physiotherapy such as AI technologies assist in improving precision, providing consistency, remote access and engagement of the patient. Real-world outcomes and limitations are discussed in relation to challenges on various cancer types, including practical limitations for breast, head and neck and colorectal cancer. Adoption of such technologies, data and AI ethics, integration in the healthcare system and privacy with AI in healthcare are explored. Multimodal data fusion, emotionally intelligent robots, brain–computer interfaces (BCIs), integration with virtual reality and in-home care are also discussed.

20.1. Introduction

Cancer treatments (surgery, chemotherapy, radiation) can result in permanent physical, neurological and psychological impairments. Cancer treatments can result in complications such as muscle weakness, fatigue, neuropathy, lymphedema and reduced mobility, negatively affecting one's quality of life. Cancer survivors need physiotherapy, as it helps them recover lost functionality, stamina and daily living activities (Silver et al. 2013). However, there are challenges in the delivery of

Chapter written by Rashmi GUDUR and Faisal Hussain HUSSAIN.

physiotherapy, which include the availability of each session's variability, the engagement and adherence of the patient and the absence of ongoing personalization and data. Cancer rehabilitation can benefit from the use of AI and robotics.

Cancer rehabilitation can use AI robotics because robotics can perform repetitive and precise tasks while AI can evaluate and integrate with the patient. Robotic exoskeletons are not the only AI-integrated devices designed for rehabilitation. AI-enabled virtual physiotherapists provide adaptive home-based rehab, enabling patients to rehabilitate anytime and anywhere (Reddy 2025).

For the rehabilitation of cancer patients, rehabilitation robots not only provide physical aid but also motivate, track progress and remotely enable therapy to those in under-resourced and isolated areas (Kumar et al. 2026a). These systems are designed for real-time user-driven rehabilitation aligned with the principles of precision medicine. However, there are still challenges to be addressed, such as high costs, accessibility, user acceptance, the need for clinical validation and ethical issues.

This chapter outlines the emerging role of AI-integrated robotic systems in rehabilitation physiotherapy for cancer patients, exploring the potential of these systems to promote integrated recovery and clinical outcomes while describing the technologies, practical uses, clinical research, advantages and challenges and anticipated progress in the field.

20.2. Background and clinical context

Although cancer treatment can be lifesaving, it can leave some survivors with the lasting effects of cancer-related disabilities that need specialized rehabilitation. These disabilities can occur from the cancer, the treatments or the sequelae from surgery, chemotherapy, radiation therapy or hormonal therapy. Patients who are recovering from breast cancer, colorectal cancer, head and neck cancers and brain tumors have the greatest burden of functional loss. Disability sequelae may include atrophy of muscle and joints, stooped posture, unstable walking, diminished lungs and stamina, exhaustion, chronic pain and cognitive deficits (Kumar et al. 2026b). These sequelae can severely hinder one's ability to perform activities of daily living (ADLs), reduce autonomy and lower one's quality of life.

The role of physiotherapy in the rehabilitation of cancer survivors is aimed at improving movement and muscle, managing the remaining symptoms and restoring stamina to the patient. The purpose of physiotherapy is manual therapy, therapeutic exercises and balance, walking and other neuromuscular systems training. A shortage of trained professionals, logistical problems and patient compliance can, however, lead to compromising the availability and consistency of quality physiotherapy.

The common practice of rehabilitation is also not created with the purpose to react to the changing emotional and physiological condition of the patient, which weakens the effectiveness of treatment (Schollhorn et al. 2010). Cancer rehabilitation remains an unknown factor due to the variety of the required rehabilitation.

As an illustration, breast cancer survivors will normally be receiving articulation and lymphedema therapy, whereas brain tumor survivors will require help with balance, hand coordination and cognition. Fatigue, which is reported by approximately 90% of patients during the treatment process, also further restricts the exercise tolerance, and it has to be carefully adjusted in terms of intensity and duration (Hussey and Gupta 2022). These issues require the incorporation of individualized and scalable physiotherapy measures.

Artificial intelligence (AI) and robotics are one of the solutions. Robotic systems may be made to carry out upper and lower limb recovery and may interface with current protocols to deliver movement patterns of a differing strength, range and repetition. They may help either passively or actively depending on the mobility of the limbs and provide biomechanical feedback, which can be used to personalize the therapy and warn clinicians about possible failures (Wang et al. 2017).

AI and robots can be combined to provide uninterrupted robotic therapy and be able to dynamically adjust the aim and protocols. Machine learning (ML) has the capability to forecast the results, update the objectives, issue disengagement notifications and algorithmically improve workout plans. Emotion-sensitive feedback can also be given by AI to increase motivation (Fatime et al. 2025).

They can be administered as therapeutic modalities as well as to provide constant monitoring and assessment, which in this case is particularly useful in the outpatient and home-care environment, where access is restricted. The use of robot systems offers safe, monitored and low-demand exercise regimens, which keep patients undergoing intensive chemotherapy or major surgeries active.

The combination of AI-based robotics is an example of how biomedical engineering, clinical oncology and data science intersect and that adaptive integrated rehabilitation – physiotherapy matching demand and shaped to the current physiological parameters and needs of a patient – is beginning (Aziz et al. 2024). Advanced and scalable rehabilitation, effective rehabilitation, will continue to be in demand as the number of cancer survivors across the globe will continue to rise.

20.3. AI-enabled robotic rehabilitation technologies

An AI-powered robotics rehabilitation incorporates robotics, AI, biomedical sensors and data analytics. This implementation enables adjusting accurate and reliable physiotherapy treatment. Such technology is especially beneficial for patients recovering from cancer surgery, patients with neurological impairments and patients who have been hospitalized for extended periods.

This section describes the main components of AI-enabled robotic rehabilitation systems, including their design, how they operate and their clinical features.

20.3.1. *Robotic systems in rehabilitation*

Robotic rehabilitation systems can be broadly categorized into exoskeletons, end-effector devices and assistive robotic platforms:

– Exoskeletons are flexible robotic bodies that assist individuals with specific mobility functions and full-body movement support. They are especially helpful during rehabilitation of the lower limbs when the individual needs help with movement training and support muscle rehabilitation. The more advanced versions of these flexible bodies perform real-time adjustments that aid the user based on their performance metrics with the help of integrated sensors and actuators.

– Guided trajectory end-effector robots direct portions of the robot body (e.g. hand or foot). These robots are practiced for upper limb rehabilitation, notably after a stroke, or breast or head and neck cancer surgeries when arm functions are impacted.

– Artificial rehabilitation technologies such as robotic rehabilitation assistive robotic arms and rehabilitation robotic platforms focus more on passive, active-assisted or resistive movements implemented according to the patient's needs. The integration of AI makes the robotic arms and platforms customized to rehabilitation, which promotes voluntary effort by the patients.

Modern robotic systems such as EGZOTech's Luna EMG and Sidra LEG are designed to support oncological rehabilitation. These devices offer modes for both passive mobilization and active training and can be used in supine positions, making them suitable for severely weakened patients or those undergoing aggressive treatment (Green et al. 2016).

20.3.2. *Integration of AI*

The conversion of the traditional robotic therapy machines into smart, reactive machines that can learn about the patient data makes the process of rehabilitation more effective:

– ML algorithms process large datasets collected from wearable sensors, robotic actuators and patient feedback to identify patterns in recovery.

– Supervised learning is used for classifying patient states (e.g. fatigue, disengagement), while reinforcement learning allows robots to improve assistance strategies through feedback loops (Tsiakas 2018).

– The analysis of intricate time-series data from biosignals such as electromyography (EMG), electroencephalography (EEG) and motion trajectories uses advanced deep learning methodologies, and in particular, Convolutional Neural Networks (CNNs) and Recurrent Neural Networks (RNNs).

– Natural language processing (NLP) capabilities enable robots to interpret verbal cues or deliver motivational support during therapy, improving human–robot interaction.

20.3.3. *Wearable sensors and biometric monitoring*

Wearable sensors play a critical role in collecting the physiological and kinematic data that fuel AI algorithms. These include:

– EMG sensors that detect muscle activation and fatigue;

– force and pressure sensors embedded in robotic joints or platforms to quantify load, resistance and gait asymmetry;

– physiological sensors to monitor heart rate, skin conductance and breathing patterns: useful for detecting exertion or emotional stress.

AI technology has the ability to track patient progress in a way that is both continuous and unobtrusive. For example, AI recognizes when patients have difficulty with a task or if they are trying to do too much, enabling the automated modification of systems to adjust the exercise intensity and provide movement instructions (Tamantini 2020).

20.3.4. *Virtual and augmented reality integration*

Incorporating virtual reality (VR) and augmented reality (AR) environments into robotic systems has become a new trend to boost engagement and provide motivating gamification to users. VIVADIV VR for children's oncology gives a VR headset to children during and after their therapy to help them provide cancer-related therapy and enjoy their therapy. These environments are made and evaluated using AI models to adapt to individual patient profiles. These environments are adjusted to the patients by gauging their emotional state so that they are challenged enough to stay engaged without feeling overwhelmed.

For example, using AI real-time no-touch biosignal analysis, the I-TROPHYTS AI-powered robotic platform adapts therapy for children and adolescents with cancer and neurological disorders. It integrates robotic arms, VR screens and AI classifiers to detect patient arousal and fatigue to adjust VIVADIV IV therapy VR gamification to the patient's therapy emotional state, modifying the cognitive and physical demands of the tasks to their therapy tasks (Le et al. 2018).

20.4. Oncology-specific applications

The application of AI-powered robotic assistance in physiotherapy is particularly impactful in oncology, where patients experience diverse and often complex rehabilitation needs. Cancer is not restricted to the primal site, but it tends to bring about systemic issues such as neuromuscular deficits, postural instability, musculoskeletal pain, lymphatic dysfunction and psychological distress. The tailored AI-based robotic systems by these cancer-specific impairments can offer specific interventions aimed at enhancing recovery, particularly where the conventional rehabilitation fails.

20.4.1. *Breast cancer*

Breast cancer survivors experience restricted mobility of shoulders, lymphedema, pain and weakness of upper limbs, especially following mastectomy, axillary node removal or radiation (Kumar et al. 2026b). These problems disrupt normal activities such as grooming, dressing or lifting. Robotic upper-limb rehabilitation systems, either end-effector or exoskeleton systems, can aid the enhancement of the range of motion, muscle strength and motor control:

– EGZO Tech's Luna EMG has been used for active-assisted arm movements, promoting neuromuscular recovery with real-time biofeedback.

– AI algorithms can monitor fatigue levels and tailor movement repetition to avoid overexertion, a common issue in post-chemotherapy patients.

– Emotion-aware robotic systems can detect frustration or pain through facial recognition or speech patterns and suggest rest or emotional support accordingly.

In combination with wearable EMG and lymphedema monitoring devices, these systems also provide early warning signals of recurrence or complications, allowing proactive intervention.

20.4.2. *Head and neck cancer*

Head and neck cancer survivors also experience dysphagia, speech difficulties, cervical stiffness and balance due to the involvement of nerves and fibrosis. While traditional therapy focuses on speech-language pathology and manual stretching, AI-enabled robotic interventions expand these options:

– Robotic neck and jaw movement platforms, combined with real-time AI analysis of muscle engagement, assist in regaining mobility and preventing fibrosis.

– Integration with speech recognition software allows for adaptive speech exercises, adjusting intensity based on voice clarity and patient effort.

– Balance training robots with embedded AI (such as dynamic platforms or VR gait systems) can correct postural imbalances, especially for patients affected by cranial nerve palsies or cerebellar involvement (Nizamis et al. 2021).

20.4.3. *Colorectal and prostate cancer*

Recovering from colorectal or prostate cancer surgery can be difficult, particularly with regard to gait problems, deconditioning, urinary incontinence, pelvic floor dysfunction and dysfunction. However, robotic gait training systems and lower limb exoskeletons can be valuable tools in rehabilitation.

– Sidra LEG, another EGZOTech solution, supports passive and active movement of the lower limbs, particularly useful in bed-bound or weak patients.

– AI models use gait symmetry, joint angle and pressure sensor data to adjust support in real time, improving motor relearning.

– AI-enabled pelvic floor biofeedback systems assist in retraining the muscles used to control continence by tracking contraction signals in real time and offering appropriate audio-visual cues.

These applications help restore dignity, autonomy and social engagement for male and female cancer survivors who struggle with incontinence and, as a result, may feel embarrassed or isolated (Green 2024).

20.4.4. *Neurological cancers*

Individuals diagnosed with tumors of the central nervous system such as gliomas, medulloblastomas or tumors of the spinal cord may experience motor and coordination difficulties, spasticity and cognitive deficits (Walker et al. 2016). These problems need comprehensive and prolonged rehabilitation, which should be physically and cognitively stimulating. Robotic systems enhanced with AI are specifically designed for such individuals.

– Exoskeletons for paraplegic or hemiplegic patients offer guided ambulation while AI assesses weight shift, stride length and effort.

– Cognitive-motor dual-task training platforms (e.g. walking while solving arithmetic tasks in VR) simultaneously stimulate brain and body recovery.

– Emotion-aware robotic platforms detect signs of confusion or cognitive fatigue through gaze tracking and facial expression analysis, adapting the pace or complexity of the session.

20.4.5. *Pediatric oncology*

Patients within this age group are not easy to rehabilitate. The effects of chemotherapy and radiation will cause kids to experience ongoing pain and will likely trigger psychosocial issues. Treatment fatigue will likely develop and treatment fatigue will likely develop.

Robots like NAO and Pepper provide social assistance and encourage kids to adhere to play therapy. Robots that people call “Pepper” and “NAO” engage kids conversationally and motivate them to complete therapeutic exercises and reward them with games and activities. Robots improve compliance with therapy, the range of motion and therapy retention. Safety and adjustability of robotic systems are achievable through considerate child-sized anthropometrics. Supported robotic trainers, reward systems and activity games will monitor knee and shoulder joints and motivate kids to complete tasks using set parameter adjustments. These trainers will adjust the grip, power and range of motion to match the therapeutic goals. Tech tools that monitor sustained AI used for functional outcomes will improve kids with rehabilitation after cancer (Gonzalez et al. 2021).

Robotic system/platform	AI integration features	Primary rehabilitation use	Cancer type/application	Key benefits
EGZO Tech Luna EMG	EMG-based machine learning algorithms for adaptive motion control	Upper limb mobility training	Breast cancer (post-mastectomy)	Personalized motion guidance; muscle activation feedback
Sidra LEG	Gait analysis using real-time AI feedback loops	Lower limb and gait rehabilitation	Colorectal; prostate cancers	Improved gait symmetry; adjustable intensity
I-TROPHYTS platform	CNN-based biosignal analysis and fatigue detection	Dual motor-cognitive training with VR	Neurological and pediatric cancers	Emotion-aware feedback; adaptive VR therapy
NAO/Pepper (Social Robots)	NLP and sentiment recognition	Pediatric physiotherapy motivation	Pediatric oncology	Enhanced engagement; reduced therapy dropout
AI-powered pelvic floor trainer	Supervised learning for muscle contraction detection	Pelvic floor and continence rehabilitation	Colorectal and prostate cancers	Real-time feedback; privacy-focused training
BCI-integrated robotic arm	EEG-based reinforcement learning	Neurological motor recovery	Brain and spinal tumors	Thought-driven control; neuroplasticity stimulation

Table 20.1. *Overview of AI-enabled robotic systems in cancer physiotherapy*

20.5. Advantages and benefits

The use of AI technologies for robotic assistance in cancer rehabilitation physiotherapy brings many advantages that, for the most part, surpass the usual rehabilitation physiotherapy procedures. This is particularly important considering the challenges that cancer patients deal with in the recovery phase – physical, cognitive and emotional. These technologies are changing the factors that most affect the delivery and experience of physiotherapy, from improved personalization to greater access.

20.5.1. *Personalization and adaptivity*

One of the most transformative benefits of AI is the ability to provide personalized rehabilitation. Standardized therapy protocols are modified manually after getting some feedback from the patient and therapist. In the case of AI-based robotic rehabilitation, the therapy robot is able to adapt to the patient in real time based on sensor, wearable and user interaction data (Deshmukh 2023). It is dynamic in terms of therapy intensity, duration, resistance and difficulty.

As an example, when a patient exhibits fatigue, sluggish reaction, diminished movement or alterations in heart rate, the robot will reduce the intensity of an exercise or recommend a rest. In contrast, the occurrence of the progress instigates an increased test, which is physiologically and emotionally adequate in the therapy. This constant modification provides the appropriate amount of stimulation to every patient.

20.5.2. *Precision and consistency*

Robotics provide highly accurate and repeatable movements, which are required in motor relearning and neuroplasticity. Exercises administered by humans also change every time, whereas a robot duplicates the same variety, speed and path, with a perfect copy (Krebs et al. 2022). This accuracy is particularly useful when mobilizing passive or highly controlled joint movements during the initial rehabilitation.

Quantitative monitoring is made possible by the same accuracy. Measures, such as joint angle and applied force, asymmetry of gait and muscle activity, are taken in real time and used to monitor progress, assess outcomes and design therapy. Insights made by data can assist clinicians in making well-informed choices and preventing the emergence of complications or setbacks.

20.5.3. *Improved patient engagement and motivation*

Involvement is a familiar obstacle in cancer rehabilitation, specifically among those patients who are grappling with fatigue, depression or pain. Gamification, VR and social interfaces are common features of AI-enhanced robots to increase the motivation and make the therapy fun. Goal-oriented activities, interactive games and customized avatars do not focus on pain, but still therapeutic performance is achieved.

Emotional robots such as NAO or Pepper are used in pediatric and geriatric populations to provide encouragement, humor and praise with the help of speech and facial recognition and have been shown to considerably enhance adherence and well-being (Jeong 2017).

20.5.4. *Accessibility and remote care*

Remote monitoring and telerehabilitation can be facilitated by AI-enabled robots, including cloud-connected ones, and this will increase the reach of physiotherapy (Goyal et al. 2025). Chemotherapy patients or surgery patients cannot travel a lot. Robotic home stations can be used as a secure, efficient and convenient substitute.

Physiotherapists are able to track the performance metrics, make changes and provide assistance without physically being present. This promotes continuity of care and alleviates system burden and exposure of the patient to a hospital environment – critical to immunocompromised patients.

20.5.5. *Staff support and clinical efficiency*

Robotic systems also assist the healthcare helping hands by increasing the capacity of clinical services. Robots can be used in a high-volume or short-staffed environment to provide monotonous exercises, allowing a therapist to address more complicated assessments, manual therapies or psychosocial interventions. AI-generated progress reports support documentation by decreasing the administration load and enhancing efficiency.

20.6. Implementation challenges

Robotic assistance assisted by AI may enhance cancer physiotherapy, but its implementation has multiple obstacles to overcome. These are the technology, clinical, economic, regulatory and the human factors. We need to learn how to address these barriers in order to implement these systems safely and equitably in an oncology setting.

20.6.1. *High cost and limited accessibility*

The cost of purchasing, installing and maintaining robotic rehab devices is a huge challenge. Multisensor systems, complex exoskeletons and AI-based units demand a great amount of money, which small hospitals, community clinics or home care services can hardly afford. In addition to the cost of purchase, the institutions have to pay for maintenance costs as well as software updates and trained personnel.

In most of the low- and middle-income nations, complex rehab equipment is virtually non-existent, contributing to gaps in care. The centers that are tech-savvy offer more recovery opportunities, and resource-constrained settings are based on the traditional approaches (Riviello et al. 2011). There is unequal insurance coverage towards robotic physiotherapy, and the reimbursement policies are still in their formation stage.

20.6.2. *Complexity of device use and setup*

The rehabilitation systems that are based on robots, particularly AI-based systems, are complex. Their installation, calibration and operation require special training; thus, they are normally restricted to well-equipped facilities that have qualified biomedical personnel. Low-digital patients might have difficulties with the device interfaces, particularly at home where no professionals are available (Lindberg et al. 2013).

Devices must be easy to operate, user-friendly and customized to the needs of individual patients, even those with cognitive impairment, hearing or sight loss, or amputated limbs. Failure to meet these standards reduces the rate of adoption and adherence to therapy.

20.6.3. *Clinical validation and evidence gaps*

There is limited research indicating that robotic rehabilitation is useful in cancer physiotherapy, although there is still no strong clinical evidence. Smaller groups, smaller durations and single centers are characteristic of many trials (Schröder et al. 2021). To demonstrate the long-term results, safety and costs, we require big, multicenter, randomized controlled trials.

The institutions interpret these systems differently without having standardized clinical guidelines to apply them and measure them. Moreover, their non-integration with electronic health records (EHRs) reduces their applicability in the wider oncology processes.

20.6.4. *Ethical and data privacy concerns*

AI is associated with various ethical concerns. The data collected by robots regarding patients, such as movement traces, physiological indicators, moods and voice logs, is personal. They have to be stored, transmitted and processed to abide by privacy laws including the Health Insurance Portability and Accountability Act (HIPAA) and General Data Protection Regulation (GDPR) (Chiu 2023).

Bias may also happen when AI models are trained on non-diverse datasets. For instance, rehabilitation models developed using data from one demographic group may underperform in others, exacerbating disparities. Transparent model design, diverse training datasets and regular audits are essential to mitigate bias and maintain trust.

20.6.5. *Patient acceptance and psychological barriers*

Some patients enjoy robotic rehabilitation, while others do not. Common concerns are robotic patients being depersonalized, machines being used to make decisions and being observed. For many cancer patients being monitored, stressed and fatigued, robotic devices can be intimidating (Gudur et al. 2023). Robots must be designed with human-centered interfaces, incorporating emotionally responsive AI, communication and personalization. Integrating virtual therapists, encouraging AI and customizable interaction can alleviate discomfort.

20.7. Future directions

Using AI-powered robotics for cancer physiotherapy is still new. Over time, advancements in the field will increase the sophistication, availability and ease of incorporation of these systems into the oncology-care continuum that is focusing on patients (Elliott 2024).

This section describes the most important technological advancements, incorporation into practice, expansion of research and the establishment of frameworks that will define the robotic rehabilitation of cancer patients for the future.

20.7.1. *Multimodal data fusion for precision rehabilitation*

In the future, robotic systems will depend more on multimodal data fusion and the integration of various data sources – such as EHRs, wearable sensors, imaging studies, patient-reported outcomes and genomic profiles – into a single dataset for each patient. AI will use the integrated dataset to devise personalized rehabilitation strategies based on a patient's tumor type, treatment phase, functional level, comorbid conditions and psychosocial factors (Rasa 2024).

Being able to devise adaptive therapy stems will ensure that therapy goals and the patient's physiology will be in sync, even when the patient's progress or regression and overall physiological status change. For instance, ML algorithms will analyze and interpret tumor burden, inflammatory markers and fatigue patterns to adjust therapy load.

20.7.2. *Emotionally intelligent and socially aware robots*

Research is looking into the development of emotionally aware robotic systems for rehabilitation therapy. Such systems could analyze a patient's emotional state during a therapy session through facial expressions, tone of voice and other cues. Robots might vary their tone, language and activity upon sensing the emotional distress, disengagement or loss of interest in the patient.

Future socially conscious robots will go beyond this, offering physical assistance and also take the role of emotional support by serving as therapeutic companions to lonely or helpless patients. This applies particularly to pediatric oncology patients and elderly survivors who might experience loneliness or experience cognitive impairment.

20.7.3. *Brain–computer interfaces (BCIs)*

A promising future direction in neuro-rehabilitation is the use of Brain–Computer Interface (BCIs). These systems can support severely neurologically impaired cancer survivors such as those with brain tumors, spinal cord injuries, or chemotherapy-induced complications by detecting neural signals (e.g. EEG) and translating them into commands for motorized assistive devices, or by enabling participation in combined mental and physical therapeutic activities.

With AI, future BCI-robotic systems will be able to learn the brain pattern of each individual; this will enhance accuracy, reduce lag time and allow thought-directed therapy. It is still in its experimental phase, but initial outcomes are positive.

20.7.4. *Expanded home-based and community rehabilitation*

With the decentralization of healthcare, home-based robotic rehabilitation platforms are going to increase in application. Small, interchangeable and inexpensive gadgets, possibly ones with haptics and remote interaction with AI, will enable patients to keep doing their therapy themselves or remotely (Zhang et al. 2024).

These systems will be connected to telehealth so that physiotherapists will monitor the progress, provide feedback and remotely change settings. In the next editions, virtual coaching avatars, live video feedback and automatic progress-based goal setting could be available.

Shared robotics might also be used by community centers and survivor groups and form group rehabilitation with peer support, which would increase motivation and mental health.

20.7.5. *Integration with AR and VR*

Combining AI robotics and augmented reality (AR) with VR will establish immersive rehab spaces. Patients would be able to perform such exercises in gamified simulations that would replicate real-life challenges, such as climbing stairs, cooking, or walking in a crowded street, but remain safe and controlled (Mane and Yadav 2025).

AR/VR-powered robots can address physical and cognitive rehabilitation, particularly for patients who have chemo-brain, depression or post-traumatic stress disorder (PTSD). The introduction of AI allows such environments to change according to the progress and preferences of individual patients in terms of their complexity, pace and feedback style.

20.8. Conclusion

AI-driven robotic assistance is a groundbreaking innovation for physiotherapy for cancer patients. With improved cancer survival rates, the need for individualized and effective rehabilitation programs also increases. The integration of AI and robotic technologies in rehabilitation offers a precision, personalization and scalability unparalleled by conventional rehabilitation. In this chapter, we describe modern rehabilitation technologies and their various components from AI algorithms to sensors, robotic integration and clinical uses. AI robotic systems go beyond mechanical assistance and automate therapy by adapting to the patient's unique needs. Real-time therapy adaptive feedback systems improve rehabilitation outcomes by setting responsive goals and changing them to meet the patients' emotional and physiological needs. AI-driven robotic systems improve patient satisfaction with therapy outcomes that consistently adjust to the patient's needs. The benefits of this approach are the nearly goal-oriented systems that in rehabilitation therapy aim to improve personalization and patient satisfaction by consistently tracking and executing the disengaged exercises and adaptive systems. These systems are vital for cancer patients who are disengaged from therapy due to fatigue, pain, emotional distress, immunosuppression and other barriers that conventional therapy systems are not designed to comprehend.

20.9. References

Aziz, R., Jawed, F., Khan, S.A., Sundus, H. (2024). Wearable IoT devices in rehabilitation: Enabling personalized precision medicine. In *Medical Robotics and AI-Assisted Diagnostics for a High-Tech Healthcare Industry*, Khang, A. (ed.). IGI Global, Hershey, PA.

Bastos-Filho, T. (ed.) (2020). *Introduction to Non-invasive EEG-Based Brain-Computer Interfaces for Assistive Technologies*. CRC Press, Boca Raton, FL.

Chiu M. (2023). Building simulacra: Gamified physical therapy for VR. Master's Thesis, Northeastern University, Boston, MA.

Deshmukh, P. (2023). The integration of artificial intelligence and wearable technology in personalized physical therapy programs for enhanced recovery outcomes. *International Journal of Artificial Intelligence, Data Science, and Machine Learning*, 4(3), 1–9.

Elliott, A. (2024). *Algorithms of Anxiety: Fear in the Digital Age*. John Wiley & Sons, Hoboken, NJ.

Fatime, O.D., Waanders, I., Lips, D.J., Witteveen, A., Nane, G.F., van Keulen, M., John, A. (2025). Unsupervised change point detection for early complication identification in post-surgical oncology patients. In *14th Supporting Health by Technology Conference*, Twente.

Gonzalez, A., Garcia, L., Kilby, J., McNair, P. (2021). Robotic devices for paediatric rehabilitation: A review of design features. *Biomedical Engineering Online*, 20(1), 89.

Goyal, N.K., Dandotiya, M., Kumari, M., Kaloria, S., Anushya, A. (2025). AI-powered haptic devices for patient rehabilitation. In *Integrating AI With Haptic Systems for Smarter Healthcare Solutions*, Kalam, A., Bossoufi, B., Somwanshi, D.K. et al. (eds). IGI Global, Hershey, PA.

Green, R. (2024). Experiences and management of urinary incontinence following treatment for prostate cancer: Disrupted embodied practices and adapting to maintain masculinity. *Health*, 28(4), 489–506.

Green, M., Marzano, V., Leditschke, I.A., Mitchell, I., Bissett, B. (2016). Mobilization of intensive care patients: A multidisciplinary practical guide for clinicians. *Journal of Multidisciplinary Healthcare*, 9, 247–256.

Gudur, A., Pati, P., Garg, P., Sharma, N. (2023). Radiomics feature selection for lung cancer subtyping and prognosis prediction: A comparative study of ant colony optimization and simulated annealing. *International Journal of Intelligent Systems and Applications in Engineering*, 12(3s), 553–565.

Hussey, C. and Gupta, A. (2022). Exercise interventions to combat cancer-related fatigue in cancer patients undergoing treatment: A review. *Cancer Investigation*, 40(9), 822–838.

Jeong S. (2017). The impact of social robots on young patients' socio-emotional wellbeing in a pediatric inpatient care context. Doctoral Dissertation, Massachusetts Institute of Technology, Cambridge, MA.

Krebs, H.I., Edwards, D.J., Volpe, B.T. (2022). Forging mens et manus: The MIT experience in upper extremity robotic therapy. In *Neurorehabilitation Technology*, Reinkensmeyer, D.J., Marchal-Crespo, L., Dietz, V. (eds). Springer, Cham.

Kumar, A., Batta, P., Ahuja, S., Rathore, P.S. (eds) (2026a). *AI-driven Innovations in Physiotherapy and Oncology 1*. ISTE Ltd, London and Wiley, Hoboken, NJ.

Kumar, A., Batta, P., Ahuja, S., Rathore, P.S. (eds) (2026b). *AI-driven Innovations in Physiotherapy and Oncology 2*. ISTE Ltd, London and Wiley, Hoboken, NJ.

Le, D.N., Van Le, C., Tromp, J.G., Nguyen, G.N. (eds) (2018). *Emerging Technologies for Health and Medicine: Virtual Reality, Augmented Reality, Artificial Intelligence, Internet of Things, Robotics, Industry 4.0*. Scrivener Publishing, Austin, TX.

Lindberg, B., Nilsson, C., Zotterman, D., Söderberg, S., Skär, L. (2013). Using information and communication technology in home care for communication between patients, family members, and healthcare professionals: A systematic review. *International Journal of Telemedicine and Applications*, 2013(1), 461829.

Lovelace, D.L.,˙ McDaniel, L.R., Golden, D. (2019). Long-term effects of breast cancer surgery, treatment, and survivor care. *Journal of Midwifery & Women's Health*, 64(6), 713–724.

Mane, A.S. and Yadav, T. (2025). The short-term efficacy of mulligan traction straight leg raise on low back pain associated with hamstring tightness in young adults. *Cureus*, 17(3), e80215. doi: 10.7759/cureus.80215.

Nizamis, K., Athanasiou, A., Almpani, S., Dimitrousis, C., Astaras, A. (2021). Converging robotic technologies in targeted neural rehabilitation: A review of emerging solutions and challenges. *Sensors*, 21(6), 2084.

Rasa, A.R. (2024). Artificial intelligence and its revolutionary role in physical and mental rehabilitation: A review of recent advancements. *BioMed Research International*, 2024(1), 9554590.

Reddy, K.J. (ed.) (2025). Technological innovations in rehabilitation: Artificial intelligence. In *Innovations in Neurocognitive Rehabilitation: Harnessing Technology for Effective Therapy*. Springer, Cham.

Riviello, E.D., Letchford, S., Achieng, L., Newton, M.W. (2011). Critical care in resource-poor settings: Lessons learned and future directions. *Critical Care Medicine*, 39(4), 860–867.

Schollhorn, W., Beckmann, H., Davids, K. (2010). Exploiting system fluctuations: Differential training in physical prevention and rehabilitation programs for health and exercise. *Medicina*, 46(6), 365–373.

Schröder, W., Gisbertz, S.S., Voeten, D.M., Gutschow, C.A., Fuchs, H.F., van Berge Henegouwen, M.I. (2021). Surgical therapy of esophageal adenocarcinoma–current standards and future perspectives. *Cancers*, 13(22), 5834.

Scoglio, A.A., Reilly, E.D., Gorman, J.A., Drebing, C.E. (2019). Use of social robots in mental health and well-being research: Systematic review. *Journal of Medical Internet Research*, 21(7), e13322.

Silver, J.K., Baima, J., Mayer, R.S. (2013). Impairment-driven cancer rehabilitation: An essential component of quality care and survivorship. *CA: A Cancer Journal for Clinicians*, 63(5), 295–317.

Tamantini, C. (2020). User state estimation methods for tailoring human-robot interaction. PhD Thesis, Campus Bio Medico University of Rome, Rome.

Tsiakas, K. (2018). Interactive learning and adaptation for personalized robot-assisted training. PhD Thesis, The University of Texas, Arlington, TX.

Walker, D., Bendel, A., Stiller, C., Indelicato, D., Smith, S., Murray, M., Bleyer, A. (2016). Central nervous system tumors. In *Cancer in Adolescents and Young Adults*, Bleyer, W.A. and Barr, R.D. (eds). Springer, Cham.

Wang, Q., Markopoulos, P., Yu, B., Chen, W., Timmermans, A. (2017). Interactive wearable systems for upper body rehabilitation: A systematic review. *Journal of Neuroengineering and Rehabilitation*, 14(1), 20.

Zhang, R., Zhou, Y., Zhang, J., Zhao, J. (2024). Cloud-integrated robotics: Transforming healthcare and rehabilitation for individuals with disabilities. *Proceedings of the Indian National Science Academy*, 90(3), 752–763.

List of Authors

Sachin AHUJA
Department of Engineering
Chandigarh University
Mohali
India

Sanjay BADJATE
Department of Electronics and Telecommunication Engineering
S.B. Jain Institute of Technology, Management and Research
Nagpur
India

Priya BATTA
Amity School of Engineering and Technology
Amity University Punjab
Mohali
India

Anand GUDUR
Department of Oncology
Krishna Hospital and Medical Research Centre
Karad
India

Rashmi GUDUR
Department of Oncology
Krishna Hospital and Medical Research Centre
Karad
India

Faisal Hussain HUSSAIN
Department of Mechanical Engineering
S.B. Jain Institute of Technology, Management and Research
Nagpur
India

Abhay KASHETWAR
Department of Electronics and Telecommunication Engineering
S.B. Jain Institute of Technology, Management and Research
Nagpur
India

Abhishek KUMAR
Department of Computer Science and Engineering
Chandigarh University
Mohali
India

Mrudula NIMBARTE
Department of Computer Science and Engineering
S.B. Jain Institute of Technology, Management and Research
Nagpur
India

Dhairyasheel PATIL
Department of Oncology Physiotherapy
Krishna College of Physiotherapy
Krishna Vishwa Vidyapeeth
"Deemed to be University"
Taluka-Karad
India

Rahul PETHE
Department of Electronics and Telecommunication Engineering
S.B. Jain Institute of Technology, Management and Research
Nagpur
India

Pramod Singh RATHORE
Manipal University Jaipur
Jaipur
India

Pankaj THOTE
Department of Electrical Engineering
S.B. Jain Institute of Technology, Management and Research
Nagpur
India

Himanshu WAGH
Department of Mechanical Engineering
S.B. Jain Institute of Technology, Management and Research
Nagpur
India

Trupti YADAV
Department of Oncology Physiotherapy
Krishna College of Physiotherapy
Krishna Vishwa Vidyapeeth
"Deemed to be University"
Taluka-Karad
India

Index

F

G

H

I

K

L

M

N

O

P

W

Other titles from

ISTE

in

Computer Engineering

2026

BHAMBRI Pankaj, DUTTA Pushan Kumar, KHAN Mudassir,
STAROSTKA-PATYK Marta
Next-Gen Supply Chains: AI, Automation, and Sustainability in a Disrupted World

DARCHE Philippe
Computer Memories 1: Core Concepts
Computer Memories 2: Static Random-Access Memory

MAHOUT Vincent, ROCACHER Thierry, AURIOL Guillaume
Programming of Embedded Systems: Applications to SM32F10x μControllers

2025

CHIKH Azeddine
Systemic Approach to Categorizing and Modeling Requirements

KUMAR Abhishek, BATTA Priya, AHUJA Sachin,
RATHORE Pramod Singh
AI-driven Innovations in Physiotherapy and Oncology 1
AI-driven Innovations in Physiotherapy and Oncology 2

TANEJA Ashu, KUMAR Abhishek, LIMKAR Suresh Vishnudas,
OUAISSA Mariya, OUAISSA Mariyam
Driving Innovation through AI and Digital Twin for 6G Powered Sustainable Ultra Smart Cities

TANEJA Ashu, KUMAR Abhishek, LIMKAR Suresh Vishnudas,
OUAISSA Mariya, OUAISSA Mariyam
6G Urban Innovation: AI and Digital Twin for Next-Gen Sustainable Cities

2024

CHAKRABORTY Rajdeep, GHOSH Anupam, KUMAR MANDAL Jyotsna,
CHOUDHURY Tanupriya, CHATTERJEE Prasenjit
Smart Edge Computing: An Operation Research Perspective
(International Perspectives in Decision Analytics and Operations Research Set – Volume 2)

HOMÈS Bernard
Fundamentals of Software Testing (2nd edition, revised and updated)

JERBI Abdessalem
Discrete-Event Simulation: Concepts and Production in Arena

PRADHAN Devasis, KUMAR SAHU Prasanna, MYO TUN Hla,
CHATTERJEE Prasenjit
Artificial and Cognitive Computing for Sustainable Healthcare Systems in Smart Cities
(International Perspectives in Decision Analytics and Operations Research Set – Volume 3)

2023

CHOUDHURY Tanupriya, KHANNA Abhirup, CHATTERJEE Prasenjit,
UM Jung-Sup, BHATTACHARYA Abhishek
Blockchain Applications in Healthcare: Innovations and Practices
(International Perspectives in Decision Analytics and Operations Research Set – Volume 1)

PAI G A Vijayalakshmi
A Textbook of Data Structures and Algorithms 1: Mastering Linear Data Structures
A Textbook of Data Structures and Algorithms 2: Mastering Nonlinear Data Structures
A Textbook of Data Structures and Algorithms 3: Mastering Advanced Data Structures and Algorithm Design Strategies

2022

HOMÈS Bernard
Advanced Testing of Systems-of-Systems 1: Theoretical Aspects
Advanced Testing of Systems-of-Systems 2: Practical Aspects

MEHTA Shikha, TIWARI Sanju, SIARRY Patrick, JABBAR M.A.
Tools, Languages, Methodologies for Representing Semantics on the Web of Things

SIDHOM Sahbi, KADDOUR Amira
Systems and Uses of Digital Sciences for Knowledge Organization (Digital Tools and Uses Set – Volume 9)

ZAIDOUN Ameur Salem
Computer Science Security: Concepts and Tools

2021

DARCHE Philippe
Microprocessor 4: Core Concepts – Software Aspects
Microprocessor 5: Software and Hardware Aspects of Development, Debugging and Testing – The Microcomputer

DELHAYE Jean-Loïc
Inside the World of Computing: Technologies, Uses, Challenges

HARDIN Thérèse, JAUME Mathieu, PESSAUX François,
VIGUIÉ DONZEAU-GOUGE Véronique
Concepts and Semantics of Programming Languages 1: A Semantical Approach with OCaml and Python
Concepts and Semantics of Programming Languages 2: Modular and Object-oriented Constructs with OCaml, Python, C++, Ada and Java

MKADMI Abderrazak
Archives in The Digital Age: Preservation and the Right to be Forgotten (Digital Tools and Uses Set – Volume 8)

TOKLU Yusuf Cengiz, BEKDAŞ Gebrail, NIGDELI Sinan Melih
Metaheuristics for Structural Design and Analysis
(Optimization Heuristics Set – Volume 3)

2020

DARCHE Philippe
Microprocessor 1: Prolegomenes – Calculation and Storage Functions – Models of Computation and Computer Architecture
Microprocessor 2: Core Concepts – Communication in a Digital System
Microprocessor 3: Core Concepts – Hardware Aspects

DUVAUT Patrick, DALLOZ Xavier, MENGA David, KOEHL François,
CHRIQUI Vidal, BRILL Joerg
Internet of Augmented Me, I.AM: Empowering Innovation for a New Sustainable Future

LAFFLY Dominique
TORUS 1 – Toward an Open Resource Using Services: Cloud Computing for Environmental Data
TORUS 2 – Toward an Open Resource Using Services: Cloud Computing for Environmental Data
TORUS 3 – Toward an Open Resource Using Services: Cloud Computing for Environmental Data

LAURENT Anne, LAURENT Dominique, MADERA Cédrine
Data Lakes
(Databases and Big Data Set – Volume 2)

OULHADJ Hamouche, DAACHI Boubaker, MENASRI Riad
Metaheuristics for Robotics
(Optimization Heuristics Set – Volume 2)

SADIQUI Ali
Computer Network Security

VENTRE Daniel
Artificial Intelligence, Cybersecurity and Cyber Defense

2019

BESBES Walid, DHOUIB Diala, WASSAN Niaz, MARREKCHI Emna
Solving Transport Problems: Towards Green Logistics

CLERC Maurice
Iterative Optimizers: Difficulty Measures and Benchmarks

GHLALA Riadh
Analytic SQL in SQL Server 2014/2016

TOUNSI Wiem
Cyber-Vigilance and Digital Trust: Cyber Security in the Era of Cloud Computing and IoT

2018

ANDRO Mathieu
Digital Libraries and Crowdsourcing
(Digital Tools and Uses Set – Volume 5)

ARNALDI Bruno, GUITTON Pascal, MOREAU Guillaume
Virtual Reality and Augmented Reality: Myths and Realities

CARDON Alain
Beyond Artificial Intelligence: From Human Consciousness to Artificial Consciousness

HOMAYOUNI S. Mahdi, FONTES Dalila B.M.M.
Metaheuristics for Maritime Operations
(Optimization Heuristics Set – Volume 1)

JEANSOULIN Robert
JavaScript and Open Data

PIVERT Olivier
NoSQL Data Models: Trends and Challenges
(Databases and Big Data Set – Volume 1)

SEDKAOUI Soraya
Data Analytics and Big Data

SALEH Imad, AMMI Mehdi, SZONIECKY Samuel
Challenges of the Internet of Things: Technology, Use, Ethics
(Digital Tools and Uses Set – Volume 7)

SZONIECKY Samuel
Ecosystems Knowledge: Modeling and Analysis Method for Information and Communication
(Digital Tools and Uses Set – Volume 6)

2017

BENMAMMAR Badr
Concurrent, Real-Time and Distributed Programming in Java: Threads, RTSJ and RMI

BOUHAÏ Nasreddine, SALEH Imad
Internet of Things: Evolutions and Innovations
(Digital Tools and Uses Set – Volume 4)

HÉLIODORE Frédéric, NAKIB Amir, ISMAIL Boussaad, OUCHRAA Salma, SCHMITT Laurent
Metaheuristics for Intelligent Electrical Networks
(Metaheuristics Set – Volume 10)

MA Haiping, SIMON Dan
Evolutionary Computation with Biogeography-based Optimization
(Metaheuristics Set – Volume 8)

PÉTROWSKI Alain, BEN-HAMIDA Sana
Evolutionary Algorithms
(Metaheuristics Set – Volume 9)

PAI G A Vijayalakshmi
Metaheuristics for Portfolio Optimization: An Introduction using MATLAB®
(Metaheuristics Set – Volume 11)

REYES-GARCIA Everardo
The Image-Interface: Graphical Supports for Visual Information
(Digital Tools and Uses Set – Volume 3)

REYES-GARCIA Everardo, BOUHAÏ Nasreddine
Designing Interactive Hypermedia Systems
(Digital Tools and Uses Set – Volume 2)

SZONIECKY Samuel, BOUHAÏ Nasreddine
Collective Intelligence and Digital Archives: Towards Knowledge Ecosystems
(Digital Tools and Uses Set – Volume 1)

2016

BLUM Christian, FESTA Paola
Metaheuristics for String Problems in Bio-informatics
(Metaheuristics Set – Volume 6)

DEROUSSI Laurent
Metaheuristics for Logistics
(Metaheuristics Set – Volume 4)

DHAENENS Clarisse, JOURDAN Laetitia
Metaheuristics for Big Data
(Metaheuristics Set – Volume 5)

LABADIE Nacima, PRINS Christian, PRODHON Caroline
Metaheuristics for Vehicle Routing Problems
(Metaheuristics Set – Volume 3)

LEROY Laure
Eyestrain Reduction in Stereoscopy

LUTTON Evelyne, PERROT Nathalie, TONDA Alberto
Evolutionary Algorithms for Food Science and Technology
(Metaheuristics Set – Volume 7)

MAGOULÈS Frédéric, ZHAO Hai-Xiang
Data Mining and Machine Learning in Building Energy Analysis

RIGO Michel
Advanced Graph Theory and Combinatorics

ZELM Martin, DOUMEINGTS Guy, MENDONÇA Joao Pedro
Enterprise Interoperability: in the Digitized and Networked Factory of the Future

2015

BARBIER Franck, RECOUSSINE Jean-Luc
COBOL Software Modernization: From Principles to Implementation with the BLU AGE® Method

CHEN Ken
Performance Evaluation by Simulation and Analysis with Applications to Computer Networks

CLERC Maurice
Guided Randomness in Optimization
(Metaheuristics Set – Volume 1)

DURAND Nicolas, GIANAZZA David, GOTTELAND Jean-Baptiste,
ALLIOT Jean-Marc
Metaheuristics for Air Traffic Management
(Metaheuristics Set – Volume 2)

MAGOULÈS Frédéric, ROUX François-Xavier, HOUZEAUX Guillaume
Parallel Scientific Computing

MUNEESAWANG Paisarn, YAMMEN Suchart
Visual Inspection Technology in the Hard Disk Drive Industry

2014

BOULANGER Jean-Louis
Formal Methods Applied to Industrial Complex Systems
Formal Methods Applied to Complex Systems: Implementation of the B Method

EXPOSITO Ernesto, DIOP Codé
Smart SOA Platforms in Cloud Computing Architectures

GARDI Frédéric, BENOIST Thierry, DARLAY Julien, ESTELLON Bertrand, MEGEL Romain
Mathematical Programming Solver based on Local Search

KRICHEN Saoussen, CHAOUACHI Jouhaina
Graph-related Optimization and Decision Support Systems

LARRIEU Nicolas, VARET Antoine
Rapid Prototyping of Software for Avionics Systems: Model-oriented Approaches for Complex Systems Certification

LAURAS Matthieu, ZELM Martin, ARCHIMÈDE Bernard, BÉNABEN Frédérick, DOUMEINGTS Guy
Enterprise Interoperability: Interoperability for Agility, Resilience and Plasticity of Collaborations

OUSSALAH Mourad Chabane
Software Architecture 1
Software Architecture 2

QUESNEL Flavien
Scheduling of Large-scale Virtualized Infrastructures: Toward Cooperative Management

RIGO Michel
Formal Languages, Automata and Numeration Systems 1: Introduction to Combinatorics on Words
Formal Languages, Automata and Numeration Systems 2: Applications to Recognizability and Decidability

SAINT-DIZIER Patrick
Musical Rhetoric: Foundations and Annotation Schemes

TOUATI Sid, DUPONT DE DINECHIN Benoit
Advanced Backend Code Optimization

2013

ANDRÉ Étienne, SOULAT Romain
The Inverse Method: Parametric Verification of Real-time Embedded Systems

BOULANGER Jean-Louis
Safety Management for Software-based Equipment

DELAHAYE Daniel, PUECHMOREL Stéphane
Modeling and Optimization of Air Traffic

FRANCOPOULO Gil
LMF — Lexical Markup Framework

GHEDIRA Khaled
Constraint Satisfaction Problems: CSP Formalisms and Techniques

GOVAERT Gérard, NADIF Mohamed
Co-Clustering: Models, Algorithms and Applications

LANDRAGIN Frédérick
Man–Machine Dialogue: Design and Challenges

LUCAS Laurent, LOSCOS Céline, REMION Yannick
3D Video: Models, From Capture to Diffusion

ROCHANGE Christine, UHRIG Sascha, SAINRAT Pascal
Time-Predictable Architectures

WAHBI Mohamed
Algorithms and Ordering Heuristics for Distributed Constraint Satisfaction Problems

ZELM Martin, VAN SINDEREN Marten, FERRAIRA PIRES Luis, DOUMEINGTS Guy
Enterprise Interoperability: Research and Applications in the Service-oriented Ecosystem

2012

ARBOLEDA Hugo, ROYER Jean-Claude
Model-Driven and Software Product Line Engineering

BLANCHET Gérard, DUPOUY Bertrand
Computer Architecture

BOULANGER Jean-Louis
Industrial Use of Formal Methods: Formal Verification
Formal Methods: Industrial Use from Model to the Code

CALVARY Gaëlle, DELOT Thierry, SÈDES Florence, TIGLI Jean-Yves
Computer Science and Ambient Intelligence

MAHOUT Vincent
Assembly Language Programming: ARM Cortex-M3 2.0: Organization, Innovation and Territory

MARLET Renaud
Program Specialization

SOTO Maria, SEVAUX Marc, ROSSI André, LAURENT Johann
Memory Allocation Problems in Embedded Systems: Optimization Methods

2011

BICHOT Charles-Edmond, SIARRY Patrick
Graph Partitioning

BOULANGER Jean-Louis
Static Analysis of Software: The Abstract Interpretation

CAFERRA Ricardo
Logic for Computer Science and Artificial Intelligence

KORDON Fabrice, HADDAD Serge, PAUTET Laurent, PETRUCCI Laure
Distributed Systems: Design and Algorithms
Models and Analysis in Distributed Systems

LORCA Xavier
Tree-based Graph Partitioning Constraint

RIDHA MAHJOUB A.
Progress in Combinatorial Optimization: ISCO2010

TRUCHET Charlotte, ASSAYAG Gerard
Constraint Programming in Music

VICAT-BLANC Pascale, GOGLIN Brice, GUILLIER Romaric,
SOUDAN Sébastien
Computing Networks: From Cluster to Cloud Computing

2010

AUBRUN Christophe, SIMON Daniel, SONG Ye-Qiong
Co-design Approaches for Dependable Networked Control Systems

AUDIBERT Pierre
Mathematics for Informatics and Computer Science

BABAU Jean-Philippe, BLAY-FORNARINO Mireille, CHAMPEAU Joël,
ROBERT Sylvain, SABETTA Antonino
Model Driven Engineering for Distributed Real-Time Systems: MARTE Modeling, Model Transformations and their Usages

BOULANGER Jean-Louis
Safety of Computer Architectures

SIGAUD Olivier, BUFFET Olivier
Markov Decision Processes in Artificial Intelligence

MONMARCHÉ Nicolas, GUINAND Frédérick, SIARRY Patrick
Artificial Ants

SOLNON Christine
Ant Colony Optimization and Constraint Programming

2009

FOURNIER Jean-Claude
Graph Theory and Applications: with exercises and problems

GUÉDON Jeanpierre
The Mojette Transform: theory and applications

JARD Claude, ROUX Olivier H.
Communicating Embedded Systems: Software and Design

LECOUTRE Christophe
Constraint Networks: Techniques and Algorithms

2008

BANÂTRE Michel, MARRON Pedro Jose, OLLERO Anibal, WOLITZ Adam
Cooperating Embedded Systems and Wireless Sensor Networks

MERZ Stephan, NAVET Nicolas
Modeling and Verification of Real-time Systems: Formalisms and Software Tools

PASCHOS Vangelis Th.
Combinatorial Optimization and Theoretical Computer Science: Interfaces and Perspectives

WALDNER Jean-Baptiste
Nanocomputers and Swarm Intelligence

2007

BENHAMOU Frédéric, JUSSIEN Narendra, O'SULLIVAN Barry
Trends in Constraint Programming

JUSSIEN Narendra
A to Z of Sudoku

2006

BABAU Jean-Philippe, CHAMPEAU Joël, GÉRARD Sébastien
From MDD Concepts to Experiments and Illustrations – DRES 2006

HABRIAS Henri, FRAPPIER Marc
Software Specification Methods: An Overview Using a Case Study

MURAT Cécile, PASCHOS Vangelis Th.
Probabilistic Combinatorial Optimization on Graphs

PANETTO Hervé, BOUDJLIDA Nacer
Interoperability for Enterprise Software and Applications: Proceedings of the Workshops and the Doctorial Symposium of the Second IFAC/IFIP I-ESA International Conference: EI2N, WSI, IS-TSPQ 2006

2005

GÉRARD Sébastien, BABAU Jean-Philippe, CHAMPEAU Joël
Model Driven Engineering for Distributed Real Time Embedded Systems

PANETTO Hervé
Interoperability of Enterprise Software and Applications: Workshops of the INTEROP-ESA 2005 International Conference (EI2N, WSI, ISIDI and IEHENA 2005)

Printed and bound by CPI Group (UK) Ltd, Croydon, CR0 4YY

28/07/2026

14931346-0005